Springer Series in
CHEMICAL PHYSICS

Series Editors: A.W. Castleman, Jr.   J.P. Toennies   K. Yamanouchi   W. Zinth

The purpose of this series is to provide comprehensive up-to-date monographs in both well established disciplines and emerging research areas within the broad fields of chemical physics and physical chemistry. The books deal with both fundamental science and applications, and may have either a theoretical or an experimental emphasis. They are aimed primarily at researchers and graduate students in chemical physics and related fields.

77 **Heterogeneous Kinetics**
    Theory of Ziegler-Natta-Kaminsky
    Polymerization
    By T. Keii

78 **Nuclear Fusion Research**
    Understanding Plasma-Surface
    Interactions
    Editors: R.E.H. Clark and D.H. Reiter

79 **Ultrafast Phenomena XIV**
    Editors: T. Kobayashi,
    T. Okada, T. Kobayashi,
    K.A. Nelson, S. De Silvestri

80 **X-Ray Diffraction**
    by Macromolecules
    By N. Kasai and M. Kakudo

81 **Advanced Time-Correlated Single**
    Photon Counting Techniques
    By W. Becker

82 **Transport Coefficients of Fluids**
    By B.C. Eu

83 **Quantum Dynamics**
    of Complex Molecular Systems
    Editors: D.A. Micha and I. Burghardt

84 **Progress in Ultrafast**
    Intense Laser Science I
    Editors: K. Yamanouchi, S.L. Chin,
    P. Agostini, and G. Ferrante

85 **Quantum Dynamics**
    Intense Laser Science II
    Editors: K. Yamanouchi, S.L. Chin,
    P. Agostini, and G. Ferrante

86 **Free Energy Calculations**
    Theory and Applications
    in Chemistry and Biology
    Editors: Ch. Chipot and A. Pohorille

87 **Analysis and Control of**
    Ultrafast Photoinduced Reactions
    Editors: O. Kühn and L. Wöste

88 **Ultrafast Phenomena XV**
    Editors: P. Corkum, D. Jonas,
    D. Miller, and A.M. Weiner

89 **Progress in Ultrafast Intense Laser**
    Science III
    Editors: K. Yamanouchi, S.L. Chin,
    P. Agostini, and F. Ferrante

90 **Thermodynamics**
    and Fluctuations
    Far from Equilibrium
    By J. Ross

91 **Progress in Ultrafast Intense Laser**
    Science IV
    Editors: K. Yamanouchi, A. Becker,
    R. Li, and S.L. Chin

92 **Ultrafast Phenomena XVI Proceedings**
    of the 16th International Conference,
    Palazzo dei Congressi Stresa, Italy,
    June 9–13, 2008
    Editors: P. Corkum, S. De Silvestri,
    K.A. Nelson et al.

93 **Energy Transfer Dynamics in**
    Biomaterial Systems
    Editors: I. Burghardt, V. May,
    D.A. Micha et al.

94 **Lectures on Ultrafast Intense Laser**
    Science I
    Editor: K. Yamanouchi

For other titles published in this series, go to
http://www.springer.com/series/676

Vladimir N. Pokrovskii

The Mesoscopic Theory of Polymer Dynamics

Second Edition

 Springer

Professor Vladimir N. Pokrovskii
Institute of Chemical Physics, Russian Academy of Sciences
Kosygin St 4, Moscow 117977, Russia
vpok@comtv.ru

Series Editors:
Professor A.W. Castleman, Jr.
Department of Chemistry, The Pennsylvania State University
152 Davey Laboratory, University Park, PA 16802, USA

Professor J.P. Toennies
Max-Planck-Institut für Strömungsforschung
Bunsenstrasse 10, 37073 Göttingen, Germany

Professor K. Yamanouchi
University of Tokyo, Department of Chemistry
Hongo 7-3-1, 113-0033 Tokyo, Japan

Professor W. Zinth
Universität München, Institut für Medizinische Optik
Öttingerstr. 67, 80538 München, Germany

ISSN 0172-6218
ISBN 978-94-007-9092-6   ISBN 978-90-481-2231-8 (eBook)
DOI 10.1007/978-90-481-2231-8
Springer Dordrecht Heidelberg London New York

Cover design: eStudio Calamar S.L.

Printed on acid-free paper

Springer is part of Springer Science+Business Media (www.springer.com)

Contents

PREFACE TO THE REVISED EDITION   xi

PREFACE TO THE FIRST EDITION   xiii

NOTATIONS AND CONVENTIONS   xv

1   **Introduction: Macromolecular Systems in Equilibrium** 1
   1.1   Microscopic Models of a Macromolecule 1
   1.2   Bead-and-Spring Model 3
   1.3   Normal Co-Ordinates 5
   1.4   Macromolecular Coil 6
   1.5   Excluded-Volume Effects 8
   1.6   Macromolecules in a Solvent 11
      1.6.1   Macromolecules in a Dilute Solution 12
      1.6.2   Weakly-Coupled Macromolecules 13
   1.7   Elasticity of Polymer Networks 16
   1.8   Crystalline and Glassy Systems 19

2   **Dynamics of a Macromolecule in a Viscous Liquid** 21
   2.1   Equation of Macromolecular Dynamics 21
   2.2   Intramacromolecular Hydrodynamic Interactions 22
   2.3   Resistance-Drag Coefficient of a Macromolecular Coil 24
   2.4   Effective Resistance-Drag Coefficient of a Particle 26
   2.5   Intramolecular Friction 28
   2.6   The Cerf-Zimm-Rouse Modes 31
   2.7   The Moments of Linear Modes 32
      2.7.1   Equations for the Moments of Co-ordinates 33
      2.7.2   The Slowest Relaxation Processes 33
      2.7.3   Fourier-Transforms of Moments 35

3   Dynamics of a Macromolecule in an Entangled System 37
3.1   Admitted Approximations in the Many-Chain Problem 38
    3.1.1   Dynamics of Entangled Course-Grained Chains 38
    3.1.2   Dynamics of a Probe Macromolecule 40
3.2   The General Form of Dynamic Equation 42
    3.2.1   The Linear Approximation 43
    3.2.2   A Non-Linear Approximation – Local Anisotropy 45
3.3   Molecular Interpretation of the Dissipative Terms 46
    3.3.1   Concept of Microviscoelasticity 46
    3.3.2   External Friction 48
    3.3.3   Intramolecular Friction 51
    3.3.4   Fundamental Dynamical Parameters 53
3.4   Markovian Form of Dynamic Equation 54
3.5   Reptation-Tube Model 56
3.6   Method of Numerical Simulation 59
    3.6.1   Non-Dimensional Form of Dynamic Equation 59
    3.6.2   Algorithm of Calculation 61

4   Conformational Relaxation 63
4.1   Correlation Functions for the Linear Dynamics 63
    4.1.1   Modified Cerf-Rouse Modes 63
    4.1.2   Equilibrium Correlation Functions 66
    4.1.3   One-Point Non-Equilibrium Correlation Functions 68
    4.1.4   Two-Point Non-Equilibrium Correlation Functions 69
4.2   Relaxation of Macromolecular Coil 71
    4.2.1   Correlation Functions for Isotropic Motion 71
    4.2.2   Effect of Local Anisotropy 73
    4.2.3   Transition Point 77
    4.2.4   Conformational Relaxation Times 78
4.3   Macromolecular Coil in a Flow 79
    4.3.1   Non-Equilibrium Correlation Functions 79
    4.3.2   Size and Form of the Macromolecular Coil 80

5   The Localisation Effect 83
5.1   Mobility of a Macromolecule 83
    5.1.1   A Macromolecule in a Viscous Liquid 84
    5.1.2   A Macromolecule in an Entangled System 85
5.2   Quasi-Elastic Neutron Scattering 93
    5.2.1   The Scattering Function 94
    5.2.2   An Estimation of Intermediate Length 96

6   Linear Viscoelasticity 99
  6.1 Stresses in the Flow System............................. 99
      6.1.1 The Stress Tensor 99
      6.1.2 Oscillatory Deformation 102
  6.2 Macromolecules in a Viscous Liquid......................103
      6.2.1 The Stress Tensor 103
      6.2.2 Dynamic Characteristics 105
      6.2.3 Initial Intrinsic Viscosity 107
      6.2.4 On the Effect of Internal Viscosity..................109
  6.3 Macromolecules in a Viscoelastic Liquid 111
      6.3.1 The Stress Tensor 111
      6.3.2 Dynamic Characteristics 112
  6.4 Entangled Macromolecules...............................115
      6.4.1 The Stress Tensor 116
      6.4.2 Dynamic Modulus and Relaxation Branches 118
      6.4.3 Self-Consistency of the Mesoscopic Approach 122
      6.4.4 Modulus of Elasticity and the Intermediate Length 124
      6.4.5 Concentration and Macromolecular Length
            Dependencies 125
      6.4.6 Frequency–Temperature Superposition 127
  6.5 Dilute Blends of Linear Polymers 128
      6.5.1 Relaxation of Probe Macromolecule..................129
      6.5.2 Characteristic Quantities 130
      6.5.3 Terminal Relaxation Time 132
      6.5.4 A Final Remark...................................134

7   Equations of Relaxation 135
  7.1 Normal-Modes Form of Dynamic Equation.................135
      7.1.1 Transition to the Normal Modes.....................135
      7.1.2 Anisotropy of Particle Mobility.....................137
  7.2 Equations for the Non-Equilibrium Moments 139
  7.3 Relaxation of the Macromolecular Conformation.............143
      7.3.1 Diffusive Relaxation 143
      7.3.2 Reptation Relaxation 145
  7.4 Relaxation of Orientational Variables 146
      7.4.1 Weakly Entangled Systems 148
      7.4.2 Strongly Entangled Systems 148
  7.5 Relaxation of the Segment Orientation 149
      7.5.1 Rubber Elasticity and Mean Orientation of Segments ..149
      7.5.2 Elementary Theory of Dielectric Relaxation...........151

8   Relaxation Processes in the Phenomenological Theory 155
   8.1   The Laws of Conservation of Momentum and Angular
       Momentum 155
   8.2   The Law of Conservation of Energy and the Balance of
       Entropy 158
   8.3   Thermodynamic Fluxes and Relaxation Processes 160
   8.4   The Principle of Relativity for Slow Motions 163
   8.5   Constitutive Relations for Non-Linear Viscoelastic Fluids 164
   8.6   Different Forms of Constitutive Relation 167

9   Non-Linear Effects of Viscoelasticity 171
   9.1   Dilute Polymer Solutions 171
       9.1.1   Constitutive Relations 172
       9.1.2   Non-Linear Effects in Simple Shear Flow 173
       9.1.3   Non-Steady-State Shear Flow 175
       9.1.4   Non-Linear Effects in Oscillatory Shear Motion 176
   9.2   Many-Mode Description of Entangled Systems 178
       9.2.1   Constitutive Relations 178
       9.2.2   Linear Approximation 180
       9.2.3   Steady-State Simple Shear Flow 184
   9.3   Single-Mode Description of Entangled System 186
       9.3.1   Weakly Entangled Systems 187
       9.3.2   Strongly Entangled Systems 189
       9.3.3   Vinogradov Constitutive Relation 191
       9.3.4   Relation between Shear and Elongational Viscosities ... 194
       9.3.5   Recoverable Strain 196

10   Optical Anisotropy 199
   10.1   The Relative Permittivity Tensor 199
   10.2   The Permittivity Tensor for Polymer Systems 202
       10.2.1   Dilute Solutions 203
       10.2.2   Entangled Systems 204
   10.3   Optical Birefringence 206
       10.3.1   Simple Elongation 206
       10.3.2   Simple Shear 207
       10.3.3   Oscillatory Deformation 208
   10.4   Anisotropy in a Simple Steady-State Shear Flow............. 209
       10.4.1   Dilute Solutions 209
       10.4.2   Entangled Systems 211
   10.5   Oscillatory Birefringence 211
       10.5.1   Dilute Solutions 211
       10.5.2   Entangled Systems 212

CONCLUSION   215

APPENDICES   217
   A   The Random Walk Problem 217
   B   Equilibrium Deformation of a Non-Linear Elastic Body....... 219
   C   The Tensor of Hydrodynamic Interaction 222
   D   Resistance Force of a Particle in a Viscoelastic Fluid 223
   E   Resistance Coefficient of a Particle in Non-Local Fluid 225
   F   Dynamics of Suspension of Dumbbells..................... 228
   G   Estimation of Some Series 239

REFERENCES   241

INDEX   253

Preface to the Revised Edition

I used the opportunity of this edition to correct some minor mistakes and clarify, wherever it possible, exposition of the theory in comparison with the previous edition of this book (Kluwer, Dordrecht *et cet*, 2000). It provokes enlargement of the book, though I tried to present the modern theory of thermic motion of long macromolecules in compact form. I have tried to accumulate the common heritage and to take into account different approaches in the theory of dynamics of linear polymers, at least, to understand and make clear the importance of various ideas for explanation of relaxation phenomena in linear polymers, to present recent development in the field.

The theory of non-equilibrium phenomena in polymer systems is based on the fundamental principles of statistical physics. However, the peculiarities of the structure and the behaviour of the systems necessitate the implementation of special methods and heuristic models that are different from those for gases and solids, so that polymer dynamics has appeared to be a special branch of physics now. The monograph contains discussions of the main principles of the theory of slow relaxation phenomena in linear polymers, elaborated in the last decades. The basic model of a macromolecule, which allows us a consistent explanation of different relaxation phenomena (diffusion, neutron scattering, viscoelasticity, optical birefringence), remains to be a coarse-grained or bead-spring model, considered in different environments: viscous, to describe the behaviour of dilute solutions, or viscoelastic, to describe the behaviour of both weakly and strongly entangled systems. Besides, extra features of dynamics of a chain in strongly entangled systems, namely the strong resistance to changes of conformation of macromolecule (the internal viscosity resistance due to the entanglements) and local anisotropy of mobility of particles of the chain, which provokes motion of macromolecule along its contour – the reptation motion, have to be taken into account. The dynamic transition point between weakly and strongly entangled systems is calculated as $M^* \approx 10 M_e$, where M_e is called conventionally 'the length of the macromolecule between adjacent entanglements'.

Thus, among the linear polymer systems, three types of systems, according to the ratio of the length of the macromolecule M to M_e : $M < 2M_e$ – non-entangled system, $2M_e < M < 10M_e$ – weakly entangled systems and $M > 10M_e$ – strongly entangled systems, have to be considered separately. The laws of the relaxation behaviour of the different systems are different: no reptation relaxation of macromolecules exists in the non-entangled and weakly entangled systems.

The properly formulated phenomenological dynamic equation for a single macromolecule remains to play a role of the central organising principle of the monograph. The model was designed to study systematically deviations from the Rouse dynamics when adding non-Markovian and anisotropic noise. The developed model describes underlying stochastic motion of particles of the chain and provides both the confinement of a macromolecule in a tube and easier (reptation) motion of the macromolecule along its contour – the features, which were envisaged by Edwards and de Gennes for the entangled systems. An intermediate length, which has the meaning of a tube radius and/or the length of a macromolecule between adjacent entanglements, is calculated through parameters of the model. The unified approach appeared to be useful for consistent explanation the relaxation phenomena in entangled linear polymers (polymer solutions and melts), and one can think that a consequent theory of viscoelasticity (so as other phenomena) in mesoscopic approximation can be developed on the base of the unified non-linear dynamics of a macromolecule.

It is my pleasure to acknowledge my gratitude to various people for the comments on the previous edition of the monograph and for advice how to improve it. During the work on the revision of the monograph, in September 2004, due to courtesy of Professor Kurt Kremer, I had a privilege to be a guest at the Max-Plank-Institut für Polymerforschung (Mainz, Germany) and to benefit from its excellent facilities for work. I have learnt and understood much from discussions of the relevant problems with Professor Kremer and members of the Institute, especially, with Burkhart Dünweg, Bernd Ewen, Tadeusz Pakula, Vahktang Rostiashvili and Nico van der Vegt. I thank all of them.

Any comments will be greatly appreciated.

Moscow, RUSSIA   *Vladimir N. Pokrovskii*
                                         http://www.ecodynamics.narod.ru/

Preface to the First Edition

Our brutal century of atom bombs and spaceships can also be called the century of polymers. In any case, the broad spreading of synthetic polymer materials is one of the signs of our time. A look at the various aspects of our life is enough to convince us that polymeric materials (textiles, plastics, rubbers) are as widely spread and important in our life as are other materials (metals and non-metals) derived from small molecules. Polymers have entered the life of the twentieth century as irreplaceable construction materials.

Polymers differ from other substances by the size of their molecules which, appropriately enough, are referred to as macromolecules, since they consist of thousands or tens of thousands of atoms (molecular weight up to 10^6 or more) and have a macroscopic rectilinear length (up to 10^{-4} cm). The atoms of a macromolecule are firmly held together by valence bonds, forming a single entity. In polymeric substances, the weaker van der Waals forces have an effect on the components of the macromolecules which form the system. The structure of polymeric systems is more complicated than that of low-molecular solids or liquids, but there are some common features: the atoms within a given macromolecule are ordered, but the centres of mass of the individual macromolecules and parts of them are distributed randomly. Remarkably, the mechanical response of polymeric systems combines the elasticity of a solid with the fluidity of a liquid. Indeed, their behaviour is described as viscoelastic, which is closely connected with slow (relaxation time to 1 sec or more) relaxation processes in systems.

The monograph is devoted to the description of the relaxation behaviour of very concentrated solutions or melts of linear polymers. In contrast to well-known text-books on polymer dynamics by Doi and Edwards (1986) and by Bird et al. (1987a), I exploit a mesoscopic approach, which deals with the dynamics of a single macromolecule among others and is based on some statements of a general kind. From a strictly phenomenological point of view, the mesoscopic approach is a microscopic macromolecular approach. It reveals the internal connection between phenomena and gives more details than the phenomenological approach. From a strictly microscopic point of view, it is a phenomenological one. It needs some mesoscopic parameters to be introduced and determined empirically. However, the mesoscopic approach permits us to explain the different phenomena of the dynamic behaviour of polymer melts – diffusion, neutron scattering, viscoelasticity, birefringence and others – from a macromolecular point of view and without any specific hypotheses. The mesoscopic approach constitutes a phenomenological frame within which the results of investigations of behaviour of weakly-coupled macromolecules can be considered. The resultant picture of the thermal motion of a macromolecule in the system appears to be consistent with the common ideas about the localisation of a macromolecule: the theory comes to introduce an intermediate length which has the sense of a tube diameter and/or the length of a macromolecule between adjacent entanglements. It appears to be the most important parameter of the theory, as it was envisaged by Edwards and by de Gennes. In fact, one needs no more parameters, apart from the monomer friction coefficient, to describe dynamics of polymer melts in mesoscopic approach.

The monograph contains the fundamentals of the theory and reflects the modern situation in understanding the relaxation behaviour of a polymer solutions and melts. The contents of the monograph can be related to the fields of molecular physics, fluid mechanics, polymer physics and materials science. I have tried to present topics in a self-contained way that makes the monograph a suitable reference book for professional researchers. I hope that the book will also prove to be useful to graduate students of above mentioned specialities who have some background in physics and mathematics. It would provide material for a one or two semester graduate-level course in polymer dynamics.

I should like gratefully to note that at different times Yu.A. Altukhov, V.B. Erenburg, V.L. Grebnev, Yu.K. Kokorin, N.P. Kruchinin, G.V. Pyshnograi, Yu.V. Tolstobrov, G.G. Tonkikh, A.A. Tskhai, V.S. Volkov and V.E. Zgaevskii participated in the investigations of the problems and in the discussions of the results. I thank them for their helpful collaboration. I would like to express special thanks to Mrs Marika Fenech who has done much work to change my drafts into a readable manuscript and to improve my English.

It is my great pleasure to acknowledge my indebtedness to Professor Sir Sam Edwards who has kindly read an original version of the manuscript. His comments and, especially, conversations with him in Cambridge in May 1998 were very useful for me. I am grateful to Professor A.D. Jenkins who has also read the entire manuscript and made many helpful remarks concerning language of the book.

This preface would be incomplete without words of acknowledgement to the University of Malta Department of Physics for its hospitality during the period of the completion of this book.

Madliena, MALTA   *Vladimir N. Pokrovskii*

Notations and Conventions

b — mean square distance between adjacent particles along the chain

B — coefficient of enhancement of "external" friction of a particle due to surrounding macromolecules

c — concentration of polymer in solution

D — coefficient of diffusion

e — unit vector

$\langle e_i e_j \rangle$ — tensor of mean orientation of segments

E — coefficient of enhancement of "internal" friction of a particle due to surrounding macromolecules

$G(\omega)$ — dynamic shear modulus

G_e — plateau value of the dynamic modulus

l — the length of a Kuhn segment

M — molecular weight or length of macromolecule

M_e — length of chain between adjacent entanglements in a very concentrated solution

n — number of macromolecules in volume unit

N — number of subchains for a macromolecule

$O(\omega)$ — strain-optical coefficient

p — pressure

q — centre of mass of a macromolecule

r^α — co-ordinate of a particle labelled α in the subchain model

R — end-to-end distance of a macromolecule

$s_{ij} = \langle e_i e_j \rangle - \frac{1}{3} \delta_{ij}$ — tensor of mean orientation of segments

$S(\omega)$ — dynamo-optical coefficient

S — radius of gyration of a macromolecular coil

T — temperature in units of energy ($1\ \mathrm{K} = 1.38 \times 10^{-16}$ erg)

$u^\alpha = \dot{r}^\alpha$ — velocity of a particle labelled α

u_{ik}^α — tensor of internal stresses for mode α

$\boldsymbol{v} = \boldsymbol{v}(\boldsymbol{x}, t)$ — macroscopic velocity of continuum

$x_{ik}^\alpha = \frac{\langle \rho_i^\alpha \rho_k^\alpha \rangle}{\langle \rho^\alpha \rho^\alpha \rangle_0}$ — tensor of conformation for mode α

z — number of Kuhn segments for a chain

$Z = \frac{M}{M_e}$ — length of macromolecule measured by M_e

$\gamma_{ij} = \frac{1}{2}(\nu_{ij} + \nu_{ji})$ — symmetric tensor of velocity gradients

ε_{ik} — tensor of relative permittivity

ζ — coefficient of friction of a particle of a chain

η — coefficient of shear viscosity

$\eta(\omega)$ — dynamic viscosity

$[\eta]$ — intrinsic viscosity

λ — coefficient of elongational viscosity

λ_{ik} — tensor of recoverable displacements

$2\mu T = \frac{3TN}{2\langle R^2 \rangle_0}$ — coefficient of subchain elasticity

$\nu_{js} = \frac{\partial v_j}{\partial x_s}$ — tensor of velocity gradients

ξ — intermediate length in an entangled system

$\boldsymbol{\rho}^\alpha$ — normal co-ordinate referred to mode α

σ_{ik} — stress tensor

τ — terminal viscoelasticity relaxation time or relaxation time of segment orientation

τ^* — characteristic 'monomer' relaxation time of a very concentrated solution

τ_ν^{R} — relaxation time of macromolecule mode ν for a flexible draining chain (Rouse approximation)

$\tau_\nu^\perp, \tau_\nu^\parallel$ — orientation and deformation times of relaxation of mode ν for macromolecule in viscous fluid

$\tau_\nu^\pm, \tau_\nu$ — relaxation times of mode ν for macromolecule in very concentrated solution

ϕ^α — random force acting on particle α

$\chi = \frac{\tau}{2B\tau^*}$ — fundamental dynamical parameter for entangled systems

$\psi = \frac{E}{B}$ — fundamental dynamical parameter for entangled systems

ω — frequency of oscillation

$\omega_{ij} = \frac{1}{2}(\nu_{ij} - \nu_{ji})$ — antisymmetric tensor of velocity gradients

Notation of the type of z_{ij}^{-1} means $(z^{-1})_{ij}$.

The Fourier transforms are defined as

$$f(\omega) = \int_{-\infty}^{\infty} f(s)e^{i\omega s}\,\mathrm{d}s,$$

$$f(s) = \int_{-\infty}^{\infty} f(\omega)e^{-i\omega s}\frac{\mathrm{d}\omega}{2\pi},$$

$$f[\omega] = \int_0^{\infty} f(s)e^{i\omega s}\,\mathrm{d}s.$$

Latin suffixes take values 1, 2, 3. Greek suffixes take values from 0 or 1 to N. The rule about summation with respect to twice repeated suffixes is used.

The averaging with respect to the realisation of random variable is noted by angle brackets.

The chapter number and respective formulae are shown in references to formulae.

Chapter 1
Introduction: Macromolecular Systems in Equilibrium

Abstract The general theory of equilibrium and non-equilibrium properties of polymer solutions and melts appears to be derived from the universal models of long macromolecules which can be applied to any flexible macromolecule notwithstanding the nature of its internal chemical structure. Although many universal models are useful in the explanation of the behaviour of the polymeric system, the theory that will be described in this book is based on the coarse-grained model of a flexible macromolecule, the so-called, bead-and-spring or subchain model. In the foundation of this model, one finds a simple idea to observe the dynamics of a set of representative points (beads, sites) along the macromolecule instead of observing the dynamics of all the atoms. It has been shown that each point can be considered as a Brownian particle, so the theory of Brownian motion can be applied to the motion of a macromolecule as a set of linear-connected beads. The large-scale or low-frequency properties of macromolecules and macromolecular systems can be universally described by this model, while the results do not depend on the arbitrary number of sites. In this chapter, the bead-and-spring model will be introduced and some properties of simple polymer systems in equilibrium are discussed.

1.1 Microscopic Models of a Macromolecule

One says that the microstate of a macromolecule is determined, if a sequence of atoms, the distances between atoms, valence angles, the potentials of interactions and so on are determined. The statistical theory of long chains developed in considerable detail in monographs (Birshtein and Ptitsyn 1966; Flory 1969) defines the equilibrium quantities that characterise a macromolecule in a whole as functions of the macromolecular microparameters.

To say nothing about atoms, valence angles and so on, one can notice that the length of a macromolecule is much larger than its breadth, so one can consider the macromolecule as a flexible, uniform, elastic thread with coefficient of elasticity a, which reflects the individual properties of the macromolecule

V.N. Pokrovskii, *The Mesoscopic Theory of Polymer Dynamics*,
Springer Series in Chemical Physics 95,
DOI 10.1007/978-90-481-2231-8_1, © Springer Science+Business Media B.V. 2010

(Flory 1969; Landau and Lifshitz 1969). Thermal fluctuations of the macro-molecule determine the dependence of the mean square end-to-end distance $\langle R^2 \rangle$ on the length of macromolecule M and temperature T which is, we assume, measured in energy units. If $MT \gg a$

$$\langle R^2 \rangle = \frac{2Ma}{T}. \tag{1.1}$$

The last relation shows that a long macromolecule rolls up into a coil at high temperatures. The smaller the elasticity coefficient a is, the more it coils up. Another name for the model of flexible thread is the model of persistence length or the Kratky-Porod model. The quantity a/T is called the persistence length (Birshtein and Ptitsyn 1966).

One can use another way to describe the long macromolecule. One can see that at high temperatures there is no correlation between the orientations of the different parts of the macromolecule, which are not close to each other along the chain. This means that the chain of freely-jointed rigid segments reflects the behaviour of a real macromolecule. This model carries the name of Werner Kuhn who introduced it in his pioneering works (Kuhn 1934).

The expression for the mean square end-to-end distance can be written as the mean square displacement of a Brownian particle after z steps of equal length l (Appendix A)

$$\langle R^2 \rangle = zl^2. \tag{1.2}$$

If we return to the chain, z is the number of Kuhn segments in the chain, and l is the length of the segment. To avoid uncertainty, one adds a condition which is usually $zl = M$, so that one has a definition of the length

$$l = \frac{\langle R^2 \rangle}{M}. \tag{1.3}$$

Formulae (1.2) and (1.3) determine the model of a freely-jointed segment chain, which is frequently used in polymer physics as a microscopic heuristic model (Mazars 1996, 1998, 1999). A Kuhn segment in the flexible polymers (polyethylene, polystyrene, for example) usually includes a few monomer units, so that a typical length of the Kuhn segment is about 10 Å or 10^{-7} cm and, at the number of segments $z = 10^4$, the end-to-end distance $\langle R^2 \rangle^{1/2}$ of a macromolecule is about 10^{-5} cm.

In such a way, there are two universal, (that is, irrespective of the chemical nature) methods of description of a macromolecule; either as a flexible thread or as freely-jointed segments. Either model reflects the properties of each macromolecule long enough to be flexible. A relation

$$\frac{2a}{T} = l$$

follows from the comparison of equations (1.1)–(1.3). This relation demonstrates the imperfection of either model when applied to a real macromolecule.

Indeed, it shows that the length of a segment or the elasticity coefficient depends on the temperature, which contradicts the proposed features of the models.

In any case, the mean square end-to-end distance of a long macromolecule $\langle R^2 \rangle$ is small compared to the length of the macromolecule. Whatever its chemical composition, a macromolecule which is long enough rolls up into a coil as a result of thermal motion, so that its mean square end-to-end distance becomes proportional to its molecular length

$$\langle R^2 \rangle \sim C_\infty(T)M. \tag{1.4}$$

The temperature dependence of the size of a macromolecular coil is included in the coefficient of stiffness $C_\infty(T)$ which has the meaning of the ratio of the squared length of a Kuhn segment to the squared length of the chemical bond, and can be calculated from the local chemical architecture of the chain. The results of the calculations were summarised by Birshtein and Ptitsyn (1966) and by Flory (1969).

The probability distribution function for the fixed end-to-end distance R of macromolecule can be written down on either ground. In the simplest case, it is the Gaussian distribution

$$W(R) = \left(\frac{3}{2\pi\langle R^2 \rangle} \right)^{\frac{3}{2}} \exp \left(-\frac{3R^2}{2\langle R^2 \rangle} \right). \tag{1.5}$$

There are a number of ways to calculate function (1.5). One of the methods is demonstrated in Appendix A.

We may note that function (1.5) has a non-realistic feature that R can be larger than the maximum extended length M of the chain. Though more realistic distribution functions are available (Birshtein and Ptitsyn 1966; Flory 1969), in this monograph, approximation (1.5) is sufficient for our purpose.

1.2 Bead-and-Spring Model

A macrostate of a macromolecule can be described with the help of the end-to-end distance $\boldsymbol{R}$. To give a more detailed description of the macromolecule, one should use a method introduced by the pioneering work reported by Kargin and Slonimskii (1948) and by Rouse (1953), whereby the macromolecule is divided into N subchains of length M/N. One can consider the ends of the macromolecule and the points, at which the subchains join to form the entire chain, as a particles (the beads), labelled 0 to N respectively, and their positions will be represented by $r^0, r^1, \ldots, r^N$.

One can assume that each subchain is also sufficiently long, so that it can be described in the same way as the entire macromolecule, in particular, one can introduce the end-to-end distance for a separate subchain b^2. The equilibrium probability distribution for the positions of all the particles in the macromolecule is determined by the multiplication of N distribution functions of the type (1.5)

$$W(r^0, r^1, \ldots, r^N) = C \exp(-\mu A_{\alpha\gamma} r^\alpha r^\gamma), \tag{1.6}$$

where

$$\mu = \frac{3}{2b^2} = \frac{3N}{2\langle R^2 \rangle}, \tag{1.7}$$

and the matrix $A_{\alpha\gamma}$ describes the connection of the particles into the chain and has the form

$$A = \begin{Vmatrix} 1 & -1 & 0 & \ldots & 0 \\ -1 & 2 & -1 & \ldots & 0 \\ \ldots & \ldots & \ldots & \ldots & \ldots \\ \ldots & \ldots & \ldots & \ldots & \ldots \\ 0 & 0 & 0 & \ldots & 1 \end{Vmatrix}. \tag{1.8}$$

One notes that the free energy of the macromolecule in this approach is given by

$$F(r^0, r^1, \ldots, r^N) = \mu T A_{\alpha\gamma} r^\alpha r^\gamma \tag{1.9}$$

and this determines the force on the particle in the first order in r

$$K_i^\nu = -\frac{\partial F}{\partial r_i^\nu} = -2\mu T A_{\nu\gamma} r_i^\gamma \tag{1.10}$$

where ν is the bead number.

In order the expressions (1.6)–(1.10) to be valid, every subchain of the model have to contain a great number of Kuhn segments. When it is determined in this way, the model is called the Gaussian subchain model: it can be generalised in a number of ways. When additional rigidity is taken into account, we have to add the interaction between different particles, so that matrix (1.8) is replaced, for example, by a five-diagonal matrix. It is also possible to take into account the finite extension of subunits by including in (1.9) terms of higher order in r.[1]

The Gaussian subchain model and its possible generalisations are universal models, which can be applied to every macromolecule, irrespective of

---

[1] A reasonable approximation for the force between two adjacent particles is given by the so-called FENE (finitely extendable non-linear elastic) spring force law (Bird et al. 1987a)

$$F_{\text{FENE}} = -\frac{kr}{1 - r/r_{\text{max}}} \tag{1.11}$$

with k and r_{max} denoting the elasticity coefficient and the upper limit for the extension. For the long subchains, when $r_{\text{max}} \to \infty$, the first term of expansion of the FENE force coincides with expression (1.10), so that the coefficient of elasticity ought to be $k = 2\mu T$, but often one uses the FENE force to simulate behaviour of shorter (down to Kuhn length) chains, while choosing a different, empirical coefficient of elasticity (Kremer and Grest 1990; Ahlrichs and Dünweg 1999; Paul and Smith 2004).

its chemical composition, which is long enough. It does not mean that the number of subchains N has to be very big. Indeed, at $N = 1$, the subchain model becomes the simplest model of a macromolecule: a dumbbell with two beads connected by elastic force. At large N the description can be simplified. Instead of discrete label α of the co-ordinate, a continuous label

$$s = \frac{\alpha}{N+1}, \quad 0 \le s \le 1$$

can be introduced, and the matrix A expressed by (1.8) can be represented as the operator

$$A \approx -\frac{1}{N^2}\frac{d^2}{ds^2}. \tag{1.12}$$

This allows one to rewrite expressions, considered here and later, in other forms and to fulfil analytical calculations more easily. In this monograph, however, we prefer to use the matrixes, bearing in the mind that the theory can be also applied to produce numerical calculations at small numbers N.

The Gaussian subchain model and its possible generalisations allows one to calculate, in a coarse-grained approximation, the different characteristics of a macromolecule and systems of macromolecules, playing a fundamental role in the theory of equilibrium and non-equilibrium properties of polymers. The model does not describe the local structure of the macromolecule in detail, but describes correctly the properties on a large-length scale.

1.3 Normal Co-Ordinates

The equilibrium and non-equilibrium characteristics of the macromolecular coil are calculated conveniently in terms of new co-ordinates, so-called normal co-ordinates, defined by

$$r^\beta = Q_{\beta\alpha}\rho^\alpha, \quad \rho^\alpha = Q_{\alpha\gamma}^{-1}r^\gamma, \tag{1.13}$$

such that the quadratic form in equations (1.6) and (1.9) assumes a diagonal form

$$Q_{\lambda\mu}A_{\lambda\gamma}Q_{\gamma\beta} = \lambda_\mu\delta_{\mu\beta}. \tag{1.14}$$

It can readily be seen that the determinant of the matrix given by (1.8) is zero, so that one of the eigenvalues, say λ_0, is always zero. The normal co-ordinate corresponding to the zeroth eigenvalue

$$\rho^0 = Q_{0\gamma}^{-1}r^\gamma$$

is proportional to the position vector of the centre of the mass of a macromolecular coil

$$q = \frac{1}{1+N}\sum_{\alpha=0}^{N}r^\alpha. \tag{1.15}$$

It is convenient to describe the behaviour of a macromolecule in a co-ordinate frame with the origin at the centre of the mass of the system. Thus $\rho^0 = 0$ and there are only N normal co-ordinates, numbered from 1 to N.

The transformation matrix $\mathbf{Q}$ can be chosen in a variety of ways, which allow us to put extra conditions on it. Usually, it is assumed orthogonal and normalised. In this case, it can be demonstrated (see, for example, Dean 1967) that the components of the transformation matrix and the eigenvalues are defined as

$$Q_{\alpha\gamma} = \left(\frac{2 - \delta_{0\gamma}}{N+1}\right)^{\frac{1}{2}} \cos\frac{(2\alpha+1)\gamma\pi}{2(N+1)}, \qquad \lambda_\alpha = 4\sin^2\frac{\pi\alpha}{2N}. \tag{1.16}$$

For large N and small values of α, the eigenvalues are then given by

$$\lambda_\alpha = \left(\frac{\pi\alpha}{N}\right)^2, \qquad \alpha = 0, 1, 2, \ldots, \ll N. \tag{1.17}$$

In the case of an orthogonal transformation, the relationship between the normal co-ordinate corresponding to the zeroth eigenvalue and the position of the centre of mass of the chain is

$$\rho^0 = q\sqrt{1+N}. \tag{1.18}$$

The distribution function (1.6), normalised to unity, then assumes the following form

$$W(\rho^1, \rho^2, \ldots, \rho^N) = \prod_{\gamma=1}^{N} \left(\frac{\mu\lambda_\gamma}{\pi}\right)^{\frac{3}{2}} \exp(-\mu\lambda_\gamma \rho^\gamma \rho^\gamma). \tag{1.19}$$

The probability distribution function allows us readily to calculate equilibrium moments of the normal co-ordinates

$$\langle \rho_i^\nu \rho_k^\nu \rangle = \int W \rho_i^\nu \rho_k^\nu \{d\rho\} = \frac{1}{2\mu\lambda_\nu}\delta_{ik},$$
$$\langle \rho_i^\nu \rho_k^\nu \rho_s^\nu \rho_j^\nu \rangle = \frac{1}{4(\mu\lambda_\nu)^2}(\delta_{ik}\delta_{sj} + \delta_{is}\delta_{kj} + \delta_{ij}\delta_{ks}). \tag{1.20}$$

In a case of a general transformation, relations (1.16) and (1.17) are not valid and ought to be replaced by other relations. A non-orthonormal transformation matrix was used at investigation of non-equilibrium properties of the macromolecule in a liquid when so-called hydrodynamic interaction was taking into account (Zimm 1956).

1.4 Macromolecular Coil

The subchain model gives a more detailed description of a macromolecule and allows one to introduce, in line with the end-to-end distance $\langle R^2 \rangle = Nb^2$, another characteristic of the macromolecular coil – the mean square radius of gyration

$$\langle S^2 \rangle = \frac{1}{1+N} \sum_{\alpha=0}^{N} \langle (r^\alpha - q)^2 \rangle, \quad q = \frac{1}{1+N} \sum_{\alpha=0}^{N} r^\alpha. \tag{1.21}$$

This quantity, as it is followed from the above definitions, can be also calculated as

$$\langle S^2 \rangle = \frac{1}{2(1+N)^2} \sum_{\alpha,\gamma=0}^{N} \langle (r^\alpha - r^\gamma)^2 \rangle. \tag{1.22}$$

In the normal co-ordinates (1.13), in the case of the orthogonal transformation, the mean square radius of gyration of the macromolecule (1.21) is expressed in equilibrium moments

$$\langle S^2 \rangle = \frac{1}{1+N} \sum_{\alpha=1}^{N} \langle \rho_i^\alpha \rho_i^\alpha \rangle.$$

The formulae (1.20) allow one to estimate the mean square radius of gyration of the macromolecule

$$\langle S^2 \rangle \approx \frac{1}{6} \langle R^2 \rangle = \frac{N}{6} b^2.$$

An important property of the Gaussian chain is that the distribution of the distance between any two particles of the chain is Gaussian and is similar to function (1.5). So, the mean values of the functions of the vector $r^\alpha - r^\alpha = e^{\alpha\gamma} |r^\alpha - r^\gamma|$ where α and γ are the labels of the particles of the chain, can be calculated with the help of the distribution function

$$W(r^\alpha - r^\gamma) = \left(\frac{3}{2\pi |\alpha - \gamma| b^2} \right)^{3/2} \exp \left\{ -\frac{3(r^\alpha - r^\gamma)^2}{2\langle (r^\alpha - r^\gamma)^2 \rangle} \right\},$$

which allows one to calculate averaged values of various quantities, for example,

$$\langle (r^\alpha - r^\gamma)^2 \rangle = |\alpha - \gamma| b^2,$$

$$\left\langle \frac{1}{|r^\alpha - r^\gamma|} \right\rangle_0 = \frac{1}{b} \left(\frac{6}{\pi |\alpha - \gamma|} \right)^{1/2}, \tag{1.23}$$

$$\left\langle \frac{e_i^{\alpha\gamma} e_j^{\alpha\gamma}}{|r^\alpha - r^\gamma|} \right\rangle_0 = \frac{1}{3b} \left(\frac{6}{\pi |\alpha - \gamma|} \right)^{1/2} \delta_{ij}.$$

To characterise the size and form of the macromolecular coil, one can introduce a function of density of the number of particles of the chain

$$\rho(r) = \sum_{\nu=0}^{N} \langle \delta(r - r^\nu) \rangle,$$

where r^ν is the co-ordinate of particle ν, and r is the separation from the mass centre of the coil. At equilibrium, one considers the macromolecular coil to have spherical symmetry. The effective radius of the macromolecular coil is assumed to be equal to the mean radius of inertia of the coil $\langle S^2 \rangle$ which is determined by equation (1.21). A spherical-symmetrical distribution function of the density of the macromolecular coil $\rho(r)$, where r is the vector from the centre of the coil, can now be introduced by relations

$$\int \rho(r)\mathrm{d}r = N, \qquad \frac{1}{N} \int \rho(r) r^2 \mathrm{d}r = \langle S^2 \rangle.$$

This allows us to approximate the function $\rho(r)$, for example, by a two-parameter exponential function

$$\rho(r) = \left(\frac{3}{2\pi \langle S^2 \rangle} \right)^{3/2} N \exp\left(-\frac{3}{2} \frac{r^2}{\langle S^2 \rangle} \right). \qquad (1.24)$$

1.5 Excluded-Volume Effects

One says that the above results are valid for a chain with non-interacting particles. However, the monomers in a real macromolecule interact with each another, and this ensures, above all, that parts of the molecule cannot occupy the place already occupied by other parts; i.e. the probabilities of successive steps are no longer statistically independent, as was assumed in the derivation of the above probability distribution functions and mean end-to-end distance (Flory 1953). So, considering the coarse-grained model, one has to introduce lateral forces of attractive and repulsive interactions. The potential energy of lateral interactions U depends on the differences of the position vectors of all particles of the chain and, in the simplest case, can be written as a sum of pair interactions

$$U = \frac{1}{2} \sum_{\nu=0}^{N} \sum_{\substack{\gamma=0 \\ \gamma \neq \nu}}^{N} u\left(|r^\gamma - r^\nu| \right). \qquad (1.25)$$

The presentation of potential in this form can be apparently justified only for large numbers N of subchains.

The effective potential $u(r)$ between two fictious particles of the chain can be chosen in a convenient form. For analytical calculations, the potential function is approximated (Doi and Edwards 1986) by the delta function,

$$u(r) = vT\delta(r). \qquad (1.26)$$

The parameter v has the dimension of volume and is called the excluded volume parameter. The above approximation of repulsive force can be apparently valid for a long macromolecule, when a very large number of subchains

N can be introduced. For a finite number of subchains, the potential can be approximated (Öttinger 1995) by a Gaussian function

$$u(r) = \frac{vT}{(2\pi\sigma^2)^{3/2}} \exp\left(-\frac{r^2}{2\sigma^2}\right),$$

where the parameter of interaction σ depends on the number of subchains of a macromolecule in such a way, that, at $N \to \infty$, $\sigma \to 0$, and repulsive potential turns into function (1.26). The dependence ought to be chosen in such a way, that properties of macromolecular coil do not depend on number of division of macromolecule into subchains.[2]

For the subchain model under consideration, an equilibrium distribution function that includes the particle interaction potential, takes the form

$$W = C \exp\left(-\mu A_{\alpha\gamma} r^\alpha r^\gamma - \frac{1}{T}U\right) \qquad (1.27)$$

where C is the normalisation constant. The definition of the quantity μ in (1.27) does not coincide with expression (1.7), so as the internal interactions are taken into account, but nevertheless the quantity can be expressed, on the basis of scaling speculations, through the mean end-to-end distance of a subchain as

$$\mu \sim b^{-2}.$$

The free energy of a macromolecule, instead of (1.9), is given by

$$F(r^0, r^1, \dots, r^N) = \mu T A_{\alpha\gamma} r^\alpha r^\gamma + U(r^0, r^1, \dots, r^N). \qquad (1.28)$$

However, if one is not interested in observing the variables $r^0, r^1, \dots, r^N$ at all, the independent on these parameters free energy can be defined. This quantity can be calculated, starting from expression (1.25) and (1.27), so that it depends on the parameters T, N, b, v, whereby the arbitrary quantity N cannot influence the free energy of the macromolecular coil and the explicit

---

[2] The problem of how to chose the effective potential for simulation purposes was recently discussed by Müller-Plathe (2002). At least, one parameter σ with dimension of length is usually included also in the function $u(r)$. The magnitude of interaction decreases when N increases, so that, for long chains, the potential can be presented in universal form as

$$u(r) = \frac{T}{N^\eta} v\left(\frac{r}{\sigma}\right).$$

The universality also assumes that $\sigma \sim b$. The index η can be estimated, when one calculates free energy of the coil. Specifically, the Lennard-Jones potential

$$u(r) = 4\epsilon\left[\left(\frac{\sigma}{r}\right)^{12} - \left(\frac{\sigma}{r}\right)^6 + \frac{1}{4}\right], \quad r < 2^{1/6}\sigma$$

is often used (Kremer and Grest 1990; Ahlrichs and Dünweg 1999; Paul and Smith 2004) to describe interaction between particles.

dependence on arbitrary parameter N has to be excluded. So, after dimensional considerations has been taken into account, one has to write free energy of the coil as a function of the only parameter

$$F(T) = Tg\left(\frac{v}{b^3}\right). \tag{1.29}$$

A relation between the mean end-to-end distance of the entire chain $\langle R^2 \rangle$ and the mean end-to-end distance of a subchain b can be found from simple speculation. This relation includes temperature T, mean distance b between the nearest along chain particles, excluded volume parameter v and the number of particles on the chain N. When dimensional considerations are taken into account, the relation can be written in the form

$$\langle R^2 \rangle = b^2 f\left(N, \frac{v}{b^3}\right). \tag{1.30}$$

Of course, the end-to-end distance of the entire macromolecule $\langle R^2 \rangle$ does not depend on the arbitrary number of subchains N at $N \to \infty$, when the ratio v/b^3 is constant. This means that the relation between $\langle R^2 \rangle$ and a finite number of subchains should be written in a way, which keeps the form of the relation under repeating divisions of the macromolecule, so that the mean square end-to-end distance of the macromolecule has to be written as a power function

$$\langle R^2 \rangle \sim N^{2\nu} b^2.$$

It is easy to see that this relation is valid for an arbitrary number of divisions. Thus, general consideration leads to the power dependence of the end-to-end distance of the macromolecule on its length

$$\langle R^2 \rangle \sim M^{2\nu}. \tag{1.31}$$

We can guess that the dimensions of a macromolecular coil with the excluded-volume effect are larger than those of the ideal coil, so that $\nu \geq 1/2$. However, it is necessary to fulfil a number of special and sophisticated calculations to find a specific value of power 2ν in expression (1.31) (Alkhimov 1991). The first estimates of the index (Flory 1953; Edwards 1965) were done by simple self-consistent methods. Then the mean end-to-end distance was calculated by a perturbation method, while the chain in a imaginable 4-dimensional space is considered to be non-perturbed, and deviation of dimensionality of the imaginable space from the real physical space ϵ is believed to be the small parameter of expansion. The first-order term gives (Gabay and Garel 1978) the following value of index

$$2\nu = \frac{9}{8}.$$

The answer is known to many decimal places (Alkhimov 1991).

A great deal of effort has been expended in attempts to find a distribution function for the end-to-end length of the chain (Valleau 1996). Oono et al. (1981) have shown that in the simplest approximation, the distribution function for non-dimensional quantity $R^2/\langle R^2 \rangle$ is close to Gaussian, so the above results allow one to write down an expression for the elasticity coefficient, when the excluded-volume effect is taken into account, in the form

$$\mu \sim \left(\frac{N}{M}\right)^{2\nu}.$$ (1.32)

It is an approximation; in fact, the index in (1.32) is slightly different from 2ν.

The lateral forces depend on temperature: at high temperatures the repulsion interactions between particles prevail; on the contrary, at low temperatures the attraction interactions prevail, so that there is a temperature at which the repulsion and attraction effects exactly compensate each other. This is the θ-temperature at which the second virial coefficient is equal to zero. It is convenient to consider the macromolecular coil at θ-temperature to be described by expressions for an ideal chain, those demonstrated in Sections 1.1–1.4. However, the old and more recent investigations (Grassberger and Hegger 1996; Yong et al. 1996) demonstrate that the last statement can only be a very convenient approximation. In fact, the concept of θ-temperature appears to be immensely more complex than the above picture (Flory 1953; Grossberg and Khokhlov 1994).

1.6 Macromolecules in a Solvent

The picture considered in the previous section is idealised one: the macromolecule does not exist in isolation but in a certain environment, for example, in a solution, which is dilute or concentrated in relation to the macromolecules (Des Cloizeaux and Jannink 1990). The important characteristic for the case is the number of macromolecules per unit of volume n which can be written down through the weight concentration of polymer in the system c and the molecular weight (or length) of the macromolecule M as

$$n = 6.026 \times 10^{23} \frac{c}{M} \text{ cm}^{-3}.$$ (1.33)

The mean distance between the centres of adjacent macromolecular coils $d \approx n^{-1/3}$ can be compared with the mean squared radius of gyration of the macromolecular coil $\langle S^2 \rangle$, which presents the mean dimension of the coil. Taking the definition (1.21) into account, one can see, that a non-dimensional parameter $n\langle R^2 \rangle^{3/2}$ is important for characterisation of polymer solutions. The condition

$$n\langle R^2 \rangle^{3/2} \approx 1$$ (1.34)

defines the critical molecular weight for a given concentration, or the critical concentration of the solution for a given molecular weight, at which the coils

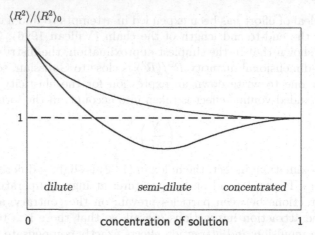

Figure 1. A macromolecular coil in a good solvent.
The curves illustrate two variants of the concentration dependence of the mean size of a macromolecular coil in solution. The example is taken of a macromolecule in a good solvent, so that at low concentrations the size of the macromolecular coil is larger than the size of ideal coil, $\langle R^2 \rangle / \langle R^2 \rangle_0 > 1$.

---

begin to overlap. However, the mean square end-to-end distance $\langle R^2 \rangle$ of the macromolecule itself depends on concentration c and molecular length M. The possible concentration dependencies of the mean square end-to-end distance of the macromolecule are depicted on Fig. 1. The increase in the concentration of the polymer from dilute to very concentrated solution can be accompanied by the mutual interpenetrating or repulsion of the macromolecular coils (Erukhimovich et al. 1976).

1.6.1 Macromolecules in a Dilute Solution

The condition for a polymer solution to be dilute can be written as

$$n\langle R^2 \rangle^{3/2} \ll 1$$

Macromolecules in dilute solutions ($c \ll 1$) can be considered as not interacting with each other, though this is not always valid (Kalashnikov 1994; Polverary and de Ven 1996).

To consider the behaviour of a single macromolecule in the solution, the interaction of the atoms of the macromolecule with the atoms of solvent molecules has to be taken into account, apart from the interactions between the different parts of the macromolecule. To find the distribution function for the chain co-ordinates, one ought to consider $N + 1$ "big" particles of chain interacting with each other and each with "small" particles of solvent. One can anticipate that after eliminating the co-ordinates of the small particles in the distribution function, the distribution function of the chain co-ordinates can be taken in the form (1.27). In this case, the energy potential of the particle U is an effective potential, while taking into account both the interaction of the atoms of the macromolecule with the atoms of solvent and the interaction of the atoms of the macromolecule with each other. The speculations and results of the previous section are valid for the considered case.

From the energy point of view polymer–solvent contacts as compared with polymer–polymer contacts are preferred for some solvents called "good" solvents in this situation. A macromolecular coil swells and enlarges its dimension in a "good" solvent. On the contrary in a "bad" solvent, a macromolecular coil decreases in its dimension and can collapse, turning into a condensed globule (Flory 1953; Grossberg and Khokhlov 1994).

The second virial coefficient of the macromolecular coil $B(T)$ depends not only on temperature but on the nature of the solvent. If one can find a solvent such that $B(T) = 0$ at a given temperature, then the solvent is called the θ-solvent. In such solvents, roughly speaking, the dimensions of the macromolecular coil are equal to those of an ideal macromolecular coil, that is the coil without particle interactions, so that relations of Sections 1.1–1.4 can be applied to this case, as a simplified description of the phenomenon.

1.6.2 Weakly-Coupled Macromolecules

In the alternative case, when the solution is very concentrated, that is the condition

$$n\langle R^2 \rangle^{3/2} \gg 1$$

is satisfied, the system of linear interacting macromolecules can exist in various physical states, depending on temperature. Transition points from the fluid state to the crystalline and/or glassy state are different for different polymers (Table 1). Further on, in this monograph, we shall consider the systems at temperatures exceeding the characteristic crystallisation and glass points, T_c and T_g, so that the system, schematically depicted in Fig. 2, can be considered to be fluid. Either macromolecule in the system can only move as freely as its macromolecular neighbours allow it to. Similar to entangled ropes, polymer chains in a concentrated system can slide past but not through each other. These topological constrains, the so-called entanglements lead to a specific interaction between the macromolecules in the system, to the formation of sites and tangles (Kholodenko and Vilgis 1998; Marcone et al. 2005). An analysis of the entangled polymer system by numeric simulations discovers a certain topological structure – primitive path mesh (Everaers et al. 2004; Kremer et al. 2005).

To describe the system in the coarse-grained approximation, the position of each macromolecule can be defined, as before, by specifying certain points along the macromolecule, spaced at distances that are equal, but not too small; as before, we shall refer to these points as particles. If one takes $N + 1$

TABLE 1. Characteristics of typical polymers

Polymer	Density in flow state, g cm^{-3}	Transition temperature, °C crystal	glass	M_e
Polystyrene	0.962	–	100	18 100
Poly-α-methylstyrene	–	–	168	13 500
Polybutadiene	0.910	−25	−110	2 200
Poly(vinyl acetate)	1.380	–	32	12 000
Poly(dimethyl siloxane)	0.960	−70	−123	8 100
Polyethylene	0.767	140	−20	5 100
Polyisoprene (natural rubber)	0.910	10	−71	5 800
Poly(methyl methacrylate)	1.380	–	100	5 900
Polyisobutylene	0.812	24	−55	8 900

points to define the position of the macromolecule, $3n(N + 1)$ co-ordinates are needed to specify the state of the entire system. Let us note that due to a great number of Kuhn segments in a separate subchain, the number density of Brownian particles is much less than the number density of the segments, so that the system of Brownian particles can be considered as dilute. The equilibrium distribution function of the system can be written as

$$W = C \exp \left(-\mu \sum_{a=1}^{n} A_{\gamma\nu} r^{a\gamma} r^{a\nu} - \frac{U}{T} \right) \tag{1.35}$$

where $r^{a\gamma}$ is the co-ordinate of the γth particle of a macromolecule labelled a or, in short, the co-ordinate of the particle $a\gamma$; the matrix A and the quantity μ are given by (1.8) and (1.32), respectively. The potential energy U associated with the "lateral" interaction between the chains depends on the differences between the co-ordinates of all the particles in the system, but in contrast to the case of a single macromolecule, described in Section 1.4, it is doubtful that the potential can be written as a sum of pair interactions.

For concentrations approaching the limiting value $(c \to 1)$, the system of macromolecular coils becomes homogeneous in space. The presence of other coils changes the potential of interaction between two particles of the same chain in such a way, that the interactions between particles of a chosen macromolecule in a highly entangled system could be neglected. This remarkable phenomenon – excluded-volume-interaction screening – was guessed by Flory (1953) and strictly confirmed by Edwards in mid sixties (Doi and Edwards 1986). This means that, for description of every macromolecule in a strongly entangled system, one can use the distribution function for ideal chains

$$W = C \exp(-\mu A_{\gamma\nu} r^{a\gamma} r^{a\nu}), \quad a = 1, 2, \ldots, n. \tag{1.36}$$

Figure 2. Schematisation of amorphous polymeric material.
Macromolecules are coupled with weak van der Waals forces. At $T > T_g$, the system is in a fluid state, and the thermal motion makes the macromolecules move besides each other. At $T < T_g$, the system is in a glassy state, large-scale conformations of the chains are frozen, and a macromolecule can change its neighbours at the deformation of the material only.

This expression could be found by integrating function (1.35) with respect to the co-ordinates of the particles of all the macromolecules apart from the particles of a chosen one.[3] In this situation, one can use the results for ideal coil, so that one can write for the parameter

$$n\langle R^2 \rangle^{3/2} \sim cM^{1/2}. \tag{1.37}$$

Free energy of the system in volume V, due to general relation (1.35), depends on the parameters n, T, V, N, b and parameters of interaction, whereby the arbitrary quantity N cannot influence the free energy of the system. So, after dimensional considerations has been taken into account, one has to write free energy for unit of volume

$$F(T) = nTg\left(\frac{v}{b^3}, n\langle R^2 \rangle^{\frac{3}{2}}\right). \tag{1.38}$$

The additional, in comparison with equation (1.29), parameter has appeared.

Discussions of dynamic phenomena in polymer melts are frequently based on assumptions about a structure of the system, which was earlier taken to be

---

[3] Experiments due to neutron scattering by the labelled macromolecules allow one to estimate the effective size of macromolecular coils in very concentrated solutions and melts of polymers (Graessley 1974; Maconachie and Richards 1978; Higgins and Benoit 1994) and confirm that the dimensions of macromolecular coils in the very concentrated system are the same as the dimensions of ideal coils. It means, indeed, that the effective interaction between particles of the chain in very concentrated solutions and melts of polymers appears changes due to the presence of other chains in correspondence with the excluded-volume-interaction screening effect. The recent discussion of the problem was given by Wittmer et al. (2007).

a network with a characteristic site lifetime and nearest-neighbour separation (Lodge 1956). Modifications of these presentations retain a certain intermediate scale, as a postulated quantity. The elaborated theory (Doi and Edwards 1986) includes this quantity as the diameter of a tube in which macromolecular displacement, i.e. reptation, is possible, but this hypothetical intermediate scale has been detected only in dynamic phenomena, and its existence should be regarded as a consequence rather than the origin of the theory. The theory that we shall consider in subsequent chapters does not rely on the assumption of an intermediate scale, but it does assume that the mean size $\langle R^2 \rangle$ and the macromolecular number density n (or concentration c) are the most significant static parameters of the system. However, an intermediate dynamical length ξ will appear in our theory later (see Section 5.1.2). The intermediate length is closely related to "the length of a macromolecule between adjacent entanglements" M_e (see examples in Table 1). The ratio M/M_e, for which we shall use a special symbol Z, appears to be a very important parameter in the theory of dynamic behaviour of entangled system.

We shall not discuss here the spatial correlation functions introduced by Daoud et al. (1975) for a more detailed description of the system structure and of the relative position of monomers, since they are relatively unimportant in our, admittedly very coarse, approximation.

1.7 Elasticity of Polymer Networks

One can imagine that macromolecules in the a dense system can be connected to each other at some points by a chemical agent. In this way, there appear polymer networks consisting of long chains, connected by the chemical bonds in an entire body. As a rule, such structures are non-regular; networks have free dangling ends, tangles; chains between adjacent junctions have different lengths, and so on. Nevertheless it is convenient to study the behaviour of a perfect network with the regular structure of chains of equal length and without any defects, as shown in Fig. 3. The mobility of the macromolecules is now restricted and the more the density of junctions ν, the more the restrictions. The mobility of the junctions about their mean positions is severely restricted too.

The main subject of the following discussion is the mechanical behaviour of networks in terms of the behaviour of the system of weakly coupled macromolecules. The network modulus of elasticity is small in comparison to the values of the elasticity modulus for low-molecular solids (Dušek and Prins 1969; Treloar 1958). Nevertheless, large (up to 1000%) recoverable deformations of the networks chains are possible.

In this section we consider the simplest approach to the thermodynamics of a deformed network, for which a tensor of displacement gradients is given by

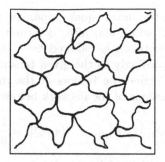

Figure 3. Schematisation of an ideal polymer network.
Every knot connects four chains by chemical bonds. At $T > T_g$, both chains and knots take part in thermal motion as Brownian particles, at $T < T_g$, the network is in a glassy state, large-scale conformations of the chains are frozen, the motion of the knots is negligible.

$$\lambda_{ik} = \frac{\partial x_i}{\partial x_k^0}, \tag{1.39}$$

where x_j^0 and x_j are the co-ordinates of a body point, before and after deformation, respectively.

In the ideal case, when one considers the network of chains of equal lengths, the stresses under the given deformation can be obtained in a very simple way. In virtue of the speculations of the previous section, free energy of the whole network can be represented as the sum of free energy of all the chains, while each of the equal chains of the network can be characterised by the same equilibrium distribution function $W(s)$, where s is the separation between adjacent junctions. In the state without deformation, the function has the form (1.5), while in a deformed state, it depends on the displacement gradients (1.39). The free energy of the whole network can be written down simply as

$$F = \nu TV\mu \int W(s)s^2 ds, \tag{1.40}$$

where ν is the number of chains in the volume unit of the network, so that νV is the number of chains in the whole network. The coefficient of elasticity μ is the same for each chain and is defined as

$$\mu = \frac{3}{2\langle s^2 \rangle_0},$$

where $\langle s^2 \rangle_0$ is the equilibrium mean square separation between adjacent junctions.

The equation (1.40) can be rewritten in the form

$$F = F_0(T, V) + \nu TV\mu(\langle s^2 \rangle - \langle s^2 \rangle_0), \tag{1.41}$$

where $\langle s^2 \rangle_0$ and $\langle s^2 \rangle$ are the mean square separations between the adjacent junctions of the network before and after deformation. The assumption of a regular structure of a network is followed by a statement that every internal length changes accordingly to the given tensor of displacement gradients (1.39). So, one can define the tensors of the mean square separations between the adjacent junctions of the network before and after deformation

$$\langle s_i s_k \rangle = \frac{1}{3} \lambda_{ij} \lambda_{kj} \langle s^2 \rangle_0, \qquad \langle s_i s_k \rangle_0 = \frac{1}{3} \delta_{ik} \langle s^2 \rangle_0.$$

It is convenient to introduce, following to Murnaghan (1954), the tensor of deformation $\Lambda_{ik} = \lambda_{si} \lambda_{sk}$. The latter is useful to write down a free energy function (1.41) of a deformed network as

$$F = F_0(T, V) + \frac{1}{2} \nu TV (\Lambda_{ii} - 3). \tag{1.42}$$

Then the relation between the free energy of body and the stress tensor (Appendix B, equation (B.7)) can be used

$$\sigma_{ik} = \frac{2}{V} \lambda_{kl} \lambda_{ij} \left(\frac{\partial F}{\partial \Lambda_{jl}} \right)_T.$$

The formula for the stress tensor of a deformed network, as follows from the above relations, is given by

$$\sigma_{ik} = -p \delta_{ik} + \nu T (\lambda_{ij} \lambda_{kj} - \delta_{ik}) \tag{1.43}$$

where p is the thermodynamic pressure, and ν is the number of chains in the volume unit of the network. Relation (1.43) defines the shear modulus of a polymer network as

$$G = \nu T. \tag{1.44}$$

Expression (1.44) is useful to estimate the number density of the active chains of the network, due to the measured temperature T and shear modulus G.

The above results, which were formulated in the early thirties (Treloar 1958), explain the main features of the elastic behaviour of polymer networks. Nevertheless, there are notable discrepancies between empirical data and the cited results. Further investigations demonstrated that the values of shear modulus and stress–strain dependence are determined substantially by topological constraints due to the proximity of the chains. The theory was improved by taking into account the discussed issue (Edwards 1967a, 1967b, 1969; Flory 1977; Erman and Flory 1978; Priss 1957, 1980, 1981). More recent developments are summarised in the work of Panyukov and Rabin (1996), where many additional relevant references could be found.

Figure 4. Schematisation of crystalline polymer material.
Crystallite sizes are of order 10^2–10^3 Å. The amorphous parts of the material are in a
fluid state at $T > T_g$ and in a glassy state at $T < T_g$.

1.8 Crystalline and Glassy Systems

In Section 1.6.2 we assumed that the temperature is high enough to consider
the van der Waals' interaction between chains to be small, so the chains (not
to consider chemical crosslinks) can move freely beside each other. To complete the picture, we will shortly describe the structure and behaviour of the
systems, described in Section 1.6.2, at lower temperatures. In contrast to the
previous cases, interchain interaction is not small, so that the mobility of the
macromolecule is very severely restricted by neighbouring macromolecules.
One can observe, instead of a fluid state, a crystalline and/or glassy state in
this case (Ferry 1980).

Due to the atomic structure of the macromolecule, some of the polymers
can be in crystalline state at a temperature $T < T_c$. The long macromolecules
are folded several times, thus creating crystallites of definite size for each
polymer (see review: Oleinik 2003). In this case the polymer materials consist
of crystalline and amorphous parts, the latter being in a rubbery state at
$T > T_g$ (Fig. 4). Crystallites are arranged in a special way so that there is
order even on large scales: so-called spherulites exist and can be discovered by
light scattering (Treloar 1958). Though the structure of the material is rather
complex, there is an important characteristic of a semi-crystalline body – the
volume fraction of the crystalline part χ $(0 < \chi < 1)$, which correlates with
the density of the material quite well.

In the temperature region between melt and glass points $T_c > T > T_g$, the
modulii of elasticity of the amorphous and crystalline parts differ very much
(Tashiro 1993), so the deformation of relatively soft and mobile amorphous
areas consisting of chain sections anchored in a crystal lattice of adjacent
crystalline region is large in comparison with the crystalline parts. So, small
(up to 10–20%) elastic deformations of the whole body can be attributed to
the deformation of the amorphous parts of the material (Zgaevskii 1977). One
can suppose that the crystallite modulus is high enough in comparison to the

modulus of the amorphous part. For intermediate values of the volume fraction of crystalline part χ, the assumption leads to the expression for modulus of semi-crystalline polymer material

$$G = \nu T \left[1 + \frac{\chi^2}{(1 - \chi)^2}\right] \tag{1.45}$$

where ν is the number density of the active chains of the amorphous part of the material. Relation (1.45), which is the generalisation of formula (1.44), is a particular case of the more general relation (Zgaevskii 1977). This model can be also used to calculate photoelastic properties of semi-crystalline polymer material (Patlazhan 1993).

At far lower temperatures $T < T_g$, polymer materials (no matter whether they are crystalline or amorphous) become rigid; one says that the polymer materials transit into the glassy state. The interaction between different macromolecules in glassy materials is not weak. The free energy of materials cannot be written down as the sum of chain free energies, as was the case for the higher temperatures considered in Sections 1.6.2 and 1.7. A macromolecule in a glassy material cannot move, the large-scale conformations of macromolecules are frozen. The structure of the material can be imagined as a very dense network capable of small elastic deformations (Laius and Kuvshinskii 1963; Shishkin et al. 1963).

Glassy materials can be deformed without fracture up to hundreds of percents. Under this so-called forced deformation, the conformations of macromolecules change but the structure of the dense network does not change. After the stresses have been removed, the sample is still in a deformed state, which is a metastable equilibrium state. The macromolecules in such a sample are frozen in a deformed state, so that on heating, when the crosslink number density decreases, forces are exerted and the sample can return to its initial form (Laius and Kuvshinskii 1963; Shishkin et al. 1963).

Chapter 2
Dynamics of a Macromolecule in a Viscous Liquid

Abstract In this chapter, the dynamics of the macromolecule moving in the uniform flow of a viscous liquid will be considered. To be accurate, one ought to consider a system consisting of a macromolecule and molecules of solvent (microscopic approach). However, since we are interested in large-scale or low-frequency dynamics of a macromolecular coil, a bead-spring model of a macromolecule can be used and molecules of solvent can be considered to constitute a continuum – a viscous liquid. This is a mesoscopic approach to the dynamics of dilute solutions of polymers. The approach provides the simplest model that appears to be rather complex, if the effects of excluded volume, hydrodynamic interaction, and internal viscosity are taken into account. Due to these effects, all the Brownian particles of the chain ought to be considered to interact with each other in a non-linear way. There is no intention to collect all the available results and methods concerning the dynamics of a macromolecule in viscous liquid in this chapter. We need to consider the results for dilute solutions mainly as a background and a preliminary step to the discussion of the dynamics of a macromolecule in very concentrated solutions and melts of polymers.

2.1 Equation of Macromolecular Dynamics

The theory of relaxation processes for a macromolecular coil is based, mainly, on the phenomenological approach to the Brownian motion of particles. Each bead of the chain is likened to a spherical Brownian particle, so that a set of the equation for motion of the macromolecule can be written as a set of coupled stochastic equations for coupled Brownian particles

$$m\frac{\mathrm{d}^2 r^\alpha}{\mathrm{d}t^2} = F^\alpha + G^\alpha + K^\alpha + \phi^\alpha, \quad \alpha = 0, 1, \ldots, N \qquad (2.1)$$

where m is the mass of a Brownian particle associated with a piece of the macromolecule of length $M/(N+1)$, r^α are the co-ordinates of the Brownian

V.N. Pokrovskii, *The Mesoscopic Theory of Polymer Dynamics*,
Springer Series in Chemical Physics 95,
DOI 10.1007/978-90-481-2231-8_2, © Springer Science+Business Media B.V. 2010

particles. Every Brownian particle is involved in thermal motion, which, as usual (Chandrasekhar 1943; Gardiner 1983; Doi and Edwards 1986), can be described by putting a stochastic force ϕ^α (for a particle labelled α) into an equation of motion of a macromolecule. The essential features of the stochastic force are connected with properties of the dissipative forces $\boldsymbol{F}^\alpha$ and $\boldsymbol{G}^\alpha$ (the fluctuation-dissipation theorem). For the linear case, the relation will be discussed in Section 2.6.

According to relations (1.10) and (1.28), the elastic forces acting on the particle are taken in the form

$$\boldsymbol{K}^\alpha = -\frac{\partial F}{\partial \boldsymbol{r}^\alpha} = -2T\mu A_{\alpha\gamma}\boldsymbol{r}^\gamma - \frac{\partial U}{\partial \boldsymbol{r}^\alpha}, \tag{2.2}$$

whereas the dissipative forces $\boldsymbol{F}^\alpha$ and $\boldsymbol{G}^\alpha$ are needed in special discussion.

To say nothing about the different equivalent forms of the theory of the Brownian motion that has been discussed by many authors (Chandrasekhar 1943; Gardiner 1983), there exist different approaches (Rouse 1953; Zimm 1956; Cerf 1958; Peterlin 1967) to the dynamics of a bead-spring chain in the flow of viscous liquid.[1] In this chapter, we shall try to formulate the theory in a unified way, embracing all the above-mentioned approaches simultaneously. Some parameters are used to characterise the motion of the particles and interaction inside the coil. This phenomenological (or, better to say, mesoscopic) approach permits the formulation of overall results regardless to the extent to which the mechanism of a particular effect is understood.

2.2 Intramacromolecular Hydrodynamic Interactions

In the study of the dynamics of the macromolecule in the subchains approximation, each particle of the chain is considered, to a first approximation, to be spherical with a radius a, so that the coefficient of resistance of the particle in a viscous liquid, according to Stokes, can be written as follows

$$\zeta_0 = 6\pi\eta_s a, \tag{2.3}$$

where η_s is viscosity coefficient of the liquid.

Each particle, moving at a velocity $\boldsymbol{u}^\gamma$, is acted upon by the hydrodynamic drag force, which has the form

$$\boldsymbol{F}^\gamma = -\zeta_0(\boldsymbol{u}^\gamma - \boldsymbol{v}^\gamma), \tag{2.4}$$

where $\boldsymbol{v}^\gamma$ is the velocity of the liquid of the point, at which the given particle is present, the velocity corresponding to the situation where no account is taken of the particle. When an assembly of particles is considered, the velocity $\boldsymbol{v}^\gamma$

---

[1] See Ferry (1990) for a short history of development.

is, generally speaking, determined by the motion of all the particles. As can be seen from formulae for forces applied to points (see Appendix C), we have

$$v_j^\gamma = \nu_{jl} r_l^\gamma - \sum_{\beta=0}^{N} H_{jl}^{\gamma\beta} F_l^\beta, \tag{2.5}$$

where $\nu_{jl} r_l^\gamma$ is velocity of the flow in the point r_l^γ in absence of the particles in the liquid, and $\nu_{ij} = \frac{\partial v_i}{\partial x_j}$ is a tensor of velocity gradients[2] of the uniform flow.[3] The components of the hydrodynamic interaction tensor $H_{jl}^{\gamma\beta}$ for $\gamma = \beta$ are zero, while, in the case $\gamma \neq \beta$, they are determined by the relation

$$H_{ik}^{\gamma\beta} = \frac{1}{8\pi\eta_s |r^\gamma - r^\beta|} \left(\delta_{ik} + e_i^{\gamma\beta} e_k^{\gamma\beta} \right) \tag{2.6}$$

where $e_i^{\gamma\beta} = (r_i^\gamma - r_i^\beta)/|r^\gamma - r^\beta|$.

A system of equations for the drag forces follows from equations (2.4) and (2.5)

$$F_j^\alpha = -\zeta_0 (u_j^\alpha - \nu_{jl} r_l^\alpha) - \zeta_0 \sum_\gamma H_{jl}^{\alpha\gamma} F_l^\gamma. \tag{2.7}$$

A solution of equations (2.7) can be written in the form

$$F_j^\alpha = -\zeta_0 B_{jl}^{\alpha\gamma} (u_l^\gamma - \nu_{li} r_i^\gamma) \tag{2.8}$$

where the matrix of hydrodynamic resistance $B_{jl}^{\alpha\gamma}$ is introduced as the matrix inverse to the matrix

$$\delta_{jl}\delta_{\alpha\gamma} + \zeta_0 H_{jl}^{\alpha\gamma}.$$

For small perturbations, the solution of equations (2.7) assumes the following form, to the first approximation

$$F_j^\alpha = -\zeta_0 \sum_\gamma \left[\left(\delta_{\alpha\gamma}\delta_{jl} - \zeta_0 H_{jl}^{\alpha\gamma} \right) (u_l^\gamma - \nu_{li} r_i^\gamma) \right]. \tag{2.9}$$

This equation shows that the resistance-drag force for a certain particle depends on the relative velocities of all the particles of the macromolecule and also on the relative distance between the particles. This expression determines an approximate matrix of hydrodynamic resistance

---

[2] Note that henceforth it will be convenient to use the following notation for the symmetric and antisymmetric tensors of velocity gradients

$$\gamma_{ij} = \frac{1}{2}(\nu_{ij} + \nu_{ji}), \qquad \omega_{ij} = \frac{1}{2}(\nu_{ij} - \nu_{ji}).$$

[3] The dependence of the velocity gradients on the co-ordinates leads to possible migration of macromolecules in a flow (Aubert and Tirell 1980; Brunn 1984) – the effect, which is not discussed in this monograph.

$$B_{jl}^{\alpha\gamma} = \delta_{\alpha\gamma}\delta_{jl} - \zeta_0 H_{jl}^{\alpha\gamma}. \tag{2.10}$$

The exact components of the matrix of hydrodynamic resistance for a two-particle chain are shown in Appendix F.

We note that the values of the hydrodynamic interaction tensor (2.6) averaged beforehand with the aid of some kind of distribution function, are frequently used to estimate the influence of the hydrodynamic interaction, as was suggested by Kirkwood and Riseman (1948).[4] For example, after averaging with respect to the equilibrium distribution function for the ideal coil and taking the relation (1.23) into account, the hydrodynamic interaction tensor (2.6) assumes the following form

$$\langle \zeta_0 H_{ij}^{\alpha\gamma} \rangle = 2h|\alpha - \gamma|^{-1/2}\delta_{ij}.$$

The non-dimensional hydrodynamic interaction parameter appears here

$$h = \frac{\zeta_0\sqrt{6/\pi}}{12\pi\eta_s b} \approx \frac{a}{b} \tag{2.11}$$

where a is the radius of a fictious particle associated with a subchain of length M/N and b is the mean square distance between neighbouring particles along the chain.

One can expect that the parameter of hydrodynamic interaction (2.11) behaves universally for subsequent division of the chain. One can reasonably guess that the quantity (2.11) does not depend on the length of the macromolecule and on the number of subchains. In this case, the hydrodynamic radius of the particle for the Gaussian chain

$$a \sim \left(\frac{M}{N}\right)^{1/2}.$$

The dependence of the friction coefficient of the particle is similar. If the excluded-volume effect is taken into account, the more general relation (2.19) is valid.

2.3 Resistance-Drag Coefficient of a Macromolecular Coil

To calculate the resistance coefficient for the macromolecular coil, we have to determine, first of all, the velocity of the coil, which is the velocity of the mass centre of the macromolecular coil

$$v = \frac{1}{1+N}\sum_{\alpha=0}^{N} u^{\alpha}$$

---

[4] A more general approach can be found in paper of Bixon and Zwanzig (1978).

and the force acting on the coil, which is the sum of forces acting on every particle of the coil

$$F = \sum_{\alpha=0}^{N} F^{\alpha}$$

where the force F^{α} is determined by equation (2.9).

We assume that the macromolecular coil moves in a non-flowing liquid and each particle has the mean velocity of the macromolecular coil as a whole. So, we can write down Stokes law for the coil

$$F_j = -\zeta_0(1+N)\left(\delta_{jl} - \frac{\zeta_0}{8\pi\eta_{\mathrm{s}}}\frac{1}{N+1}\sum_{\alpha=0}^{N}\sum_{\gamma=0}^{N}\frac{\delta_{jl}+e_j^{\alpha\gamma}e_l^{\alpha\gamma}}{|r^{\alpha}-r^{\gamma}|}\right)v_l. \quad (2.12)$$

After the preliminary averaging of the right-hand side of relation (2.12) with respect to the equilibrium distribution function has been done (see relations (1.23)), we have

$$F = -\zeta_{\mathrm{M}}v, \qquad \zeta_{\mathrm{M}} = \zeta_0(1+N)\left(1 - \frac{2h}{N+1}\sum_{\alpha=0}^{N}\sum_{\gamma=0}^{N}\frac{1}{|\alpha-\gamma|^{1/2}}\right).$$

One can easily estimate the asymptotic behaviour of the sum as

$$\frac{1}{N+1}\sum_{\alpha=0}^{N}\sum_{\gamma=0}^{N}\frac{1}{|\alpha-\gamma|^{1/2}} \approx 2.47\,N^{1/2},$$

so that the above equations are followed the asymptotic expression for the friction coefficient

$$\zeta_{\mathrm{M}} = \zeta_0 N\left(1 - 4.94\,hN^{1/2}\right). \quad (2.13)$$

This expression, with accuracy up to the first-order terms in the power of h, practically coincides with expression derived by Kirkwood and Riseman (1948); they have the numerical coefficient 5.33 instead of 4.94, though their way of calculation was more accurate.

It is understandable that the resistance coefficient decreases as the hydrodynamic interaction increases. However, if one uses the bead-spring model of a macromolecule, the resistance coefficient of the whole macromolecule cannot depend on the arbitrary number of subchains N.[5] To ensure this, one has to consider that the product $hN^{1/2}$ does not depend on N which implies that the coefficient of hydrodynamic interaction changes with N as $h \sim N^{-1/2}$ which means, in this situation, that coefficient of resistance of a particle always remains to be proportional to the length of the subchain. All this is valid,

---

[5] Kirkwood and Riseman (1948) did not encounter this problem, because they used the bead-rod or, in other words, pearl-necklace model of macromolecule (Kramers 1946), in which N is a number of Kuhn's stiff segments, so that N present the length of the macromolecule.

when the hydrodynamic interaction is weak and we consider the first-order corrections to the resistance coefficient.[6]

In the general case of arbitrary values of h, simple speculations appeared to be useful to determine the dependence of the resistance coefficient on the length of the macromolecule (Gennes 1979). The excluded-volume effects (see Section 1.5) can also be taken into account and one considers the resistance coefficient to be a function of two non-dimensional parameters h and v/b^3. For a macromolecule, consisting of N smaller subchains, the friction coefficient can be written as

$$\zeta_M = \zeta\, Z\left(N, \frac{v}{b^3}, h\right),$$

where ζ is the friction coefficient of a particle of the chain. To obtain the dependence of the resistance coefficient on the length of the macromolecule, we compare the resistance coefficients for the two different presentations of a macromolecule. One can assume, as was done in Section 1.5 for a similar consideration, that, at $N \to \infty$, the quantities v/b^3 and h do not depend on the number of divisions of the macromolecule into subchains. The requirement of the universality of the representation of the resistance coefficient is followed by the asymptotic (long macromolecules) expression for the dependence of the resistance coefficient on the length of the macromolecule.

$$\zeta_M \sim M^{(z-2)\nu}, \quad 0 < (z-2)\nu \le 1. \tag{2.14}$$

Here ν is the index introduced in relation (1.31), whereas z is a new index, so-called dynamic index. To calculate the index in the power function, it is necessary to use special methods (Al-Naomi et al. 1978; Baldwin and Helfand 1990; Öttinger 1989b, 1990), which gives values from 3 to 4 for the parameter z.

2.4 Effective Resistance-Drag Coefficient of a Particle

One may note that, in linear approximation with respect to the velocity of a particle (see, for example, equations (2.4) and (2.9)), the expression for forces are determined by small velocities of the particles and of the flow. The force, acting on a particle in the flow, does not depend on the specific choice of hydrodynamic interaction and can be written in the following general form

$$F_j^\alpha = -\zeta_0 B_{ji}^{\alpha\gamma}(u_i^\gamma - \nu_{il} r_l^\gamma). \tag{2.15}$$

The resistance matrix depends on co-ordinates of all particles, in non-linear manner. The situation is illustrated in Appendix F for the case of two particles. To avoid the non-linear problem, one uses the preliminary averaging of the hydrodynamic resistance matrix (Kirkwood and Riseman 1948; Zimm

---

[6] The development of the theory of translational mobility of a macromolecule can be found in papers of Dünweg et al. (2002) and Liu and Dünweg (2003).

1956). If one averages with respect to the equilibrium distribution function, the matrix takes the form

$$B_{ji}^{\alpha\gamma} = B_{\alpha\gamma}\delta_{ji} \tag{2.16}$$

where matrix $B_{\alpha\gamma}$ does not depend on the co-ordinates and assumes, under the conditions of weak hydrodynamic interaction, the following form, according to equations (2.10) and (2.11),

$$B_{\alpha\gamma} = \delta_{\alpha\gamma} - 2h|\alpha - \gamma|^{-1/2}. \tag{2.17}$$

This is the first term of expansion in powers of the parameter of hydrodynamic interaction.

When normal co-ordinates, defined by equations (1.13), are employed, it is possible to make use of the arbitrariness of the transform matrix to define matrix Q in such a way that matrix B in the right-hand side of equation (2.16) assumes a diagonal form after transformation. The problem of the simultaneous adjustment of the symmetrical matrices A and B to a diagonal form does have a solution. Since matrix A is defined non-negatively and B is defined positively, it is possible to find a transformation such that B is transformed into a unit matrix (with accuracy to constant multiplier), and A into a diagonal matrix. Therefore, one can write simultaneously the equations

$$\begin{aligned} Q_{\alpha\lambda}A_{\alpha\gamma}Q_{\gamma\nu} &= \lambda_{\nu}\delta_{\lambda\nu}, \\ \zeta_0 Q_{\alpha\lambda}B_{\alpha\gamma}Q_{\gamma\nu} &= \zeta\delta_{\lambda\nu}. \end{aligned} \tag{2.18}$$

One ought to introduce the effective coefficient of friction of the particle ζ into relation (2.18) to ensure the physical dimensionality of the friction coefficient. Eigenvalues λ_{μ} are now defined not by equations (1.16), but by more general expression that will be discussed in Section 2.6.

The dependence of the effective friction coefficient on the length of the macromolecule is of special interest. In a case when the hydrodynamic interaction of the particles of the macromolecule may be neglected, i.e. when the coil is, as it were, free-draining, the coefficient of resistance of the latter is proportional to the length of the macromolecule and the coefficient of friction of the particle associated with length M/N is proportional to this length

$$\zeta \sim \frac{M}{N}.$$

The mutual influence of the particles leads to their shielding within the coil and the overall coefficient of the resistance of the coil proves to be smaller than that for a free-draining coil. The requirement of covariance in relation to successive subdivisions of the macromolecule into subchains gives rise, according to formula (2.14), to the following power dependence for large values of N

$$\zeta \sim \left(\frac{M}{N}\right)^{(z-2)\nu}. \tag{2.19}$$

In order to calculate the power exponents, a calculation based on specific representations, similar the case, when volume-effects are taken into account, is necessary. One notes that these results are valid for infinitely long chains.

2.5 Intramolecular Friction

On the deformation of the macromolecule, i.e. when the particles constituting the chain are involved in relative motion, an additional dissipation of energy takes place and intramolecular friction forces appear. In the simplest case of a chain with two particles (a dumbbell), the force associated with the internal viscosity depends on the relative velocity of the ends of the dumbbell $u^1 - u^0$ and is proportional, according to Kuhn and Kuhn (1945) to

$$-(u_j^1 - u_j^0)e_je_i \tag{2.20}$$

where e is a unit vector in the direction of the vector connecting the particles of the dumbbell and κ is the phenomenological internal friction coefficient.

When a multi-particle model of the macromolecule (Slonimskii–Kargin–Rouse model) is considered, one must assume that the force acting on each particle is determined by the difference between the velocities of all the particles $u^\gamma - u^\beta$. These quantities must be introduced in such a way that dissipative forces do not appear on the rotation of the macromolecular coil as a whole, whereupon $u_j^\alpha = \Omega_{jl} r_l^\alpha$. Thus, in terms of a linear approximation with respect to velocities, the internal friction force must be formulated as follows

$$G_i^\alpha = - \sum_{\beta \neq \alpha} C_{\alpha\beta}(u_j^\alpha - u_j^\beta)e_j^{\alpha\beta}e_i^{\alpha\beta}, \tag{2.21}$$

where $e_j^{\alpha\beta} = (r_j^\alpha - r_j^\beta)/|r^\alpha - r^\beta|$. Matrix $C_{\alpha\beta}$ is symmetrical, the components of the matrix are non-negative and may depend on the distance between the particles. The diagonal components of the matrix are equal to zero.

The internal friction force can also be written in the form

$$G_i^\alpha = -G_{ij}^{\alpha\gamma} u_j^\gamma, \tag{2.22}$$

where the matrix

$$G_{ij}^{\alpha\beta} = \delta_{\alpha\beta} \sum_{\gamma \neq \alpha} C_{\alpha\gamma} e_i^{\alpha\gamma} e_j^{\alpha\gamma} - C_{\alpha\beta} e_i^{\alpha\beta} e_j^{\alpha\beta} \tag{2.23}$$

has been introduced.

The written matrix is symmetrical with respect to the upper and lower indices. Expression (2.23) defines the general form of a matrix of internal friction, which allows the force to remain unchanged on the rotation of the macromolecular coil as a whole. In contrast to matrix $C_{\alpha\beta}$, matrix (2.23) has non-zero diagonal components, which are depicted by the first term in (2.23). Since the components of matrix $C_{\alpha\beta}$ are non-negative, the diagonal components of matrix $G_{\alpha\beta}$ exceed the non-diagonal ones and can be considered to be approximately diagonal to the indices α and β.

Expression (2.22) for an internal friction force is non-linear with respect to the co-ordinates. To avoid the non-linearity, some simpler forms for internal friction force were used (Cerf 1958). One can introduce a preliminary-averaged matrix of internal viscosity

$$\langle G_{ik}^{\alpha\gamma} \rangle = G_{\alpha\gamma}\delta_{ik},$$

where $G_{\alpha\gamma}$ is now a symmetrical numerical matrix which retains the main features of matrix (2.23), so that, instead of equation (2.22), we obtain the following expression for the force

$$G_i^\alpha = -G_{\alpha\gamma}u_i^\gamma.$$

The equation clearly does not satisfy the requirement that the internal viscosity force disappears when the coil is rotated as a whole. By ensuring linearisation of the internal friction force according to Cerf's procedure, equation (2.22) may be modified and written thus

$$G_j^\alpha = -G_{\alpha\gamma}(u_i^\gamma - \Omega_{il}r_l^\gamma). \tag{2.24}$$

The speed of rotation of the macromolecular coil in a flow Ω_{jl} is determined by the velocity gradients

$$\Omega_{jl} = \omega_{jl} + A_{jlsk}\gamma_{sk}.$$

When linear effects are considered, matrix A_{jlsk} can be determined by considering the average rotation of the coil subjected to equilibrium averaging. Since the coil is spherical at equilibrium, it follows from symmetry conditions that

$$\Omega_{jl} = \omega_{jl}$$

to within first-order terms, so that the internal friction force can be written as

$$G_j^\alpha = -G_{\alpha\gamma}(u_j^\gamma - \omega_{jl}r_l^\gamma). \tag{2.25}$$

In terms of the normal co-ordinates introduced by equation (1.13), the matrix of the internal friction can be written as follows

$$Q_{\alpha\lambda}G_{\alpha\gamma}Q_{\gamma\mu} = -\zeta\varphi_\alpha\delta_{\lambda\mu}$$

and for the internal friction force, we have

$$G_j^\alpha = -\zeta\varphi_\alpha(\dot{\rho}_j^\alpha - \omega_{jl}\rho_l^\alpha) \tag{2.26}$$

where ζ is the effective coefficient of friction, φ_α is an internal viscosity coefficient of mode α. It is noteworthy that the representation of the force in the form of equation (2.26) is possible only for weak intramolecular friction, $\varphi_\alpha \ll 1$.

The characteristics $\varphi_\alpha = \varphi_\alpha(M, \alpha)$ of the intramolecular friction forces in equations (2.26), introduced here as phenomenological quantities, should not depend on the method of subdivision of the macromolecule into subchains and, by virtue of the nature of the transformation, should be a function of the ratio α/M. One may expect that φ_α is a monotonically increasing function of the number of the mode α. This dependence can be fitted by

$$\varphi_\alpha = \varphi_1 \alpha^\theta \sim \left(\frac{\alpha}{M}\right)^\theta, \quad \varphi_1 \sim M^{-\theta}, \qquad (2.27)$$

where θ is a positive number and φ_1 is a measure of the internal viscosity.[7] For the considered subchain model, the internal rigidity cannot reach infinity, so it is better to use the following approximation

$$\varphi_\alpha = \frac{\varphi_1 \varphi_\infty \alpha^\theta}{\varphi_\infty + \varphi_1 \alpha^\theta}.$$

The internal viscosity force is defined phenomenologically by equations (2.26) formulated above. Various internal-friction mechanisms, discussed in a number of studies (Adelman and Freed 1977; Dasbach et al. 1992; Gennes 1977; Kuhn and Kuhn 1945; MacInnes 1977a, 1977b; Peterlin 1972; Rabin and Öttinger 1990) are possible. Investigation of various models should lead to the determination of matrices $C_{\alpha\beta}$ and $G_{\alpha\beta}$ and the dependence of the internal friction coefficients on the chain length and on the parameters of the macromolecule.

The significance and importance of the internal viscosity can be elucidated by comparing the consequences of the theory with experimental data, which will be discussed further on. However, here one should note that the phenomenological characteristics of the intramolecular friction prove to depend not only on the characteristics of the macromolecule, as might have been expected, but also on the properties of the liquid in which the macromolecule is present (Schrag 1991).

The internal viscosity of the macromolecule is a consequence of the intramolecular relaxation processes occurring on the deformation of the macromolecule at a finite rate. The very introduction of the internal viscosity is possible only insofar as the deformation times are large, compared with the relaxation times of the intramolecular processes. If the deformation frequencies are of the same order of magnitude as the reciprocal of the relaxation time, these relaxation processes must be taken explicitly into account and the internal viscosity force have to be written, instead of (2.26) as

$$G_j^\alpha = -\zeta \int_0^\infty \varphi_\alpha(s)(\dot{\rho}_j^\alpha - \omega_{jl}\rho_l^\alpha)_{t-s}\, \mathrm{d}s. \qquad (2.28)$$

---

[7] To satisfy empirical relations in viscoelasticity and optical anisotropy of dilute solutions of polymers (see Sections 6.2.3 and 10.4.1), one has to assume that $\theta = z\nu - 1$.

This relation, at $\varphi_\alpha(s) \sim \delta(s)$, is equivalent to relation (2.26).

2.6 The Cerf-Zimm-Rouse Modes

Now one can return to the equation (2.1) for the dynamics of the macro-molecule in the flow of a viscous liquid. The dissipative forces acting on the particles of the chain have generally non-linear forms, but the assumptions, when these force can be written in linear approximation, were discussed in the previous sections, so that we are able to write, in terms of the normal co-ordinates introduced previously and by taking into account all the considerations described above, the dynamic equation

$$Q_{\gamma\alpha}Q_{\gamma\mu}m\frac{\mathrm{d}^2\rho_i^\mu}{\mathrm{d}t^2} = -\zeta(\dot\rho_i^\alpha - \nu_{ij}\rho_j^\alpha) - \zeta\varphi_\alpha(\dot\rho_i^\alpha - \omega_{ij}\rho_j^\alpha) - 2\mu T\lambda_\alpha\rho_i^\alpha + \xi_i^\alpha,$$

$$\xi_i^\alpha = Q_{\gamma\alpha}\phi_i^\gamma, \quad \alpha = 0, 1, 2, \ldots, N \tag{2.29}$$

The transformation matrix $Q_{\alpha\nu}$ is not, generally speaking, orthogonal and the left-hand side of the equation formulated therefore includes the derivatives of all the co-ordinates, but we shall not dwell on this factor, bearing in mind, that in the limit $m \to 0$ in which we are interested, the left-hand side of the equation vanishes.

At the above limit, equation (2.29) at $\alpha = 0$ is the equation of motion for the centre of the mass of the macromolecule – a diffusion mode. At $\alpha = 1, 2, \ldots \ll N$, equation (2.29) defines the independent relaxation modes of the macromolecule.

It is convenient here to introduce two sets of relaxation times

$$\tau_\alpha^\perp = \frac{\zeta}{4T\mu\lambda_\alpha}, \qquad \tau_\alpha^\| = (1 + \varphi_\alpha)\tau_\alpha^\perp, \quad \alpha = 1, 2, \ldots, \ll N \tag{2.30}$$

as relaxation times of the mean dimensions of the macromolecular coil (see Section 2.7.2), whereas every mode is characterised by two relaxation times: orientational and deformational. These terms are justified, when one considers the dynamics of dumbbells with arbitrary big internal viscosity (see Appendix F).

The behaviour of modes with small numbers should be independent on the arbitrary number of subdivisions N. This means that the relaxation times should not depend on N. Since the dependence of quantities μ and ζ on the number of subdivisions was elucidated previously (equations (1.32) and (2.19)), the above requirement immediately leads to the expression

$$\lambda_\alpha \sim \left(\frac{\alpha}{N}\right)^{z\nu}, \quad \alpha = 1, 2, \ldots, \ll N,$$

so that for the relaxation times one has

$$\tau_\alpha^\perp \sim \left(\frac{\alpha}{M}\right)^{-z\nu}.$$

The situation of a freely-draining macromolecule without excluded-volume effects and internal viscosity, when $z\nu = 2$, and the above eigenvalues reduce to (1.17), is especially simple. In this case, equation (2.29) describes Rouse modes, and it is convenient to use the largest orientation relaxation time

$$\tau_1 = \frac{\zeta N \langle R^2 \rangle}{6\pi^2 T} = \frac{\zeta N^2}{4\pi^2 \mu T} \sim M^2, \qquad (2.31)$$

where $\langle R^2 \rangle$ is the end-to-end distance, as a characteristic (Rouse) relaxation time of a macromolecule.

The random force ξ_i^γ in the dynamic equations (2.29) is determined by its average moments and is specified from the condition that the equilibrium moments of the co-ordinates and velocities are known beforehand (Chandrasekhar 1943). In the linearised version, with $\varphi_\alpha \ll 1$, this requirement determines the relation

$$\langle \xi_i^\alpha(t) \xi_j^\gamma(t') \rangle = 2T\zeta(1 + \varphi_\alpha)\delta_{\alpha\gamma}\delta_{ij}\delta(t - t') \qquad (2.32)$$

which is valid to within first-order terms in the velocity gradients. Here and henceforth the angular brackets indicate averaging with respect to the assembly of realisations of the random force.

Let us notice that the eigenvalues λ_α in equation (2.29) are considered constant here and henceforth. The same applies to φ_α. However, the introduced dissipative matrices are, generally speaking, functions of invariants $\rho^\alpha \rho^\alpha$ or of mean values $\langle \rho^\alpha \rho^\alpha \rangle$. The latter are functions of the velocity gradients, the expansion of which begins with a second-order term. It will be necessary to take this into account when discussing the non-linear results of the calculations.

2.7 The Moments of Linear Modes

In this section we refer to the stochastic equation (2.29) to calculate the mode moments, that is, the averaged values of the products of the normal co-ordinates and their velocities. It is convenient in this section to omit the label of mode and to rewrite the dynamic equation for the relaxation mode in the form of two linear equations

$$\frac{d\rho_i}{dt} = \psi_i,$$

$$m\frac{d\psi_i}{dt} = -\zeta(\psi_i - \nu_{ij}\rho_j) - \zeta\varphi(\psi_i - \omega_{ij}\rho_j) - 2T\mu\lambda\rho_i + \xi_i. \qquad (2.33)$$

2.7.1 Equations for the Moments of Co-ordinates

To calculate second-order moments of co-ordinates and velocities, one can start with the rates of change of quantities that can be written as follows

$$\frac{d\langle \rho_i \rho_k \rangle}{dt} = \left\langle \rho_i \frac{d\rho_k}{dt} \right\rangle + \left\langle \rho_k \frac{d\rho_i}{dt} \right\rangle,$$

$$\frac{d\langle \psi_i \psi_k \rangle}{dt} = \left\langle \psi_i \frac{d\psi_k}{dt} \right\rangle + \left\langle \psi_k \frac{d\psi_i}{dt} \right\rangle,$$

$$\frac{d\langle \rho_i \psi_k \rangle}{dt} = \left\langle \rho_i \frac{d\psi_k}{dt} \right\rangle + \left\langle \psi_k \frac{d\rho_i}{dt} \right\rangle,$$

while it is assumed that the equilibrium values of the moments are given by

$$\langle \rho_i \rho_k \rangle_0 = \frac{1}{2\mu\lambda} \delta_{ik}, \quad \langle \psi_i \psi_k \rangle_0 = \frac{T}{m} \delta_{ik}, \quad \langle \rho_i \psi_k \rangle_0 = 0.$$

Then, one can use equations (2.33) to obtain equations for the moments. After one has determined the averaged values of the products of the variables and the random force, the equations for the moments take the form

$$\frac{d\langle \rho_i \rho_k \rangle}{dt} = \langle \rho_i \psi_k \rangle + \langle \rho_k \psi_i \rangle, \tag{2.34}$$

$$\frac{d\langle \psi_i \psi_k \rangle}{dt} = \frac{2T\mu\lambda}{m} (\langle \rho_i \psi_k \rangle + \langle \rho_k \psi_i \rangle)$$

$$+ \frac{\varsigma}{m} \left[2\frac{T}{m}\delta_{ik} - 2\langle \psi_i \psi_k \rangle + \nu_{ij}\langle \rho_j \psi_k \rangle + \nu_{kj}\langle \rho_j \psi_k \rangle \right.$$

$$+ \varphi \left(2\frac{T}{m}\delta_{ik} - 2\langle \psi_i \psi_k \rangle + \omega_{ij}\langle \rho_j \psi_k \rangle + \omega_{kj}\langle \rho_j \psi_i \rangle \right) \right], \tag{2.35}$$

$$\frac{d\langle \rho_i \psi_k \rangle}{dt} = \langle \psi_i \psi_k \rangle - \frac{2T\mu\lambda}{m}\langle \rho_i \rho_k \rangle$$

$$- \frac{\varsigma}{m} \left[\langle \rho_i \psi_k \rangle - \nu_{kj}\langle \rho_j \rho_i \rangle + \varphi(\langle \rho_i \psi_k \rangle - \omega_{kj}\langle \rho_j \rho_i \rangle) \right]. \tag{2.36}$$

It is easy to see that, at zeroth velocity gradients, the right-hand sides of the above equations are identically equal to zero.

2.7.2 The Slowest Relaxation Processes

The set of equations (2.34)–(2.36) for the second-order moments of co-ordinates and velocities can be simplified, if we consider the situation when the distribution of velocities corresponds to equilibrium, that is, we put $m \to 0$. In this case, equation (2.35) is followed by relation

$$\langle \rho_i \psi_k \rangle = -\frac{1}{2\tau^{\parallel}} \left(\langle \rho_i \rho_k \rangle - \frac{1}{2\mu\lambda}\delta_{ik} \right) + \nu_{kj}\langle \rho_j \rho_i \rangle - \varphi\gamma_{kj}\langle \rho_j \rho_i \rangle \tag{2.37}$$

where the relaxation times are given by relations (also by formulae (2.30))

$$\tau^{\parallel} = (1 + \varphi)\tau^{\perp}, \qquad \tau^{\perp} = \frac{\zeta}{4T\mu\lambda}. \tag{2.38}$$

Now, one can use equations (2.34) to obtain relaxation equations for the moments of co-ordinates

$$\frac{\mathrm{d}\langle\rho_i\rho_k\rangle}{\mathrm{d}t} - \nu_{ij}\langle\rho_j\rho_k\rangle - \nu_{kj}\langle\rho_j\rho_i\rangle$$

$$= -\frac{1}{\tau^{\parallel}}\left(\langle\rho_i\rho_k\rangle - \frac{1}{2\mu\lambda}\delta_{ik}\right) - \varphi(\gamma_{ij}\langle\rho_j\rho_k\rangle + \gamma_{kj}\langle\rho_j\rho_i\rangle). \tag{2.39}$$

The relaxation time $\tau^{\parallel}$ refers to the deformation processes. Indeed, by carrying out a direct summation of equation (2.39) with identical indices, one finds

$$\frac{\mathrm{d}\langle\rho^2\rangle}{\mathrm{d}t} = -\frac{1}{\tau^{\parallel}}\left(\langle\rho^2\rangle - \frac{3}{2\mu\lambda}\right) + 2(1 - \varphi)\gamma_{ij}\langle\rho_j\rho_i\rangle. \tag{2.40}$$

This equation describes only the deformation of the macromolecular coil and therefore $\tau^{\parallel}$ is a relaxation time of the deformation process. It can be shown (see Appendix F) that the orientation relaxation process is characterised by the relaxation time $\tau^{\perp}$.

Explicit expressions for the moments will be necessary later to calculate the physical quantities. In the non-steady-state case, the second-order moments of co-ordinates are calculated as solutions of equations (2.39). To find the solutions, we multiply equation (2.39) by $\exp(\frac{t}{\tau^{\parallel}})$ and integrate over time from t to ∞. After some transformation, we obtain

$$\langle\rho_i\rho_k\rangle = \frac{1}{2\mu\lambda}\delta_{ik} + \int_0^{\infty} \exp\left(-\frac{s}{\tau^{\parallel}}\right)$$

$$\times \left[\nu_{ij}\langle\rho_j\rho_k\rangle + \nu_{kj}\langle\rho_j\rho_i\rangle - \varphi(\gamma_{ij}\langle\rho_j\rho_k\rangle + \gamma_{kj}\langle\rho_j\rho_i\rangle)\right]_{t-s}\mathrm{d}s.$$

The moments and velocity gradients in the integrand are taken at the point of time $t - s$.

Now we can use the equilibrium moments to find the first terms of the expansion of the moments as a series of repeated integrals

$$\langle\rho_i\rho_k\rangle = \frac{1}{2\mu\lambda}\left\{\delta_{ik} + 2(1 - \varphi)\int_0^{\infty} \exp\left(-\frac{s}{\tau^{\parallel}}\right)\gamma_{ik}(t - s)\mathrm{d}s\right\}. \tag{2.41}$$

The iteration procedure can be continued.

In the steady-state case, the expansion assumes the form

$$\langle\rho_i\rho_k\rangle = \frac{1}{2\mu\lambda}\{\delta_{ik} + 2\tau^{\perp}\gamma_{ik}$$

$$+ 2(\tau^{\perp})^2\left[2\gamma_{ij}\gamma_{jk} + (1 + \varphi)(\omega_{ij}\gamma_{jk} + \omega_{kj}\gamma_{ji})\right]\}. \tag{2.42}$$

We may note that, in the approximation of the preliminary averaging, which was used, the expressions for the moments are valid only to within second-order terms with respect to the velocity gradients.

2.7.3 Fourier-Transforms of Moments

One can calculate the mode moments in different way. One can pass from equations (2.34)–(2.36) to the set of algebraic equations, introducing the Fourier-transforms of the unknown functions

$$\langle \rho_i \rho_k \rangle = \int_{-\infty}^{\infty} R_{ik}(\omega) e^{-i\omega t} \frac{d\omega}{2\pi},$$

$$\langle \rho_i \psi_k \rangle = \int_{-\infty}^{\infty} Y_{ik}(\omega) e^{-i\omega t} \frac{d\omega}{2\pi},$$

$$\langle \psi_i \psi_k \rangle = \int_{-\infty}^{\infty} Z_{ik}(\omega) e^{-i\omega t} \frac{d\omega}{2\pi}.$$

The solution of the resulting set of equations can be written accurately, within the first order terms with respect to the velocity gradients, as

$$R_{ik}(\omega) = \frac{1}{2\mu\lambda} \left[\delta_{ik}\delta(\omega) + 2C\tau^\perp \gamma_{ik}(\omega) \right],$$

$$Y_{ik}(\omega) = \frac{1}{2\mu\lambda} \left[-\omega_{ik}(\omega) - i\omega C\tau^\perp \gamma_{ik}(\omega) \right],$$

$$Z_{ik}(\omega) = \frac{T}{m} \left[\delta_{ik}\delta(\omega) + \frac{2i\omega\tau_m C}{2(1+\varphi) - i\omega\tau_m} \tau^\perp \gamma_{ik}(\omega) \right], \qquad (2.43)$$

$$C = \frac{2(1+\varphi - i\omega\tau_m)}{[1 - i\omega\tau^\perp(1+\varphi) - \tau_m\tau^\perp\omega^2][2(1+\varphi) - i\omega\tau_m] - i\omega\tau_m}.$$

The solution contains two characteristic relaxation times

$$\tau^\perp = \frac{\varsigma}{4T\mu\lambda}, \qquad \tau_m = \frac{m}{\varsigma}.$$

The first relaxation time is much bigger than the second one within the limits of applicability of the subchain model. So, the terms multiplied by the quantity $\omega\tau_m$ in relations (2.43) can be neglected, and expressions can be written down in the simpler form

$$R_{ik}(\omega) = \frac{1}{2\mu\lambda} \left[\delta_{ik}\delta(\omega) + \frac{2\tau^\perp}{1 - i\omega\tau^\parallel} \gamma_{ik}(\omega) \right],$$

$$Y_{ik}(\omega) = \frac{1}{2\mu\lambda} \left[-\omega_{ik}(\omega) + \frac{-i\omega\tau^\perp}{1 - i\omega\tau^\parallel} \gamma_{ik}(\omega) \right], \qquad (2.44)$$

$$Z_{ik}(\omega) = \frac{T}{m} \delta_{ik}\delta(\omega). \qquad (2.45)$$

It can easily be seen that the first expression from equations (2.44) corresponds to expressions (2.41) and (2.42).

Chapter 3
Dynamics of a Macromolecule in an Entangled System

Abstract In this chapter, a system of entangled macromolecules in fluid state, that is a concentrated solution or a melt of polymer, will be considered. Every macromolecule in the investigated system can move among the others macromolecules, exchanging neighbours and remaining the integrity of each individual macromolecule unaffected. It allows introducing the mesoscopic approximation, which deals with the motion of a single macromolecule in an effective medium, created by the neighbouring macromolecules. One can note that the tradition of the mesoscopic approach begins with the first work on concentrated polymer solutions (Ferry et al. in J. Appl. Phys. 26:259–362, 1955), in which some specifying hypotheses about the environment of the probe macromolecule were formulated. Some earliest approaches to the problem are developed by Edwards with collaborators (Scolnick and Kolinski in Adv. Chem. Phys. 78:223–278, 1990). One of the hypotheses ascribes the properties of a relaxing medium to the environment of a probe macromolecule (Edwards and Grant in J. Phys. A: Math. Nucl. Gen. 6:1169–1185, 1973). This idea was developed later into the theory, based on the non-Markovian stochastic equation. An alternative hypothesis was an assumption about the tube and reptation motion of macromolecules (Doi and Edwards in The Theory of Polymer Dynamics (Oxford University Press, Oxford), 1986). Now, one can see that both the first and the second hypothesis reflect the reality, and the theory, which will be exposed here, can be considered as a reconciliation of the alternative approaches. In this chapter, a unified model of macromolecular dynamics can be formulated as the Rouse model of a chain of coupled Brownian particles in the presence of a random dynamic force. It came to a consistent theory of the phenomena and constitutes a phenomenological frame within which both the results of empirical investigations and the results of microscopic, many-chains approaches can be considered. The mesoscopic approach reveals the internal connection between phenomena and provides more details than a strictly phenomenological approach.

V.N. Pokrovskii, *The Mesoscopic Theory of Polymer Dynamics*,
Springer Series in Chemical Physics 95,
DOI 10.1007/978-90-481-2231-8_3, © Springer Science+Business Media B.V. 2010

3.1 Admitted Approximations in the Many-Chain Problem

To say nothing about the truly atomistic models, every flexible macromolecule can be universally presented as consisting of z freely jointed rigid segments (Kuhn segments, see Section 1.1) – this is considered as a microscopic approach. So, the system of entangled macromolecules can be imagined as consisting of nz interacting segments, every z of them being connected in chain, and the basis heuristic model, kinetics of which has to be investigated, is a system of interacting rigid segments connected in chains. In other words, it is a system of interacting Kuhn-Kramers chains. The system is dense, the interactions are strong, and it seems to be a rather complex problem, which has not solved yet,[1] so that one has to look for more coarse approximations to describe dynamics of this system. One can use the coarse-grained Gaussian model for every macromolecule (bead-and-spring model, see Section 1.2) to describe the behaviour of the system. This is a heuristic model easier to consider. To formulate an equation for the large-scale stochastic dynamics of the entangled system as dynamics of interacting chains of Brownian particles, one can consider a system of interacting segments (atoms) and follow Zwanzig-Mori method (Zwanzig 1961; Mori 1965), described, for example, in monographs (Hansen and McDonald 1986; Boon and Yip 1980). There is no available solution of the problem; nevertheless, one can easily imagine a general form of the anticipated results.

3.1.1 Dynamics of Entangled Course-Grained Chains

When one is interested in slow modes of motion of the system, each macromolecule of the system can be schematically described in a coarse-grained way as consisting of $N + 1$ linearly-coupled Brownian particles, and we shall be able to look at the system as a suspension of $n(N + 1)$ interacting Brownian particles. An anticipated result for dynamic equation of the chains in equilibrium situation can be presented as a system of stochastic non-Markovian equations

$$m\frac{\mathrm{d}^2 r^{a\alpha}}{\mathrm{d}t^2} = -\int_0^\infty B_{a\alpha,b\beta}(s)\dot{r}^{b\beta}(t-s)\,\mathrm{d}s - \frac{\partial U}{\partial r^{a\alpha}} + \phi^{a\alpha}(t),$$
$$a = 1, 2, \ldots, \quad \alpha = 0, 1, 2, \ldots, N, \tag{3.1}$$

where $r^{a\alpha}$ is a co-ordinate of a particle (where a is the label of the macromolecule to which the Brownian particle belongs, and α is the label of the particle in the macromolecule), m is the mass of a Brownian particle associated with a section of the macromolecule of length $M/(1+N)$. The potential $U(r^{a\alpha})$ depicts interaction of a particle $r^{a\alpha}$ with particles of its own and

---

[1] Curtiss and Bird (1981a and 1981b) have posed and considered such a problem.

the other macromolecules. The integral term on the right is the friction force (external and internal resistance), determined through the memory matrix $B_{a\alpha,b\beta}(s)$, by all the Brownian particles in the system. The properly derived memory function must be expressed in terms of interaction between segments and includes relaxation time which, generally speaking, cannot be neglected.

One can think that this situation, described by equations (3.1), can be visualised as a picture of interacting (and connected in chains) Brownian particles suspended in anisotropic viscoelastic 'segment liquid'. Introduction of macroscopic concepts is unavoidable consequence of transition from microscopic to mesoscopic approach, or better to say, from the microscopic model of interacting Kuhn-Kramers chains to mesoscopic model of interacting chains of Brownian particles.

Up to now no specific results for memory function and effective potential in equations (3.1) are available,[2] so that, to simplify the system (3.1), one has to make some suggestions, two of which are intensively exploited.

Course-Grained Interacting Chains in Non-Relaxing Medium

The situation looks simpler, if one assumes that relaxation times of the surrounding can be neglected, and one obtains for the collective motion of the entire set of macromolecules, considered as a set of Brownian particles, a system of stochastic Markovian equations

$$m\frac{d^2 r^{a\alpha}}{dt^2} = -\zeta \dot{r}^{a\alpha} - 2\mu T A_{\alpha\gamma} r^{a\gamma} - \frac{\partial U}{\partial r^{a\alpha}} + \phi^{a\alpha}(t),$$
$$a = 1, 2, \ldots, \quad \alpha = 0, 1, 2, \ldots, N, \tag{3.2}$$

In other words, it is assumed here that the particles are surrounded by a isotropic viscous (not viscoelastic) liquid, and ζ is a friction coefficient of the particle in viscous liquid. The second term represents the elastic force due to the nearest Brownian particles along the chain, and the third term is the direct short-ranged interaction (excluded volume effects, see Section 1.5) between all the Brownian particles. The last term represents the random thermal force defined through multiple interparticle interactions. The hydrodynamic interaction and intramolecular friction forces (internal viscosity or kinetic stiffness), which arise when the macromolecular coil is deformed (see Sections 2.2 and 2.4), are omitted here.

When this approximation is valid? The empirical estimation of the relaxation time of the medium shows that, for the systems of short macromolecules ($M < 2M_e$), the relaxation time of the medium indeed can be neglected, so that the approximation is valid for these systems. For the systems of longer

---

[2] The methods used by Karl Freed with associates (Chang and Freed 1993; Tang et al. 1995; Guenza and Freed 1996; Kostov and Freed 1997) for calculation time-correlation functions of single macromolecules can be apparently useful in a more complicated case of many interacting macromolecules.

macromolecules ($M > 2M_e$), when the entanglements exist, there seems to be no evidence that the relaxation times are equal to zero.

Course-Grained Non-Interacting Chains in Relaxing Medium

As an alternative to the approximation of non-relaxing medium (equations (3.2)), one can suppose that the effect of the direct interactions between Brownian particles is less than the effect of the effective relaxing medium. It is supported by the fact, that the number density of the Brownian particles in the coarse-grained approximation is much less than the number density of segments, so that the Brownian particles make up a weakly interacting system. If the effect of direct interactions of coarse-grained chains with each other can be neglected at all, the system (3.1) appears to be a collection of n independent equations, everyone of which describes effective dynamics of a chain of fictional Brownian particles. It is an amazing possibility: *the system of entangled interacting macromolecules can be considered as a collection of non-interacting chains of Brownian particles suspended in a liquid, which is made up of interacting segments.* The chains of Brownian particles appear to behave independently, though the system is closely packed. This is something of a paradox which, nevertheless, is confirmed in the following chapters.

Thus, one can choose from the two possibilities to simplify the system (3.1). We are convinced, that the approximation of independent chains appears to be a very good initial approximation. The situation appears to be similar to a situation in dilute solutions discussed in the previous chapter. However, in contrast to the case of dilute solutions, the correlation times of the surrounding medium cannot be neglected for entangled systems. The initial phase of the theory might be found to be rather formal but the justification for every theory regarding physics eventually rests on the agreement between deductions made from it and experiments, and on the simplicity and consistency of the formalism. Comparison with experiment will be discussed in Chapters 5, 6, 9 and 10.

3.1.2 Dynamics of a Probe Macromolecule

The behaviour of a single macromolecule appears to be crucial in the discussion of properties of entangled polymers. It is not difficult to imagine the result of eliminating all co-ordinates, apart of the course-grained co-ordinates of the only chosen single chain in the system of entangled macromolecules, as schematised in Fig. 5. Whatever the way one chooses, to start from dynamics of interacting rigid Kuhn segments or from dynamics of interacting Brownian particles (equations (3.1) or (3.2)), an anticipated result in linear approximation and for the equilibrium situation can be written as

$$m\frac{\mathrm{d}^2 \boldsymbol{r}^\alpha}{\mathrm{d}t^2} = -\int_0^\infty \Gamma_{\alpha\gamma}(s)\dot{\boldsymbol{r}}^\gamma(t-s)\mathrm{d}s - 2\mu T A_{\alpha\gamma}\boldsymbol{r}^\gamma(t) + \boldsymbol{\phi}^\alpha(t),$$
$$\alpha = 0, 1, 2, \ldots, N, \tag{3.3}$$

Figure 5. Macromolecule in an entangled system.
Entangled macromolecules, connected with weak van der Waals forces, make up a very
concentrated polymer solution or a polymer melt. Mesoscopic approach considers the
coarse-grained dynamics of a single macromolecule. The surrounding macromolecules are
considered a reacting medium.

---

where the memory tensor function $\Gamma_{\alpha\gamma}(s)$ is connected with the correlation
function of the random force $\phi^{\alpha}(t)$

$$\langle \phi^{\alpha}(t)\phi^{\gamma}(t')\rangle = 6T\Gamma_{\alpha\gamma}(t-t').$$

In terms of our previous discussion, the result for the memory function $\Gamma_{\alpha\gamma}(s)$
has to retain traces from, generally speaking, two consequent steps: the transition from the microscopic picture of interacting segments to a picture of
a system of interacting coarse-grained chains (equation (3.1) or (3.2)) and
transition from one of these systems to the single-chain equation (3.3).

The results for the memory function $\Gamma_{\alpha\gamma}(s)$ are available for the case,
when the system (3.2) is chosen as a starting point of derivation of a dynamic equation for a single chain in the system of entangled macromolecules
(Schweizer 1989a, 1989b; Vilgis and Genz 1994; Guenza 1999; Rostiashvili et
al. 1999; Fatkullin et al. 2000). It means that the segment carrier liquid is
assumed viscous. Schweizer (1989a, 1989b) employed the Mori-Zwanzig pro-
jector operator techniques to the problem. Another method of derivation of
the same equation was presented by Vilgis and Genz (1994) and Rostiashvili
et al. (1999). After having succeeded in eliminating all variables from the
set of stochastic equations, other than those that refer to the chosen macromolecule, and some approximations, the scholars came to equations (3.3) and
evaluated, which is the most essential, the memory function $\Gamma_{\alpha\gamma}(s)$ through
the intermolecular correlation functions and structural dynamic factor of the
system of interacting Brownian particles. Their results allow us to estimate
the contribution into the memory function from the second step of derivation
and allow us to judge about importance of this contribution – the analysis
that yet has to be done.

3.2 The General Form of Dynamic Equation

The most advanced theories of relaxation phenomena in a system of entangled macromolecules is based on the dynamics of a single macromolecule. Dynamics of the tagged macromolecule is simplified by the assumption that the neighbouring macromolecules can be described as a uniform structureless medium and all important interactions can be reduced to intramolecular interactions. The dynamic equation for a macromolecule can be written as a modification of equation (2.1) for dynamics of macromolecule in viscous liquid

$$m\frac{\mathrm{d}^2 r_i^\alpha}{\mathrm{d}t^2} = -\zeta(\dot{r}_i^\alpha - \nu_{ij}r_j^\alpha) + F_i^\alpha + G_i^\alpha - 2T\mu A_{\alpha\gamma}r_i^\gamma + \phi_i^\alpha,$$
$$\alpha = 0, 1, 2, \ldots, N, \tag{3.4}$$

where m is the mass of a Brownian particle associated with a piece of the macromolecule of length $M/(N+1)$, r^α are the co-ordinates of the Brownian particles, the label $\alpha = 0, 1, \ldots, N$ being the label of the particle in the macromolecule. The external resistance experienced by a moving particle are divided into two terms, namely, the resistance due to the 'monomer' liquid, represented by a coefficient ζ, and the reaction of the neighbouring chains F_i^α. The equation assumes one more dissipative term: internal resistance force G_i^α which obeys the requirement

$$\sum_{\alpha=0}^{N} G_i^\alpha = 0. \tag{3.5}$$

Indeed, the intramacromolecular forces, both dissipative and elastic, do not affect motion of the coil as a whole.

The fourth term on the right hand side of (3.4) represents the elastic forces on each Brownian particle due to its neighbours along the chain; the forces ensure the integrity of the macromolecule. Note that this term in equation (3.4) can be taken to be identical to the similar term in equation for dynamic of a single macromolecule due to a remarkable phenomenon – screening of intramolecular interactions, which was already discussed in Section 1.6.2. The last term on the right hand side of (3.4) represents a stochastic thermal force. The correlation function of the stochastic forces $\langle \phi_i^\alpha(t)\phi_k^\gamma(t') \rangle$ is connected with the dissipative forces (the fluctuation-dissipation theorem). The relation will be discussed later, in Section 3.4 for the case, when the equation (3.4) is linear in velocities, but can be non-linear in co-ordinates.

In virtue of the results, described in the previous section, it is natural to present the extra dissipative terms in equation (3.4) in an integral forms. One has to require the resistance forces to be independent of the rotation of the co-ordinate system with constant angular velocity and, assuming also the proper covariance and linearity in the velocities of particles, determines (Pokrovskii and Volkov 1978a) the general form of the terms

$$F_i^\alpha = -\int_0^\infty B_{ik}^{\alpha\gamma}(s)(\dot{r}_k^\gamma - \nu_{kj}r_j^\gamma)_{t-s}\mathrm{d}s \qquad (3.6)$$

$$G_i^\alpha = -\int_0^\infty G_{ik}^{\alpha\gamma}(s)(\dot{r}_k^\gamma - \omega_{kj}r_j^\gamma)_{t-s}\mathrm{d}s \qquad (3.7)$$

where $v_i^\alpha = \nu_{ik}r_k^\alpha$ is the mean velocity of the medium, while ν_{ik} is the tensor of velocity gradients in the point where particle α is located. The resulting dynamic equation for a chain looks like a stochastic equation with memory function terms – generalised Langevin equation.[3] The presence of the top and bottom labels in the symbols of memory functions shows the influence of the other particles of the chain and the anisotropy of the medium on the motion of the considered particle. Strictly speaking, we should also write these terms in the form of non-local (in co-ordinates) expressions, since the agitation propagates directly through the chain to a distance $\langle R^2 \rangle$ – the distance that is large in comparison to the size of the Brownian particle under consideration. However, for the sake of simplicity, this will not be shown here, although the consequences of a non-local effect will be investigated in Section 3.3.2.

One can assume that each Brownian particle of the chain is situated in a similar environment, which is approximately correct for long chains, so that we can rewrite the memory functions in (3.6) and (3.7) as

$$B_{ik}^{\alpha\gamma}(s) = H_{ik}^{\alpha\gamma}\beta(s), \qquad G_{ik}^{\alpha\gamma}(s) = G_{ik}^{\alpha\gamma}\varphi(s)$$

where $\beta(s)$ and $\varphi(s)$ are universal scalar memory functions. This allows to make the memory terms more tractable.

3.2.1 The Linear Approximation

In the simplest case, one can assume that there is neither global nor local anisotropy, which means that

$$H_{ik}^{\alpha\gamma} = H^{\alpha\gamma}\delta_{ik}, \qquad G_{ik}^{\alpha\gamma} = G^{\alpha\gamma}\delta_{ik}.$$

---

[3] Some scholars are being stuck to Markov stochastic processes and determine the dynamics of a probe macromolecule, in contrast to equation (3.4)–(3.7), as a dynamics of a chain in a certain viscous medium – modified Rouse dynamics. These theories stem from works of Ferry et al. (1955), Vinogradov et al. (1972b) and others who tried to modify the friction coefficient of Brownian particles in such a way, to make it possible to interpret results of investigation of viscoelasticity of linear polymer. An advanced example of such theories is the constraint-release theory, due to Graessley (1982) and many others (see review by Watanabe 1999), which suggests a detailed mechanism of a large-scale lateral motion of a macromolecule in an entangled system due to process of release of some constraints of the probe chain and jumps of some parts of the chain in lateral direction. To the best of our experience, Markov processes, that is the Rouse dynamics (even with modified friction coefficients), cannot adequately describe the basic stochastic motion of a macromolecule among the neighbouring macromolecules and we are going to show that non-Markov stochastic processes are adequate for this aim.

The non-diagonal terms of the matrixes $H^{\alpha\gamma}$ and $G^{\alpha\gamma}$ are connected with mutual influence of the particles of the chain. One can admit that, in accordance with the works by Edwards and Freed (1974) and Freed and Edwards (1974, 1975), the hydrodynamic interaction in the system between the particles of the chain becomes negligible, and one can introduce a diagonal matrix of external resistance, but, in virtue of relation (3.5), one cannot introduce non-zero diagonal matrix of internal resistance, so that the simplest forms of the matrixes are

$$H_{ik}^{\alpha\gamma} = \delta_{\alpha\gamma}\delta_{ik}, \qquad G_{ik}^{\alpha\gamma} = G^{\alpha\gamma}\delta_{ik}. \tag{3.8}$$

The symmetrical numerical matrix $G^{\alpha\gamma}$ represents the influence of motion of the particle γ on the motion of the particle α. The only general requirement one ought to put on matrix $G^{\alpha\gamma}$ in the last relation is the following: in normal co-ordinates it has a zero eigenvalue. The simplest forms, satisfying the requirement (3.5), can be written as

$$H_{ij}^{\alpha\gamma}u_j^\gamma = u_i^\alpha, \qquad G_{ij}^{\alpha\gamma}u_j^\gamma = \frac{1}{N}\left\{(N+1)u_i^\alpha - \sum_{\gamma=0}^{N}u_i^\gamma\right\} \tag{3.9}$$

so that the components of the numerical matrix $G^{\alpha\gamma}$ are defined as

$$G = \left\|\begin{matrix} 1 & -1/N & \dots & -1/N \\ -1/N & 1 & \dots & -1/N \\ \dots & \dots & \dots & \dots \\ \dots & \dots & \dots & \dots \\ -1/N & -1/N & \dots & 1 \end{matrix}\right\|. \tag{3.10}$$

Thus, equation (3.4) in the simplest case can be specified in the form

$$m\frac{d^2 r_i^\alpha}{dt^2} = -\zeta(\dot{r}_i^\alpha - \nu_{ij}r_j^\alpha) - \int_0^\infty \beta(s)(\dot{r}_i^\alpha - \nu_{ij}r_j^\alpha)_{t-s}ds$$

$$- G^{\alpha\gamma}\int_0^\infty \varphi(s)(\dot{r}_i^\gamma - \omega_{ij}r_j^\gamma)_{t-s}ds - 2\mu T A_{\alpha\gamma}r_i^\gamma + \phi_i^\alpha(t). \tag{3.11}$$

The equation (3.11) is the equation for the dynamics of a single macromolecule in the case of linear dependence on the co-ordinates and velocities.[4] Let us note that, if memory functions $\beta(s)$ and $\varphi(s)$ turn into δ-functions, equation (3.11) becomes identical to the equation of motion of the macromolecule in a viscous liquid, which was used in Chapter 2 to describe the dynamics of a macromolecule in this case.

---

[4] Particular cases of dynamic equation (3.11) were investigated by Ronca (1983) and by Hess (1986, 1988) who apparently did not know about previously published results. They made unsuccessful attempts to describe dynamics of macromolecule in an entangled system without the second dissipative term which is connected with the internal resistance forces. One can see in subsequent chapters that the properties of polymer melts cannot be understood correctly without this term. The importance of the internal resistance term was recognised by Pokrovskii and Volkov (1978b) after the first attempt to tackle the problem (Pokrovskii and Volkov 1978a).

3.2.2 A Non-Linear Approximation – Local Anisotropy

The simple relations (3.8)–(3.11) are valid for linear approximation when there is neither global nor local anisotropy, that is the particles have spherical forms and the medium is isotropic. The set of equations (3.11) describes basic stochastic motion of the Brownian particles of the chain and allows us to introduce essential restrictions on the motion of particles by forces of external and internal resistance. In linear approximation, the equations determine the generalised Cerf-Rouse modes. However, there are some important effects, which cannot be described without some non-linear terms in equation of dynamics of a macromolecule. The non-linearity is connected with the easier motion of the chain along its contour – the reptation motion of the macromolecule, introduced by Gennes (1971) for explanation mobility of macromolecules among other chains – the effect, which is also very important for explanation some effects of viscoelasticity in strongly entangled system. It is assumed that a macromolecule moves among other macromolecules like a snake and create an effective tube, inside which the particles of the chain are moving with increased mobility. This effect can be described by introduction of local anisotropy of mobility of a particles of the chain.

For the case, when the local anisotropy is taken into account, the relations (3.9) ought to be generalised as

$$H_{ij}^{\alpha\gamma} u_j^\gamma = u_i^\alpha - \frac{3}{2} a_e \left(e_i^\alpha e_j^\alpha - \frac{1}{3}\delta_{ij} \right) u_j^\alpha,$$

$$G_{ij}^{\alpha\gamma} u_j^\gamma = \frac{1}{N} \left\{ (N+1) \left[u_i^\alpha - \frac{3}{2} a_i \left(e_i^\alpha e_j^\alpha - \frac{1}{3}\delta_{ij} \right) u_j^\alpha \right] \right.$$

$$\left. - \sum_{\gamma=0}^N \left[u_i^\gamma - \frac{3}{2} a_i \left(e_i^\gamma e_j^\gamma - \frac{1}{3}\delta_{ij} \right) u_j^\gamma \right] \right\}, \qquad (3.12)$$

$$e_i^\alpha = \frac{r_i^{\alpha+1} - r_i^{\alpha-1}}{|r^{\alpha+1} - r^{\alpha-1}|}, \quad \alpha = 1, 2, \ldots, N-1,$$

$$e_i^0 e_j^0 = e_i^N e_j^N = \frac{1}{3}\delta_{ij},$$

where a_e and a_i are parameters of local anisotropy introduced in such a way, that positive values of the parameters correspond to increase in mobility along the contour of the chain. To describe the effect of local anisotropy, to every internal particle of the chain is ascribed the direction vector e^α, while the end particles of the chain having no direction, so that there are $N-1$ vectors for a chain. However, it is convenient further formally to ascribe the vectors e^0 and e^N for the end particles and to define products of components of the vectors as above. For the linear case, when on average $e_i^\alpha e_j^\alpha = (1/3)\delta_{ij}$, one returns from equations (3.12) to relations (3.9). The matrixes $H_{ij}^{\gamma\mu}$ and $G_{ij}^{\gamma\mu}$, as defined by equations (3.12), have the forms

$$H_{ij}^{\alpha\gamma} = \delta_{\alpha\gamma} \left[\delta_{ij} - \frac{3}{2} a_e \left(e_i^\gamma e_j^\gamma - \frac{1}{3}\delta_{ij} \right) \right],$$

$$G_{ij}^{\alpha\gamma} = G^{\alpha\gamma} \left[\delta_{ij} - \frac{3}{2} a_i \left(e_i^\gamma e_j^\gamma - \frac{1}{3}\delta_{ij} \right) \right].$$

(3.13)

In the simplest case the matrix G has the form (3.10), but can be modified.

Introduction of the local anisotropy of mobility allows us to specify the matrixes of the extra forces of external and internal resistance and to formulate dynamic equations, which will be discussed in Section 3.4. One can expect that, as a result of the introduction of the local anisotropy, mobility of a particle along the axis of a macromolecule appears to be bigger than mobility in the perpendicular direction, so that the entire macromolecule can move more easily along its contour. The local anisotropy hinders also change of the form of the macromolecular coil, and, by this way, plays a role similar to the role of the term with internal resistance in linear version of the model.

3.3 Molecular Interpretation of the Dissipative Terms

In the case, when one applies the coarse-grained approximation for the description of chains, each particle of the chain can be considered as moving in a liquid, which represents a dense system made of the interacting rigid Kuhn segments. The memory functions $\beta(s)$ and $\varphi(s)$ in equations (3.11) cannot be determined from general considerations: they could be found theoretically as correlation functions of the random force in microscopic dynamics of interacting Kuhn-Kramers chains, or, otherwise, the memory functions ought to chosen in such a way, that the final results would describe empirical facts. At the moment, we have no choice as to look for empirical memory functions.

3.3.1 Concept of Microviscoelasticity

Underlying Relaxation Process

The effective dense medium, in which a particle of the chain is moving, consists of the interacting rigid Kuhn segments and has properties of relaxing liquid; thus, the concept of microviscoelasticity, instead of the concept of microviscosity in the case of dilute solutions, can be introduced. The times of relaxation of the surrounding medium are times of relaxation of the mean orientation of segments $\langle e_i e_k \rangle$, whereas one can accept, that the rotation of a separate segment in a dense system of long linear macromolecules is determined strongly by its environment, being weakly dependent on its position in the chain. We assume that a fundamental process has a single relaxation time τ, so that the relaxation of orientation is described by equation

$$\frac{\mathrm{d}\langle e_i e_k \rangle}{\mathrm{d}t} = -\frac{1}{\tau} \left(\langle e_i e_k \rangle - \frac{1}{3}\delta_{ik} \right).$$

(3.14)

The medium surrounding of the given segment consists from both the segments of neighbouring macromolecules and the segments of the tagged macromolecule, so that the relaxation time τ depends on the lengths both of probe macromolecule M and macromolecules of environment M_0. The relaxation time τ has to be considered as a relaxation time of the mechanical (viscoelastic) reaction of the ambient medium.

So, one can write down the specification for the memory functions in equations (3.11) as

$$\beta(s) = \frac{\zeta}{\tau} B \exp\left(-\frac{s}{\tau}\right), \qquad \varphi(s) = \frac{\zeta}{\tau} E \exp\left(-\frac{s}{\tau}\right). \qquad (3.15)$$

In this formulae, ζ is a friction coefficient of a particle in a "monomer" liquid, while non-dimensional phenomenological quantities B and E are measures of the increase in the 'external' and 'internal' friction coefficients due to the neighbouring macromolecules.

Self-Consistency of the Approach

In line with the relaxation time τ of the ambient medium, it is convenient to use the non-dimensional quantity

$$\chi = \frac{\tau}{2B\tau^*}, \qquad (3.16)$$

where τ^* is the characteristic Rouse relaxation time of the macromolecule in viscous 'monomer' liquid, which is a combination of some parameters of the dynamic equations (3.11)

$$\tau^* = \frac{\zeta N^2}{4\pi^2 \mu T} \sim M^2.$$

The quantity χ is a characteristic of a macromolecule with molecular weight M in the surrounding, consisting of linear polymer with molecular weight M_0. We shall distinguish the macromolecules, even the system is a polymer melt, $M = M_0$. It is especially essential, if one consider a system with a small additive of a similar polymer – a dilute blend.

In the case of the bulk polymer, the requirement of self-consistency of the theory states that the relaxation time τ can be interpreted as a characteristic of the whole system. Properties of the system will be calculated in Sections 6.3.2 and 6.4.3, which allows one to estimate relaxation time τ and quantity χ. It will be demonstrated that, for weakly entangled systems $(2M_e < M < 10M_e)$, the quantity χ has the self-consistent value

$$\chi \approx \frac{\pi^2}{30}.$$

For strongly entangled systems ($M > 10M_e$), the requirement of self-consistency is fulfilled identically, while the quantity χ is connected with the intermediate length ξ (see Section 5.1.2, formula (5.8)), or (as we shall see in Section 6.4.4, formula (6.55)) with the length of the macromolecule between adjacent entanglement M_e, that is

$$\chi = \frac{\tau}{2B\tau^*} \approx \begin{cases} (2\xi)^2/\langle R^2 \rangle, \\ M_e/M. \end{cases}$$

In the last case, the parameter has the meaning of the ratio of "a length of macromolecule between adjacent entanglements" to the length of the macromolecule (see Section 5.1.2). The parameters τ, χ, ξ, and M_e appear to be equivalent for the strongly entangled systems. One of these parameters is used to describe polymer dynamics in either interpretation.

The parameter χ is always small for entangled systems. Due to the above written results, the self-consistent values of the quantity can be approximated, for $M/M_e > 2$, as

$$\chi \approx \frac{\pi^2}{6} \frac{1}{1+2Z}, \quad Z = \frac{M}{M_e}. \tag{3.17}$$

It will be convenient to measure the length of macromolecules in units of M_e, M_e being the length of a part of a macromolecule between 'adjacent entanglements'.

3.3.2 External Friction

If we consider very slow motion of a macromolecular coil with constant velocity, the force of internal resistance can be neglected and the resistance-drag coefficient for the external force can be written down as

$$\zeta B = \int_0^\infty \beta(s)\mathrm{d}s, \tag{3.18}$$

where the non-dimensional quantity B is a measure of the increase in the friction coefficient, due to the fact that the particle is moving among neighbouring macromolecules, perturbing them. There is a slight difference in resistance, when the particle moves along the chain or in a perpendicular direction, but, in this subsection, the anisotropy of resistance will be neglected for simplicity.

Overlapping-Coils Friction

Let us imagine, following Pokrovskii and Pyshnograi (1988), the shear motion of the system as a motion of overlapping macromolecular coils, each of which is characterised by the function (1.24) of the mean number density of Brownian particles

$$\rho(r) = \left(\frac{3}{2\pi\langle S^2\rangle_0}\right)^{3/2} N \exp\left(-\frac{3r^2}{2\langle S^2\rangle_0}\right) \tag{3.19}$$

where r is the distance from the mass centre of the macromolecule.

The motion of a Brownian particle of the chosen macromolecule agitates a volume with size of $\langle S^2\rangle_0^{1/2}$ through its adjacent chain particles. This volume is the bigger the longer the macromolecules are. Note once more that the agitation comes through the chain, not through viscous friction. In this situation, for a particle with radius $a \ll \langle S^2\rangle_0^{1/2}$, the average environment has to be considered as a non-local liquid for which the following stress tensor can be written

$$\sigma_{ij} = -p\delta_{ij} + 2 \int \eta(\boldsymbol{r} - \boldsymbol{r}')\gamma_{ik}(\boldsymbol{r}')d\boldsymbol{r}'.$$

If the influence function $\eta(\boldsymbol{r})$ is known, the resistance-drag coefficient of a Brownian particle can be calculated (see Appendix E) as

$$\zeta B = 6\pi a \int \eta(\boldsymbol{r})d\boldsymbol{r}. \tag{3.20}$$

To find the influence function $\eta(s)$, we shall consider shear deformation of the system at velocity gradient γ_{ij}, while two macromolecular coils, separated by a distance d_j, move beside each other at velocity $\gamma_{ij}d_j$. We add to the sum the contributions of every coil, apart from the chosen one, and find the density distribution of the energy dissipation for the chosen coil. The proportionality coefficient depends only on the concentration of the Brownian particles, if an assumption is made that local dissipation is determined by relative velocities of macromolecular coils,

$$\eta(\boldsymbol{r})\gamma_{ij}\gamma_{ij} \sim \sum_a \rho(\boldsymbol{r})\rho(\boldsymbol{r} - \boldsymbol{d}^a)d_l^a d_k^a \gamma_{il}\gamma_{jk}. \tag{3.21}$$

When the linear in velocity gradients approach is considered, the equilibrium density distribution function (3.19) can be used. We turn the sum into the integral and, after calculating, obtain

$$\eta(\boldsymbol{r})\gamma_{ij}\gamma_{ij} \sim nN^2 \left(\frac{3}{2\pi\langle S^2\rangle_0}\right)^{3/2}$$

$$\times \left[(\langle S^2\rangle_0 - r^2)\delta_{lk} + r_l r_k\right] \times \exp\left(-\frac{3r^2}{2\langle S^2\rangle_0}\right) \gamma_{il}\gamma_{ik}. \tag{3.22}$$

Now the friction coefficient of the Brownian particle is calculated according to (3.20)

$$\zeta B \sim nN^2.$$

It means that

$$B \sim M^2. \tag{3.23}$$

The approximation of overlapping coils is very rough, the real value of the index in the dependence of the coefficient B on the length of the macromolecules apparently ought to be bigger than the estimated one, because one has to take into account extra motions of the macromolecule among the other chains in order to disentangle itself from its neighbours, as it was speculated by Bueche (1956).

Constraint-Release Estimate

The constraint-release theory, due to Graessley (1982), Klein (1986) and many others (see review by Watanabe 1999), studied the detailed mechanism of a large-scale lateral motion of a macromolecule in an entangled system, due to process of release of some constraints of the probe chain and jumps of some parts of the chain in lateral direction. The result of this consideration (specifically, the relaxation times of the macromolecule) is equivalent to the formal assumption that a particle of the chain is moving through the environment as a particle in viscous medium. The friction coefficient of the particle can be presented as a product of the friction coefficient in 'monomer' liquid, multiplied by some measure of enhancement of the friction coefficient due to the neighbouring chains. This measure of enhancement corresponds to the above parameter B, so that, referring to the results of the calculations, one can say that the constraint-release mechanism determines the dependence of the coefficient B on the lengths of the neighbouring chains as

$$B \sim M^3. \tag{3.24}$$

Approximation of the Dependence

So, according to the alternative estimations (3.23) and (3.24), the coefficient of friction of a Brownian particle increases with increase in length of macromolecules. One has to distinguish the probe macromolecule (with molecular weight or length M) and the neighbouring macromolecules (with the length M_0), even if all of them are equal. The derived estimates of the parameter B show that the parameter depends only on the length of neighbouring macromolecules, so that the derived dependence of the enhancement coefficient on molecular weights of the macromolecules can be written as

$$B \sim Z_0^\delta Z^0, \quad Z_0 = \frac{M_0}{M_e}, \quad Z = \frac{M}{M_e}.$$

The length of macromolecules is measured in units of M_e, M_e being the length of a part of a macromolecule between 'adjacent entanglements'). For the index δ, the above estimations give the values between 2 and 3.

It is convenient to have an approximate expression for the dependence of the parameter B on molecular length, and, accepting $B = 1$ at $Z_0 = 2$, one can write the universal function

$$B \approx \left(\frac{Z_0}{2} \right)^{\delta}. \tag{3.25}$$

The formula contains only one index δ, which has been estimated theoretically ($\delta = 2$ or 3) and empirically. In the last case, note, that in virtue of equation (6.52), the parameter B can be derived directly from measurements of coefficient of viscosity η

$$B = \frac{6\eta}{\pi^2 n T \tau^*},$$

where τ^* is the characteristic Rouse relaxation time of the macromolecule in viscous 'monomer' liquid (see Section 3.3.1). According to experiment, coefficient of viscosity is proportional to $M^{3.4}$, so that the reliable empirical estimation of index δ, is $\delta = 2.4$. This value corresponds to the above theoretical estimation of index.

3.3.3 Intramolecular Friction

If deformation of the system is fast enough (that is, before relaxation of chains can occurs), one expects that macromolecules deform affinely, i.e., for every particle $\dot{r}_i^{\alpha} = \nu_{ij} r_j^{\alpha}$, where ν_{ij} is the velocity gradient, and r_j^{α} is the position in space of a particle α of a chain. Under given deformation, the external force (3.6) is equal to zero, while the intramolecular resistance force (3.7) is proportional to $\dot{r}_i^{\alpha} - \omega_{ij} r_j^{\alpha}$, where ω_{ij} is the vorticity, or $\gamma_{ij} r_j^{\alpha}$, where γ_{ij} is the symmetric part of the velocity gradient, so that this force is a force of the intramolecular resistance due to the change in shape of the macromolecular coil (kinetic stiffness). As far as we consider the coarse-grained approximation, all the neighbouring chains, or, one can say, the particles of coarse-grained chains follow the deformation affinely, and there is no apparent cause for this force. To explain the emerging of the force, we have to refer to more detailed model of macromolecule – to the chain of freely-jointed rigid segments. Apparently, small parts of macromolecules cannot follow the deformation affinely, segments can only rotate, and an extra force is needed to change the direction of a segment in the case, when the segments of the other chains present around. That is why we can say that the internal resistance force for a macromolecule in a polymer melt has to be attributed to the interaction with neighbouring chains, though in the coarse-grained approximation we forget about segments, and this force is characterised by only phenomenological coefficient of internal resistance, which, in the simplest case, can be denoted as

$$\zeta E = \int_0^{\infty} \varphi(s) \mathrm{d}s. \tag{3.26}$$

This quantity has value of zero for non-entangled systems and increase with increase in the length of macromolecules. As for external force, there is a slight difference in resistance, when the particle moves along the chain or in a perpendicular direction, but, in this subsection, this effect is neglected for simplicity.

Approximation of Internal Resistance Matrix

The force of intramolecular resistance appears, when relative motion of the particles exists, so that one can write a general expression (which is identical to expression (2.21) for a chain in a dilute solution)

$$G_{ij}^{\alpha\gamma} = \sum_{\gamma \neq \alpha} C_{\alpha\gamma}(u_j^\alpha - u_j^\gamma) e_j^{\alpha\gamma} e_i^{\alpha\gamma},$$

where $e_j^{\alpha\gamma} = (r_j^\alpha - r_j^\gamma)/|r^\alpha - r^\gamma|$. Matrix $C_{\alpha\gamma}$ is symmetrical, the components of the matrix are non-negative and may depend on the distance between the particles. The diagonal components of the matrix are equal to zero. One can also reasonably assume that components of the matrix are equal to zero, if the difference between indexes $|\alpha - \gamma|$ is less than a certain value.

One can rewrite the matrix of internal resistance in the following form

$$G_{ij}^{\alpha\gamma} = \delta_{\alpha\gamma} \sum_{\beta \neq \alpha} C_{\alpha\beta} e_i^{\alpha\beta} e_j^{\alpha\beta} - C_{\alpha\gamma} e_i^{\alpha\gamma} e_j^{\alpha\gamma}. \tag{3.27}$$

This expression defines the general form of a matrix of internal friction, which allows the force to remain unchanged on the rotation of the macromolecular coil as a whole. The written matrix is symmetrical with respect to the upper and lower indices and, in contrast to matrix $C_{\alpha\beta}$, has non-zero diagonal components, which are depicted by the first term in (3.27). In equilibrium situations, after averaging over the orientation, matrix (3.27) can be presented as

$$G_{ij}^{\alpha\gamma} = G^{\alpha\beta} \delta_{ij}, \quad G^{\alpha\beta} = \frac{1}{3}\left(\delta_{\alpha\gamma} \sum_{\beta \neq \alpha} C_{\alpha\beta} - C_{\alpha\gamma}\right).$$

Since the components of matrix $C_{\alpha\beta}$ are non-negative, the diagonal components of matrix $G^{\alpha\beta}$ exceed the non-diagonal ones and can be considered to be approximately diagonal with respect to the indices α and β. The effect is very strong for long macromolecules and reduces to zero at $M \approx 2M_e$. As an initial approximation, to express the idea of severe confinement, one can assume that the intramolecular resistance force is determined equally by all the particles of the chain, so that the matrix is reduced to the already written matrix (3.10). It is easy to find, that in normal co-ordinates (1.13), matrix (3.10) has a diagonal form with the eigenvalues

$$\varphi_\alpha = \begin{cases} 0, & \alpha = 0, \\ 1, & \alpha \neq 0. \end{cases} \tag{3.28}$$

One can directly check that, if matrix (3.10) is modified, for example, zeros are placed on a few diagonals next to the main diagonal in the matrix, the transformed matrix retains approximately its diagonal form, while eigenvalues are close to unity and decrease slightly when the index of an eigenvalue increases. So, the effect of diagonals with zeros can be neglected indeed and the above matrix does approximate the situation for large-scale motions of the chain at $N \to \infty$

Approximation of the Measure of Internal Resistance

For the systems of long macromolecules (strongly entangled systems), the requirements of universality and self-consistency allow us to write practically identical asymptotic relations (5.17) and (6.53) between the parameter χ, introduced in Section 3.3.1, and the ratio E/B, which allows us to write for this case

$$E \sim M_0^\delta M.$$

For the weakly entangled systems, one can expect, that the ratio E/B, that is the parameter of 'internal' viscosity is small. It can be demonstrated in Section 4.2.3, that transition point from weakly to strongly entangled systems occurs at $E \approx B$. To describe these facts, one can use any convenient approximate function for the measure of internal resistance, for example, the simple formula

$$E \approx B(Z_0) \frac{12\,(Z-2)^2}{Z+768}, \quad Z_0 = \frac{M_0}{M_e}, \quad Z = \frac{M}{M_e}. \tag{3.29}$$

3.3.4 Fundamental Dynamical Parameters

To describe the behaviour of a macromolecule in an entangled system, we have introduced the ratio of the relaxation times χ and two parameters B and E connected with the external and the internal resistance, respectively. These parameters play a fundamental role in the description of the dynamical behaviour of polymer systems, so that it is worthwhile to discuss them once more and to consider their dependencies on the concentration of polymer in the system.

Equations (3.17), (3.25) and (3.29) define the dependence of the parameters on the length of a macromolecule due to empirical evidence. The above-written relations are applicable to all linear polymers, whatever their chemical structure is. One can also define these quantities as functions of concentration. Indeed, one can see that the parameters χ, B and E can be written as functions of a single argument. Actually, since the above kinetic restrictions on the motion of a macromolecule are related to the geometry of the system, the only parameters in this case are the number of macromolecules per unit volume n and the mean square end-to-end distance $\langle R^2 \rangle$, while (see formulae (1.4) and (1.33))

$$n \sim \frac{c}{M}, \quad \langle R^2 \rangle \sim C_\infty(T)M.$$

We shall not pay attention to the optional slight dependence of the last quantity on the concentration (see Section 1.6). The non-dimensional quantities

χ, B and E can therefore be regarded as universal and independent of the chemical structure of the polymer functions of the non-dimensional parameter

$$n\langle R^2\rangle^{3/2} \sim c\, C_\infty^{3/2} M^{1/2}.$$

Now the dependencies of the phenomenological parameters χ, B and E on the concentration of polymer c can also be given. From the above relations, it follows, for example, that for the strongly entangled systems

$$\chi \sim c^{-2} C_\infty^{-3} M^{-1}, \qquad B \sim c^{2\delta} C_\infty^{3\delta} M_0^{\delta}, \qquad E \sim c^{2(1+\delta)} C_\infty^{3(1+\delta)} M_0^{\delta} M. \quad (3.30)$$

In these formulae, however, the coefficients of proportionality must be estimated empirically.

Note that the above estimates of the coefficients B and E are valid for linear macromolecule. The principles of the theory can be applied to macromolecules of a different architecture: to macromolecule as a ring, a brush, a star, or something else. One can expect that the enhancement coefficient for the friction coefficient of a Brownian particle can be also introduced, but for macromolecule of complex architecture, index δ in (3.25) can be specific in each case. Moreover, dependence on the molecular weights cannot be a power function at all. Of course, the choice of memory functions is eventually justified by empirical facts discussed in later chapters, so we consider the memory functions (3.15) to be empirical, but to give a rather good description for the case $c^2 M \to \infty$ for linear macromolecules. In other situations, the memory functions (3.15) ought to be chosen in different ways.

3.4 Markovian Form of Dynamic Equation

Now one can return to dynamic equation (3.4) of a macromolecule in very concentrated solutions and melts of polymers, which can be rewritten in the form

$$\frac{dr_i^\alpha}{dt} = u_i^\alpha,$$

$$m\frac{du_i^\alpha}{dt} = -\zeta(u_i^\alpha - \nu_{ij} r_j^\alpha) + F_i^\alpha + G_i^\alpha - 2\mu T A_{\alpha\gamma} r_i^\gamma + \phi_i^\alpha(t). \quad (3.31)$$

Due to the preceding analysis, the extra forces of external and internal resistance F_i^α and G_i^α can be specified as

$$F_i^\alpha = -B\frac{\varsigma}{\tau}\int_0^\infty \exp\left(-\frac{s}{\tau}\right) H_{ij}^{\alpha\gamma}(u_j^\gamma - \nu_{jl} r_l^\gamma)_{t-s} ds, \quad (3.32)$$

$$G_i^\alpha = -E\frac{\varsigma}{\tau}\int_0^\infty \exp\left(-\frac{s}{\tau}\right) G_{ij}^{\alpha\gamma}(u_j^\gamma - \omega_{jl} r_l^\gamma)_{t-s} ds. \quad (3.33)$$

In these formulae, ζ is a friction coefficient of a particle in a "monomer" liquid, B and E are phenomenological parameters discussed in the previous sections, while matrixes $H_{ij}^{\alpha\gamma}$ and $G_{ij}^{\alpha\gamma}$ are numerical matrixes defined by relations (3.8) and (3.10) in linear approximation and by relation (3.13) in approximation of local anisotropy.

Expressions (3.32) and (3.33) are solutions of equations which are written below in the simplest covariant form (see Section 8.4 and Appendix D)

$$\tau\left(\frac{dF_i^\alpha}{dt} - \omega_{il}F_l^\alpha\right) + F_i^\alpha = -\zeta BH_{ij}^{\alpha\gamma}(u_j^\gamma - \nu_{jl}r_l^\gamma),$$ (3.34)

$$\tau\left(\frac{dG_i^\alpha}{dt} - \omega_{il}G_l^\alpha\right) + G_i^\alpha = -\zeta EG_{ij}^{\alpha\gamma}(u_j^\gamma - \omega_{jl}r_l^\gamma).$$ (3.35)

The properties of the stochastic forces in the system of equations (3.31)–(3.35) are determined by the corresponding correlation functions which, usually (Chandrasekhar 1943), are found from the requirement that, at equilibrium, the set of equations must lead to well-known results. This condition leads to connection of the coefficients of friction with random-force correlation functions – the dissipation-fluctuation theorem. In the case under consideration, when matrixes $H_{ij}^{\alpha\gamma}$ and $G_{ij}^{\alpha\gamma}$ depend on the co-ordinates but not on the velocities of particles, the correlation functions of the stochastic forces in the system of equations (3.31) can be easily determined, according to the general rule (Dünweg 2003), as

$$\langle\phi_i^\alpha(t)\phi_k^\gamma(t')\rangle = T\zeta\left[2\delta_{\alpha\gamma}\delta_{ik}\delta(t-t') + \frac{1}{\tau}(BH_{ij}^{\alpha\gamma} + EG_{ij}^{\alpha\gamma})\exp\left(-\frac{t-t'}{\tau}\right)\right].$$ (3.36)

The random process in equation (3.31) can be conveniently represented as the sum of two independent processes

$$\phi_i^\alpha(t) = \bar{\phi}_i^\alpha(t) + \tilde{\phi}_i^\alpha(t),$$

so that, introducing the variable $\Phi_i^\alpha = F_i^\alpha + G_i^\alpha + \tilde{\phi}_i^\alpha(t)$, the system of equations (3.31), (3.34) and (3.35) can be written as

$$\frac{dr_i^\alpha}{dt} = u_i^\alpha,$$

$$m\frac{du_i^\alpha}{dt} = -\zeta(u_i^\alpha - \nu_{ij}r_j^\alpha) + \Phi_i^\alpha - 2\mu TA_{\alpha\gamma}r_i^\gamma + \bar{\phi}_i^\alpha(t),$$ (3.37)

$$\tau\frac{d\Phi_i^\alpha}{dt} = -\Phi_i^\alpha - \zeta BH_{ij}^{\alpha\gamma}(u_j^\gamma - \nu_{jl}r_l^\gamma) - \zeta EG_{ij}^{\alpha\gamma}(u_j^\gamma - \omega_{jl}r_l^\gamma) + \sigma_i^u(t).$$

The first two of the above equations represent Langevin equation for the Rouse chain in presence of extra random force Φ_i^α. One can note that dynamics of a polymer chain in random fields was studied extensively (Baumgärter and Muthukumar 1996; Ebert et al. 1996), as a possible mode of motion of a macromolecule in entangled system. Note also that the two top equation from (3.37) (at $m = 0$) are identical to the Langevin equation, which was formulated (Migliorini et al. 2003) for investigating the behaviour of polymer chain in a random static field. The equation was investigated numerically by Milchev et al. (2004). However, in contrast to the cited works, the force Φ_i^α in equations (3.37) for chain in entangled system is not static and can be specially defined according to the third equation.

The random process in the last stochastic equation from set (3.37) is related to the above introduced random process by equation

$$\sigma_i^\gamma = \tilde{\phi}_i^\gamma + \tau \left(\frac{d}{dt} \tilde{\phi}_i^\gamma - \omega_{il} \tilde{\phi}_l^\gamma \right), \tag{3.38}$$

which can be looked upon as the equation for the random force $\tilde{\phi}_i^\gamma$ for the given random quantity σ_i^α. Then, if the relation

$$\langle \sigma_i^\gamma(t) \sigma_j^\mu(t') \rangle = 2\zeta (B\,H_{ij}^{\gamma\mu} + E\,G_{ij}^{\gamma\mu})\,\delta(t - t') \tag{3.39}$$

is satisfied, the random force correlator satisfies the following relation

$$\langle \tilde{\phi}_i^\gamma(t) \tilde{\phi}_j^\mu(t') \rangle = \frac{T\zeta}{\tau} \left(B\,H_{ij}^{\gamma\mu} + E\,G_{ij}^{\gamma\mu} \right) \exp\left(-\frac{t - t'}{\tau} \right). \tag{3.40}$$

This relation, in line with relation

$$\langle \bar{\phi}_i^\gamma(t) \bar{\phi}_j^\mu(t') \rangle = 2T\zeta \delta_{\gamma\mu} \delta_{ij} \delta(t - t'), \tag{3.41}$$

return us to the random-force correlation function (3.36).

The set of stochastic equations given by (3.37) is equivalent (in the linear case) to equations (3.11) with the memory functions defined in Section 3.3, but, in contrast to equations (3.11), set (3.37) is written as a set of Markov stochastic equations. This enables us to determine the variables that describe the collective motion of the set of macromolecules. In this particular approximation, the interaction between neighbouring macromolecules ensures that the phase variables of the elementary motion are co-ordinates, velocities, and some other vector variables – the extra forces. This set of phase variables describes the dynamics of the entire set of entangled macromolecules. Note that the Markovian representation of the equation of macromolecular dynamics cannot be made for any arbitrary case, but only for some simple approximations of the memory functions. We are considering the case with a single relaxation time, but generalisation for a case with a few relaxation times is possible.

3.5 Reptation-Tube Model

The system of dynamic equations (3.37) for a chain of Brownian particles with local anisotropy of mobility appears to be rather complicated for direct analysis, and one ought to use numerical methods, described in the next Section, to be convinced that equations (3.37) really describe the observed effects. Not to explore non-linear equations, one can exaggerate anisotropy of mobility, assuming that unbounded lateral motion of particles is completely suppressed due to the presence of many neighbouring coils. By this way, one comes to a very elegant linear model of reptating macromolecule proposed by Doi and Edwards (1978) (see also Doi and Edwards 1986).

Following Doi and Edwards (1978), we shall consider a bead-spring model consisting of $Z = M/M_e$ subchains and assume that the distance between adjacent particles along the chain is constant and equal to a certain intermediate length ξ, which is considered to be the radius of 'a tube', so that the number of particles is not arbitrary, but satisfies the condition

$$Z\xi^2 = \langle R^2 \rangle. \tag{3.42}$$

The states of the macromolecule will be considered in points of time in a time interval Δt, so that the stochastic motion of Brownian particles of the chain can be described by the equation for the particle co-ordinates

$$r^0(t + \Delta t) = \frac{1 + \phi(t)}{2} r^1(t) + \frac{1 - \phi(t)}{2} [r^0(t) + v(t)],$$

$$r^\nu(t + \Delta t) = \frac{1 + \phi(t)}{2} r^{\nu+1}(t) + \frac{1 - \phi(t)}{2} r^{\nu-1}(t), \quad \nu = 1, 2, \ldots, Z - 1, \tag{3.43}$$

$$r^Z(t + \Delta t) = \frac{1 + \phi(t)}{2} [r^Z(t) + v(t)] + \frac{1 - \phi(t)}{2} r^{Z-1}(t)$$

where $\phi(t)$ is a random quantity, which takes the values $+1$ or -1, and $v(t)$ is a vector of constant length ξ and random direction, so that

$$\langle \phi(t)\phi(u) \rangle = \delta_{tu}, \quad \langle \phi(t) \rangle = 0,$$
$$\langle v(t)v(u) \rangle = \delta_{tu}\xi^2, \quad \langle v(t) \rangle = 0. \tag{3.44}$$

The set of equations (3.43) describes the stochastic motion of a chain along its contour. The "head" and the "tail" particles of the chain can choose random directions. Any other particle follows the neighbouring particles in front or behind. The smaller the time interval Δt is the quicker moves the chain. Clearly, the time interval cannot be an arbitrary quantity and is specified by the requirement that the squared displacement of the entire chain by diffusion for the interval Δt is equal to ξ^2, so that

$$\xi^2 = 2D_0\Delta t = \frac{2T}{\zeta Z}\Delta t \tag{3.45}$$

where $D_0 = T/(\zeta Z)$ is the diffusion coefficient of the macromolecule in a monomeric viscous liquid (see Section 5.1.1 for explanation). Note that we follow the original Doi-Edwards model in which diffusion of the chain is considered to be one-dimensional.

The model described by equations (3.42)–(3.45) is valid for equilibrium situations. For chain in a flow, one ought to define displacements of the particles under flow and to consider the average values (3.44) to depend on the velocity gradient (Doi and Edwards 1986). McLeish and Milner (1999) considered mechanism of reptation motion of branched macromolecules of different architecture.

It is convenient to rewrite equations (3.43) in more compact form, taking also definition of Δt into account,

$$\frac{r^{\alpha}(t+\Delta t) - r^{\alpha}(t)}{\Delta t} = -\frac{T}{\zeta \xi^2 Z^3} A_{\alpha\gamma} r^{\gamma}(t) + \sigma^{\alpha}(t), \quad \alpha = 0, 1, 2, \ldots, Z, \quad (3.46)$$

where the stochastic force is defined as

$$\sigma^{\alpha}(t) = \frac{1}{2} \phi(t) \times \begin{cases} r^1(t) - r^0(t) + v(t), & \alpha = 0, \\ r^{\alpha+1}(t) - r^{\alpha-1}(t), & \alpha = 1, 2, \ldots, N-1, \\ r^{Z-1}(t) - r^Z(t) + v(t), & \alpha = Z. \end{cases} \quad (3.47)$$

To obtain relation (3.46), one has to take into account that motion of the particles of the chain ought to be considered to be coherent. Now, it is not difficult to pass from equation (3.46) to the normal-mode equation

$$\frac{d\rho^{\alpha}}{dt} = -\frac{\pi^2 T \alpha^2}{\zeta \xi^2 Z^3} \rho^{\alpha} + Q_{\gamma\alpha} \sigma^{\gamma}(t), \quad \alpha = 0, 1, 2, \ldots, \ll Z. \quad (3.48)$$

These equations describe the reptation normal relaxation modes, which can be compared with the Rouse modes of the chain in a viscous liquid, described by equation (2.29). In contrast to equation (2.29) the stochastic forces (3.47) depend on the co-ordinates of particles, equation (3.48) describes anisotropic motion of beads along the contour of a macromolecule.

It is instructive to compare the system of equations (3.46) and (3.47) with the system (3.37). One can see that both the radius of the tube and the positions of the particles in the Doi-Edwards model are, in fact, mean quantities from the point of view of a model of underlying stochastic motion described by equations (3.37). The intermediate length ξ emerges at analysis of system (3.37) and can be expressed through the other parameters of the theory (see details in Chapter 5). The mean value of position of the particles can be also calculated to get a complete justification of the above model. The direct introduction of the mean quantities to describe dynamics of macromolecule led to an oversimplified, mechanistic model, which, nevertheless, allows one to make correct estimates of conformational relaxation times and coefficient of diffusion of a macromolecule in strongly entangled systems (see Sections 4.2.2 and 5.1.2). However, attempts to use this model to formulate the theory of viscoelasticity of entangled systems encounted some difficulties (for details, see Section 6.4, especially the footnote on p. 133) and were unsuccessful.

There were different generalisations of the reptation-tube model, aimed to soften the borders of the tube and to take into account the underlying stochastic dynamics. It seems that the correct expansion of the Doi-Edwards model, including the underlying stochastic motion and specific movement of the chain along its contour – the reptation mobility as a particular mode of motion, is presented by equations (3.37), (3.39) and (3.41). In any case, the introduction of local anisotropy of mobility of a particle of chain, as described by these equations, allows one to get the same effects on the relaxation times and mobility of macromolecule, which are determined by the Doi-Edwards model.

3.6 Method of Numerical Simulation

One can consider equations (3.37), (3.39) and (3.41) to be a basic system of equations for description of dynamics of entangled systems. The system can be investigated analytically in linear approximation as will be demonstrated in the ensuing chapters. However, to study these non-linear equations in complete form, one has to use numerical methods of simulation of the stochastic processes for the particle coordinates.

3.6.1 Non-Dimensional Form of Dynamic Equation

It is convenient to use the time scale τ^*, which is called the Rouse characteristic relaxation time and is a combination of parameters of the theory

$$\tau^* = \frac{\zeta N \langle R^2 \rangle}{6\pi^2 T} = \frac{\zeta N^2}{4\pi^2 \mu T} \sim M^2, \tag{3.49}$$

and the length scale $R = \sqrt{\langle R^2 \rangle}$, where $\langle R^2 \rangle$ is the end-to-end distance, to define non-dimensional variables as

$$\frac{\mathrm{d}}{\mathrm{d}t} = \frac{1}{\tau^*} \frac{\mathrm{d}}{\mathrm{d}s},$$
$$r_i^\alpha = R\, R_i^\alpha,$$
$$u_i^\alpha = \frac{R}{\tau^*} U_i^\alpha,$$
$$F_i^\alpha + G_i^\alpha + \tilde{\phi}_i^\alpha(t) = \frac{\zeta R}{\tau^*}\, \Phi_i^\alpha,$$

so that the system of equations of dynamics of macromolecule can be written in the form

$$\frac{dR_i^\alpha}{ds} = U_i^\alpha,$$

$$\frac{m}{\zeta\tau^*}\frac{dU_i^\alpha}{ds} = -U_i^\alpha + \Phi_i^\alpha - \frac{N^2}{2\pi^2}A_{\alpha\gamma}R_i^\gamma + \bar{f}_i^\alpha(s),$$

$$2\chi B\frac{d\Phi_i^\alpha}{ds} = -\Phi_i^\alpha - BH_{ij}^{\alpha\gamma}U_j^\gamma - EG_{ij}^{\alpha\gamma}U_j^\gamma + \tilde{f}_i^\alpha(s), \quad \chi = \frac{\tau}{2\tau^*B},$$

$$\bar{f}_i^\alpha(s) = \frac{\tau^*}{\zeta R}\bar{\phi}_i^\alpha(\tau^*s), \qquad \tilde{f}_i^\alpha(s) = \chi B\frac{d\tilde{\phi}_i^\alpha}{ds} + \tilde{\phi}_i^\alpha(s),$$

$$\langle\bar{f}_i^\gamma(s)\bar{f}_j^\mu(s')\rangle = \left(\frac{\tau^*}{\zeta R}\right)^2\langle\bar{\phi}_i^\gamma(\tau^*s)\bar{\phi}_j^\mu(\tau^*s')\rangle = \frac{N}{3\pi^2}\delta_{\gamma\mu}\delta_{ij}\delta(s-s'), \quad (3.50)$$

$$\langle\tilde{f}_i^\gamma(s)\tilde{f}_j^\mu(s')\rangle = \frac{N}{3\pi^2}(BH_{ij}^{\gamma\mu} + EG_{ij}^{\gamma\mu})\delta(s-s'). \tag{3.51}$$

The inertial effects can be neglected ($m = 0$), so that the above system of equations can be written as

$$\frac{dR_i^\alpha}{ds} = U_i^\alpha,$$

$$\frac{d\Phi_i^\alpha}{ds} = \frac{1}{2\chi B}\left(-\Phi_i^\alpha - BH_{ij}^{\alpha\gamma}U_j^\gamma - EG_{ij}^{\alpha\gamma}U_j^\gamma + \tilde{f}_i^\alpha(s)\right),$$

$$U_i^\alpha = \Phi_j^\alpha - \frac{1}{2}\frac{N^2}{\pi^2}A_{\alpha\gamma}R_j^\gamma + \bar{f}_j^\alpha(s). \tag{3.52}$$

Relations (3.50) and (3.51) are being satisfied, if the random processes are given as

$$\bar{f}_i^\gamma(s) = \left(\frac{N}{3\pi^2}\right)^{1/2}\bar{g}_i^\gamma(s), \tag{3.53}$$

$$\tilde{f}_i^\gamma(s) = \left(\frac{N}{3\pi^2}B\right)^{1/2}\left\{(A_e\delta_{ij} + C_e e_i^\gamma e_j^\gamma)\tilde{g}_{ej}^\gamma(s) + \frac{1}{N}\sqrt{\frac{E}{B}}\right.$$

$$\left.\times\left[(1+N)(A_i\delta_{ij} + C_i e_i^\gamma e_j^\gamma)\tilde{g}_{ij}^\gamma(s) - \sum_{\alpha=0}^N(A_i\delta_{ij} + C_i e_i^\alpha e_j^\alpha)\tilde{g}_{ij}^\alpha(s)\right]\right\},$$

$$A_e = \sqrt{1+a_e/2}, \quad C_e = -A_e + \sqrt{1-a_e},$$

$$A_i = \sqrt{(1+a_i)/2}, \quad C_i = -A_i + \sqrt{1-a_i}, \tag{3.54}$$

where $\bar{g}_i^\gamma(s)$, $\tilde{g}_{ej}^\gamma(s)$ and $\tilde{g}_{ij}^\gamma(s)$ are independent Gaussian random processes with dispersion equal to unity.

Dynamics of a single macromolecule in an entangled system is defined by the system of non-linear equations (3.52)–(3.54), containing some phenomenological parameters, which will be identified later.

3.6.2 Algorithm of Calculation

We use the simplest method (Eyler method, with the step of integration h)
to write the algorithm for numerical solution of equations (3.52)

$$R_i^\alpha(s+h) = R_i^\alpha(s) + hU_i^\alpha,$$

$$\Phi_i^\alpha(s+h) = \Phi_i^\alpha(s) + \frac{h}{2\chi}\left(-\frac{1}{B}\Phi_i^\alpha - H_{ij}^{\alpha\gamma}U_j^\gamma - \psi G_{ij}^{\alpha\gamma}U_j^\gamma + \Delta\bar{f}_i^\alpha(s)\right), \quad (3.55)$$

$$U_i^\alpha = \Phi_j^\alpha - \frac{N^2}{2\pi^2}A_{\alpha\gamma}R_j^\gamma + \Delta\tilde{f}_j^\alpha(s).$$

The random forces are defined here as

$$\Delta\bar{f}_j^\alpha(s) = \frac{1}{h}\int_s^{s+h}\bar{f}_j^\alpha(u)\,du, \qquad \Delta\tilde{f}_j^\alpha(s) = \frac{1}{hB}\int_s^{s+h}\tilde{f}_j^\alpha(u)\,du.$$

Calculating the dispersions of the random processes $\Delta\bar{f}_j^\alpha(s)$ and $\Delta\tilde{f}_i^\alpha(s)$, one
has to take into account that relations (3.50) and (3.51) for the random forces
are written for the continuous time, so that in the discrete approach one has

$$\langle\Delta\bar{f}_i^\gamma(s)\,\Delta\bar{f}_j^\mu(s')\rangle = \frac{N}{3\pi^2h}\delta_{\gamma\mu}\delta_{ij}\delta(s-s'), \qquad (3.56)$$

$$\langle\Delta\tilde{f}_i^\gamma(s)\,\Delta\tilde{f}_j^\mu(s')\rangle = \frac{N}{3\pi^2hB}(H_{ij}^{\gamma\mu} + \psi G_{ij}^{\gamma\mu})\,\delta(s-s'). \qquad (3.57)$$

It is easy to see that, the expressions for the random forces have to be similar
to relations (3.53) and (3.54), that is

$$\Delta\bar{f}_i^\gamma(s) = \left(\frac{N}{3\pi^2h}\right)^{1/2}\bar{g}_i^\gamma(s), \qquad (3.58)$$

$$\Delta\tilde{f}_i^\gamma(s) = \left(\frac{N}{3\pi^2hB}\right)^{1/2}\left\{(A_e\delta_{ij} + C_e e_i^\gamma e_j^\gamma)\,\tilde{g}_{ej}^\gamma(s)\right.$$

$$+ \frac{1}{N}\sqrt{\frac{E}{B}}\times\left[(1+N)(A_i\delta_{ij} + C_i e_i^\gamma e_j^\gamma)\,\tilde{g}_{ij}^\gamma(s)\right.$$

$$\left.\left. - \sum_{\alpha=0}^{N}(A_i\delta_{ij} + C_i e_i^\alpha e_j^\alpha)\,\tilde{g}_{ij}^\alpha(s)\right]\right\}. \qquad (3.59)$$

To solve the system of equations (3.55), initial values of co-ordinates and
an extra random force have to be chosen. We accept that

$$R_i^0(0) = 0, \qquad R_i^\alpha(0) = R_i^{\alpha-1}(0) + \frac{1}{\sqrt{N}}g_i^\alpha, \quad \alpha = 1, 2, \ldots, N, \qquad (3.60)$$

where g_i^α is a Gaussian random process with dispersion equal to unity. Initial
values of the extra random force can be chosen as

$$\Phi_i^\alpha(0) = 0, \quad \alpha = 0, 1, 2, \ldots, N, \ i = 1, 2, 3. \tag{3.61}$$

As a result of calculation, one has the positions of the particles

$$R_i^\alpha(s), \quad \alpha = 1, 2, \ldots, N, \ i = 1, 2, 3, \ s = 0, h, 2h, 3h, \ldots,$$

which allows one to calculate mean values of different quantities. It is convenient also to use the normal co-ordinates defined by equation (1.13).

Note that steady-state situations are investigated, so that the end-to-end distance $\langle R^2 \rangle$ and the mean gyration radius

$$S^2 = \frac{1}{1+N} \sum_{\alpha=0}^{N} \sum_{i=1}^{3} (R_i^\alpha - Q_i)^2, \quad Q_i = \frac{1}{1+N} \sum_{\alpha=0}^{N} R_i^\alpha \tag{3.62}$$

must be constant on average. The mean kinetic energy for one degree of freedom also must be constant

$$\frac{1}{3(1+N)} \sum_{\alpha=0}^{N} \sum_{i=1}^{3} U_i^\alpha U_i^\alpha \sim const. \tag{3.63}$$

The above conditions allow us to monitor whether the fluctuation-dissipation relations are valid during calculations.

Chapter 4
Conformational Relaxation

Abstract The fundamental model of macromolecular dynamics in an entangled system, which was formulated in the previous chapter, imitates the basic isotropic stochastic motion of the particles of the chain among the neighbouring chains, and includes a special non-linear effect – reptation motion of the macromolecule. The system of equations allows one to find correlation functions of coordinates and calculate conformational relaxation times of macromolecular coils. In the analytic investigation, some approximations of the fundamental system will be considered: instead of a single non-linear equation, we shall consider two particular cases: linear mesoscopic equation for weakly entangled systems and the original non-amended reptation-tube model for strongly entangled systems. We consider these two models as complementary models and combine the results, unless an analysis of unified non-linear model is available. The numerical investigation of complete non-linear model allows us to calculate the times of relaxation of the macromolecular coil, while the transition point between two modes of relaxation (diffusive and reptation) is evaluated as $10M_e$. Both for the weakly and strongly entangled systems, in contrast to relaxation behaviour of the macromolecule in a viscous liquid, two relaxation branches emerge as characteristics of the relaxation behaviour of a macromolecular coil in a system of entangled macromolecules.

4.1 Correlation Functions for the Linear Dynamics

4.1.1 Modified Cerf-Rouse Modes

One has to refer to dynamic equations (3.11) of the macromolecule to find independent modes of motion. The matrices A and G in these equations are defined by equations (1.8) and (3.10), and, in the normal co-ordinates (1.13), simultaneously have diagonal forms

V.N. Pokrovskii, *The Mesoscopic Theory of Polymer Dynamics*,
Springer Series in Chemical Physics 95,
DOI 10.1007/978-90-481-2231-8_4, © Springer Science+Business Media B.V. 2010

$$Q_{\alpha\lambda}A_{\alpha\gamma}Q_{\gamma\nu} = \lambda_\nu \delta_{\lambda\nu},$$

$$Q_{\alpha\lambda}G_{\alpha\gamma}Q_{\gamma\nu} = \varphi_\nu \delta_{\lambda\nu},$$

where the eigenvalues λ_ν and φ_ν are defined by equation (1.17) and (3.28), correspondingly.

The zeroth eigenvalues of matrix A and G are zero, so that without any approximation, one can write an equation for diffusion mode

$$m\frac{d^2\rho_i^0}{dt^2} = -\int_0^\infty \beta(s)(\dot{\rho}_i^0 - \nu_{ij}\rho_j^0)_{t-s}ds + \xi_i^0. \tag{4.1}$$

The intermolecular forces are naturally absent from the equation for the zeroth mode, because this mode presents the motion of the mass centre of the coil.

In accordance to approximate form (3.10), the other eigenvalues of the matrix G are constant and equal to unity, so that the set of equations for relaxation modes of the macromolecule now assumes the form

$$m\frac{d^2\rho_i^\nu}{dt^2} = -\int_0^\infty \beta(s)(\dot{\rho}_i^\nu - \nu_{ij}\rho_j^\nu)_{t-s}ds$$

$$-\int_0^\infty \varphi(s)(\dot{\rho}_i^\nu - \omega_{ij}\rho_j^\nu)_{t-s}ds - 2\mu T\lambda_\nu\rho_i^\nu + \xi_i^\nu, \quad \nu = 1, 2, \ldots, N. \tag{4.2}$$

Let us note, that the matrixes A and G are approximations of the real situation; though, in any case, the zeroth eigenvalues of the matrixes must be zero and equation (4.1) for diffusive mode is valid, the other eigenvalues of matrix G depends, generally speaking, on the mode label. In fact, the written equations for the relaxation modes are implementation of the statements that the motion of a single macromolecule can be separated from others, and the motion of a single macromolecule can be expanded into an independent motion of modes.

We shall now start with the formal representation of the solution of equations (4.2), which, first of all, is conveniently written in the form

$$\rho_i^\nu(t) = \int_0^\infty \{\chi_\nu(s)\xi_i^\nu(t-s) + [\mu_\nu(s)\nu_{il}(t-s) + \pi_\nu(s)\omega_{il}(t-s)]\rho_l^\nu(t-s)\}ds, \tag{4.3}$$

where functions χ_α, μ_α and π_α are determined by their Fourier one-side transforms

$$\chi_\nu[\omega] = (2T\mu\lambda_\nu - m\omega^2 - i\omega B[\omega])^{-1},$$

$$\mu_\nu[\omega] = \beta[\omega]\chi_\nu[\omega], \qquad \pi_\nu[\omega] = \varphi[\omega]\chi_\nu[\omega]. \tag{4.4}$$

Functions χ_α, μ_α and π_α always vanish for $s \to 0$ and $s \to \infty$, if $m \neq 0$. Within the limits of applicability of the subchain model, the inertial effects have to be omitted, i.e. we can believe that $m = 0$, but this limit can change the values of functions (4.4) as functions of time s at limiting cases for $s \to 0$ and $s \to \infty$. To avoid any discrepancies, the results ought to be calculated at $m \neq 0$. Then the limiting values at $m \to 0$ can be obtained.

The expression for the velocity of the normal co-ordinate follows from equation (4.3). Differentiating (4.3) with respect to time, and integrating by parts, we find, by using the above shown properties of the integrands, that

$$\dot{\rho}_i^\nu(t) = \int_0^\infty \left\{ \dot{\chi}_\nu(s)\xi_i^\nu(t-s) + \left[\dot{\mu}_\nu(s)\nu_{il}(t-s) + \dot{\pi}_\nu(s)\omega_{il}(t-s) \right] \rho_l^\nu(t-s) \right\} \mathrm{d}s.$$
$$(4.5)$$

Iteration of (4.3) and (4.5) can be used to expand the normal co-ordinates and their velocities into a power series of small velocity gradients of the medium. We can write down the zero-order approximation

$$\rho_{i0}^\nu(t) = \int_0^\infty \chi_\nu(s)\xi_i^\nu(t-s)\mathrm{d}s,$$

$$\dot{\rho}_{i0}^\nu(t) = \int_0^\infty \dot{\chi}_\nu(s)\xi_i^\nu(t-s)\mathrm{d}s$$
$$(4.6)$$

and the first-order approximation

$$\rho_i^\nu(t) = \rho_{i0}^\nu(t) + \int_0^\infty \left[\mu_\nu(s)\nu_{il}(t-s) + \pi_\nu(s)\omega_{il}(t-s) \right] \rho_{l0}^\nu(t-s)\mathrm{d}s,$$
$$(4.7)$$

$$\dot{\rho}_i^\nu(t) = \dot{\rho}_{i0}^\nu(t) + \int_0^\infty \left[\dot{\mu}_\nu(s)\nu_{il}(t-s) + \dot{\pi}_\nu(s)\omega_{il}(t-s) \right] \rho_{l0}^\nu(t-s)\mathrm{d}s.$$

Now, we have to discuss in some details the properties of the stochastic force $\xi_i^\alpha(t)$, defined so that $\langle \xi_i^\alpha(t) \rangle = 0$. The second-order moment

$$K_{ij}^{\alpha\gamma}(t,t') = \langle \xi_i^\alpha(t)\xi_j^\gamma(t') \rangle$$
$$(4.8)$$

depends on the velocity gradients and can be expanded into a power series of this quantity. The first-order term cannot, in general, satisfy the conditions of symmetry under interchange of the arguments of function (4.8), and must therefore be discarded. This means that, to within first-order terms in the velocity gradients, the correlation function has the same form as in the equilibrium, i.e. time-independent, case

$$K_{ij}^{\alpha\gamma}(t,t') = K_\alpha(t-t')\delta_{\alpha\gamma}\delta_{ij}.$$

The random force correlator is determined by the rule that, at equilibrium, the moments of the velocities and the co-ordinates must be known. In our simple case, the Fourier transform of the correlator is determined as follows

$$K(\omega) = \int_{-\infty}^\infty K(s)e^{i\omega s}\mathrm{d}s = 2T\mathrm{Re}B[\omega]$$
$$(4.9)$$

where the one-sided Fourier transform of a function is indicated by square brackets

$$B[\omega] = \beta[\omega] + \varphi[\omega].$$

4.1.2 Equilibrium Correlation Functions

The expansion of normal co-ordinates and their velocities (4.7) allows us to calculate various moments of co-ordinates and velocities, which are needed to determine physical quantities, first of all of second-order moments. For simplicity, we shall omit the label of the normal co-ordinates at calculation of the moments.

First, we shall consider moments at zero-velocity gradients; in other words, the equilibrium moments that depend on just one argument

$$\langle \rho_i(t)\rho_k(t-s)\rangle_0 = M(s)\delta_{ik},$$
$$\langle \dot{\rho}_i(t)\dot{\rho}_k(t-s)\rangle_0 = L(s)\delta_{ik}, \tag{4.10}$$
$$\langle \rho_i(t)\dot{\rho}_k(t-s)\rangle_0 = S(s)\delta_{ik}.$$

The angle brackets denote the averaging over the ensemble of the realisation of the random forces in the equations of motion (4.1) and (4.2).

It is easy to see that the equilibrium moments (4.10) satisfy the following relations

$$M(s) = M(-s), \qquad \frac{\mathrm{d}M(s)}{\mathrm{d}s} = -S(s) = S(-s),$$
$$-\frac{\mathrm{d}^2 M(s)}{\mathrm{d}s^2} = L(s) = L(-s).$$

Thus, we obtain relations between the Fourier transforms of moments

$$L(\omega) = -\omega^2 M(\omega), \qquad S(\omega) = -i\omega M(\omega) \tag{4.11}$$

and, taking into account the symmetry properties of the moments (4.10), the Fourier transforms can be represented by the one-sided Fourier transforms

$$M(\omega) = M[\omega] + M[-\omega],$$
$$L(\omega) = L[\omega] + L[-\omega],$$
$$S(\omega) = S[\omega] - S[-\omega].$$

We use relation (4.4) to write an expression for the moment of the normal co-ordinate

$$M(u) = \int_0^\infty \int_0^\infty \chi(s)\chi(v)K(u-s-v)\mathrm{d}s\,\mathrm{d}v, \tag{4.12}$$

where the correlation function of the random forces is defined by relation (4.9). Multiplying (4.12) by $e^{i\omega u}$ and integrating with respect to u from $-\infty$ to ∞, we find

$$M(\omega) = \frac{T(B[\omega] + B[-\omega])}{(2T\mu\lambda - m\omega^2 - i\omega B[\omega])(2T\mu\lambda - m\omega^2 + i\omega B[-\omega])}. \qquad (4.13)$$

The last expression can be represented as a sum of two terms, which is, generally speaking, ambiguous. But as far as we know expressions for the moment at $t = 0$, the expansion is not ambiguous

$$M(\omega) = \frac{1}{2\mu\lambda} \left(\frac{B[\omega] - im\omega}{2T\mu\lambda - m\omega^2 - i\omega B[\omega]} + \frac{B[-\omega] + im\omega}{2T\mu\lambda - m\omega^2 + i\omega B[-\omega]} \right).$$

Comparing with the above presentation, we have the one-sided Fourier transform of moment

$$M[\omega] = \frac{1}{2\mu\lambda} \frac{B[\omega] - im\omega}{2T\mu\lambda - m\omega^2 - i\omega B[\omega]}. \qquad (4.14)$$

The first term of the expansion of the quantity in a power series of $(-i\omega)^{-1}$ has the form

$$M[\omega] = \frac{1}{2\mu\lambda} \frac{1}{-i\omega}.$$

It is followed by

$$\lim_{s \to 0} M(s) = \frac{1}{2\mu\lambda}.$$

Note that the case, when $m = 0$, gives the correct results only for $B(\omega) \neq 0$ at $\omega \to \infty$. Otherwise, to get the correct results, it is essential to maintain the order in which the limit is approached, hence we ought to take $m = 0$ after calculation.

The other moments of velocities and co-ordinates can be determined according to relations (4.11) or can be obtained after multiplying some expressions from (4.7) and averaging the result. In either way, we obtain an expression for the Fourier transform of the equilibrium moment of velocities

$$L(\omega) = T \left(\frac{i\omega}{2T\mu\lambda - m\omega^2 - i\omega B[\omega]} + \frac{-i\omega}{2T\mu\lambda - m\omega^2 + i\omega B[-\omega]} \right).$$

The last relation is followed by

$$L[\omega] = T \frac{i\omega}{2T\mu\lambda - m\omega^2 - i\omega B[\omega]}. \qquad (4.15)$$

It gives the correct result for the limiting value of the velocity moment

$$\lim_{s \to 0} L(s) = \frac{T}{m}.$$

In a similar way, the expression for the Fourier transform of the equilibrium moment of co-ordinate and velocity can be found

$$S[\omega] = \frac{T}{2T\mu\lambda - m\omega^2 - i\omega B[\omega]}.\tag{4.16}$$

The limiting value follows from (4.16)

$$\lim_{s\to 0} S(s) = 0.$$

4.1.3 One-Point Non-Equilibrium Correlation Functions

We turn to the non-equilibrium moments of co-ordinates and velocities of linear macromolecules. As a first step, we shall consider one-point second-order moments. The expressions for co-ordinates and velocities (4.7) with the same arguments can be used to make up proper combinations, and by averaging over the ensemble of realisation of random forces, we find the moments with accuracy to the first-order terms in velocity gradients. Then, by taking into account the properties of equilibrium moments and the antisymmetry of tensor ω_{il}, we find that

$$\langle \rho_i(t)\rho_k(t)\rangle = \frac{1}{2\mu\lambda}\delta_{ik} + 2\int_0^\infty \mu(s)M(s)\gamma_{ik}(t-s)\mathrm{d}s,\tag{4.17}$$

$$\langle \dot\rho_i(t)\dot\rho_k(t)\rangle = \frac{T}{m}\delta_{ik} + 2\int_0^\infty \dot\mu(s)\dot M(s)\gamma_{ik}(t-s)\mathrm{d}s,\tag{4.18}$$

$$\langle \rho_i(t)\dot\rho_k(t)\rangle = \int_0^\infty [\mu(s)S(s)\nu_{ik}(t-s) + \dot\mu(s)M(s)\nu_{ki}(t-s)$$

$$+ \pi(s)S(s)\omega_{ik}(t-s) + \dot\pi(s)M(s)\omega_{ki}(t-s)]\,\mathrm{d}s.$$

The last expression can be simplified in the case, when the inertial forces acting on the Brownian particles are unimportant, that is $m = 0$. This is the only case that is of interest for application. In this case, by taking expressions (4.4) and (4.14) into account, we find the auxiliary relation

$$\pi(s) + \mu(s) + R(s) = 2\mu\lambda M(s).$$

which contains a function $R(x)$ of a non-negative argument x – the function of instant relaxation.[1]

---

[1] To keep the correct expression for correlation functions when limit $m = 0$ is approached, it is convenient to use the function of a non-negative argument

$$R(t) = \lim_{\tau\to 0} e^{-t/\tau} = \begin{cases} 1, & t = 0, \\ 0, & t > 0. \end{cases}$$

So as the moments (4.17) and (4.18) are expressed in terms of the functions $M(s)$ and $\mu(s)$, it is convenient to express the third moment in these functions as well. After calculating, we obtain

$$\langle \rho_i(t)\dot{\rho}_k(t)\rangle = \frac{1}{2\mu\lambda}\omega_{ki} + \int_0^\infty \left[\mu(s)\dot{M}(s) + \dot{\mu}(s)M(s)\right]\gamma_{ik}(t-s)\mathrm{d}s. \quad (4.19)$$

We see that the non-equilibrium moments are expressed in terms of the equilibrium moment of co-ordinate $M(s)$ and its derivative, which were determined in the previous section by their Fourier transforms.

4.1.4 Two-Point Non-Equilibrium Correlation Functions

Now, we turn to the calculation of two-point moments. We take the quantities defined by expression (4.7) and average the products of $\rho_i(t)$ and $\rho_k(t-s)$, $\rho_i(t)$ and $\dot{\rho}_k(t-s)$, respectively. By taking into account the properties of the equilibrium moments, we find that

$$\langle \rho_i(t)\rho_k(t-s)\rangle$$
$$= M(s)\delta_{ik} + \int_0^\infty \Big\{\mu(u)\Big[M(u-s)\nu_{ik}(t-u) + M(u+s)\nu_{ki}(t-s-u)\Big]$$
$$+ \pi(u)\Big[M(u-s)\omega_{ik}(t-u) + M(u+s)\omega_{ki}(t-s-u)\Big]\Big\}\mathrm{d}u,$$

$$\langle \rho_i(t)\dot{\rho}_k(t-s)\rangle$$
$$= S(s)\delta_{ki} + \int_0^\infty \big[\mu(u)S(u-s)\nu_{ik}(t-u) + \dot{\mu}(u)M(u+s)\nu_{ki}(t-s-u)$$
$$+ \pi(u)S(u-s)\omega_{ik}(t-u) + \dot{\pi}(u)M(u+s)\omega_{ki}(t-s-u)\big]\,\mathrm{d}u.$$

These expressions can be written in a simplified form when inertia effects are not taken into consideration.

$$\langle \rho_i(t)\rho_k(t-s)\rangle$$
$$= M(s)\delta_{ik} + \int_0^\infty \mu(u)\Big[M(u-s)\gamma_{ik}(t-u) + M(u+s)\gamma_{ik}(t-s-u)\Big]\mathrm{d}u$$
$$+ 2\mu\lambda\int_0^\infty M(u)\Big[M(u-s)\omega_{ik}(t-u) + M(u+s)\omega_{ki}(t-s-u)\Big]\mathrm{d}u,$$
$$\langle \rho_i(t)\dot{\rho}_k(t-s)\rangle$$
$$= S(s)\delta_{ik} + M(s)\omega_{ki}(t-s)$$

The derivative of the function of instant relaxation is expressed in the delta-function

$$\dot{R}(t) = -2\delta(t).$$

$$+ \int_0^\infty \left[\mu(u)\dot{M}(u-s)\gamma_{ik}(t-u) + \dot{\mu}(u)M(u+s)\gamma_{ik}(t-s-u) \right] du$$

$$+ 2\pi\lambda \int_0^\infty \left[M(u)\dot{M}(u-s)\omega_{ik}(t-u) \right.$$

$$\left. + \dot{M}(u)M(u+s)\omega_{ki}(t-s-u) \right] du.$$

These expressions ought to be transformed to eliminate the dependence on the antisymmetrical tensor of the velocity gradients. We can use the new variable $v = s - u$ to rewrite some of the integrals in the above expressions. So, for example,

$$\int_0^\infty M(u)M(u-s)\omega_{ik}(t-u)du = \int_{-s}^\infty M(v+s)M(v)\omega_{ik}(t-s-v)dv$$

$$= \int_0^\infty M(u+s)M(u)\omega_{ik}(t-s-u)du + \int_{-s}^0 M(u+s)M(u)\omega_{ik}(t-s-u)du.$$

Similar transformations allow us to find new expressions for the considered moments

$$\langle \rho_i(t)\rho_k(t-s) \rangle = M(s)\delta_{ik}$$

$$+ \int_0^\infty [\mu(u+s)M(u) + \mu(u)M(u+s)]\gamma_{ki}(t-s-u)du$$

$$+ \int_{-s}^0 \mu(u+s)M(u)\gamma_{ik}(t-s-u)du$$

$$+ 2\mu\lambda \int_{-s}^0 M(u+s)M(u)\omega_{ik}(t-s-u)du,$$

$$\langle \rho_i(t)\dot{\rho}_k(t-s) \rangle = S(s)\delta_{ik} + M(s)\omega_{ki}(t-s)$$

$$+ \int_0^\infty [\mu(u+s)\dot{M}(u) + \dot{\mu}(u)M(u+s)]\gamma_{ik}(t-s-u)du$$

$$+ \int_{-s}^0 \mu(u+s)\dot{M}(u)\gamma_{ik}(t-s-u)du$$

$$+ 2\mu\lambda \int_{-s}^0 M(u+s)\dot{M}(u)\omega_{ik}(t-s-u)du.$$

We see that the last integrals in the previous formulae can be omitted, so the final expressions for the moments take the final form

$$\langle \rho_i(t)\rho_k(t-s)\rangle = M(s)\delta_{ik}$$
$$+ \int_0^\infty [\mu(u+s)M(u) + \mu(u)M(u+s)]\gamma_{ki}(t-s-u)\mathrm{d}u,$$

$$(4.20)$$

$$\langle \rho_i(t)\dot{\rho}_k(t-s)\rangle = S(s)\delta_{ik} + M(s)\omega_{ki}(t-s)$$
$$+ \int_0^\infty [\mu(u+s)\dot{M}(u) + \dot{\mu}(u)M(u+s)]\gamma_{ik}(t-s-u)\mathrm{d}u.$$

$$(4.21)$$

Naturally, the expressions (4.17) and (4.19) for one-point moments follow, at $s = 0$, from formulae (4.20) and (4.21), respectively.

4.2 Relaxation of Macromolecular Coil

The results discussed in the previous section are valid in linear approximation for any concrete representations of the memory functions $\beta(s)$ and $\varphi(s)$. To calculate relaxation times for macromolecular coil, one has to specify the memory functions and include the effect of local anisotropy.

4.2.1 Correlation Functions for Isotropic Motion

In accordance with equations (3.15), the memory functions $\beta(s)$ and $\varphi(s)$ in the dynamic equations are given by their one-sided transforms

$$\beta[\omega] = \zeta\left(1 + \frac{B}{1 - i\omega\tau}\right), \qquad \varphi[\omega] = \frac{\zeta E}{1 - i\omega\tau}. \qquad (4.22)$$

In this case the theory, apart from the characteristic Rouse relaxation time τ^*, contains three more parameters, namely: the relaxation time τ of the medium, the measure B of the increase in the resistance of the particle when it moves among the chains, and the measure of internal viscosity E associated with resistance to the deformation of the coil due to the present of ambient macro-molecules.

We use expression (4.22) to specify the quantities (4.4) and calculate equilibrium correlation functions (4.14) for the case, when $m = 0$,

$$\mu_\nu[\omega] = \frac{2(1 + B - i\omega\tau)\tau_\nu^{\mathrm{R}}}{(1 - i\omega 2\tau_\nu^+)(1 - i\omega 2\tau_\nu^-)}, \qquad (4.23)$$

$$M_\nu[\omega] = \frac{1}{2\mu\lambda}\frac{2(1 + B + E - i\omega\tau)\tau_\nu^{\mathrm{R}}}{(1 - i\omega 2\tau_\nu^+)(1 - i\omega 2\tau_\nu^-)}. \qquad (4.24)$$

Here, τ_ν^{R} are the relaxation times of the macromolecule in a monomer viscous fluid – Rouse relaxation times

$$\tau_\nu^{\mathrm{R}} = \frac{\zeta}{4T\mu\lambda_\nu} = \frac{\tau^*}{\nu^2}, \quad \tau^* = \frac{\zeta b^2 N^2}{6\pi^2 T} \tag{4.25}$$

and symbols for the new sets of relaxation times are introduced

$$2\tau_\nu^\pm = \tau_\nu \pm \left(\tau_\nu^2 - 2\tau\tau_\nu^{\mathrm{R}}\right)^{1/2},$$

$$\tau_\nu = \frac{\tau}{2} + \tau_\nu^{\mathrm{R}}(1 + B + E) = \tau^* B\left(\chi + \frac{1}{\nu^2}(1 + \psi)\right), \tag{4.26}$$

where $\psi = E/B$ and, in accordance with definition (3.16), $\chi = \tau/(2\tau^* B)$. The following relations are valid for introduced relaxation times

$$2\tau\tau_\nu^{\mathrm{R}} = 4\tau_\nu^+ \tau_\nu^-, \qquad \tau_\nu^+ - \tau_\nu^- = \left(\tau_\nu^2 - 2\tau\tau_\nu^{\mathrm{R}}\right)^{1/2}.$$

The dynamic equations determine the two relaxation branches, while one of them contains the small relaxation times τ_α^-, the other – the large ones τ_α^+ which practically for long macromolecules coincide with the relaxation time τ_α. Further on, it is convenient to consider asymptotic formulae for small and large mode numbers separately, so that for these branches, one has approximations

$$\tau_\alpha^+ = (B + E)\tau_\alpha^{\mathrm{R}}, \qquad \tau_\alpha^- = \frac{\tau}{2(B + E)}, \qquad \alpha^2 \ll \frac{1 + \psi}{\chi},$$

$$\tau_\alpha^+ = \frac{\tau}{2}, \qquad\qquad \tau_\alpha^- = \tau_\alpha^{\mathrm{R}}, \qquad \alpha^2 \gg \frac{1 + \psi}{\chi}. \tag{4.27}$$

To determine the functions $\mu(s)$ and $M(s)$ from equations (4.23) and (4.24), one can use the reciprocal Laplace transform. Before calculating, we remind the reader that the correct results can be obtained when the mass is retained in expressions (4.4) and (4.14). This changes expressions (4.23) and (4.24). However, it is easier to operate with limiting (at $m \to 0$) expressions. The final results can be improved by adding terms that contain function $R(t)$, described in a footnote on one of the previous pages. We can also find, by simple alternative calculations, that

$$\mu_\nu(t) = T_\nu^+ \exp\left(-\frac{t}{2\tau_\nu^+}\right) - T_\nu^- \exp\left(-\frac{t}{2\tau_\nu^-}\right) - R(t), \tag{4.28}$$

$$M_\nu(t) = \frac{1}{2\mu\lambda_\nu}\left[S_\nu^+ \exp\left(-\frac{t}{2\tau_\nu^+}\right) - S_\nu^- \exp\left(-\frac{t}{2\tau_\nu^-}\right)\right], \tag{4.29}$$

where

$$T_\nu^\pm = \frac{\tau_\nu^{\mathrm{R}}(1 + B) - \tau_\nu^\mp}{\tau_\nu^+ - \tau_\nu^-}, \qquad S_\nu^\pm = \frac{\tau_\nu^{\mathrm{R}}(1 + B + E) - \tau_\nu^\mp}{\tau_\nu^+ - \tau_\nu^-}. \tag{4.30}$$

Equation (4.29) defines a correlation function due to the diffusive mechanism of relaxation. One can see that time dependence of the equilibrium correlation functions of normal co-ordinates is determined by relaxation processes.

Some simplifications can be achieved for the large values of B. It appears to be valid the following relation

$$\tau \tau_\nu^R \ll \tau_\nu^2$$

and the relaxation times (4.26) can be written in the form

$$\tau_\nu^+ = \tau_\nu - \frac{\tau \tau_\nu^R}{2\tau_\nu}, \qquad \tau_\nu^- = \frac{\tau \tau_\nu^R}{2\tau_\nu}. \tag{4.31}$$

In the limiting case of very large values of the parameter B when $\zeta \to 0$ (but $\zeta B \neq 0$, $\zeta E \neq 0$), we find that

$$\tau_\nu^+ \to \tau_\nu, \qquad \tau_\nu^- \to 0.$$

In this limiting case, expressions (4.28) and (4.29) can be written as

$$\mu_\nu(t) = \frac{B\tau_\nu^R}{\tau_\nu} \exp\left(-\frac{t}{2\tau_\nu}\right) - \frac{B\tau_\nu^R}{\tau_\nu} R(t), \tag{4.32}$$

$$M_\nu(t) = \frac{1}{2\mu\lambda_\nu} \left[\frac{(B+E)\tau_\nu^R}{\tau_\nu} \exp\left(-\frac{t}{2\tau_\nu}\right) + \frac{\tau}{2\tau_\nu} R(t) \right]. \tag{4.33}$$

Let us note that formula (4.33) is a generalisation of the equilibrium correlation function of the normal co-ordinates of the macromolecule in a viscous liquid

$$M_\nu(t) = \frac{1}{2\mu\lambda_\nu} \exp\left(-\frac{t}{2\tau_\nu^R}\right). \tag{4.34}$$

There is a great difference between the relaxation behaviour of the system of entangled macromolecules and the relaxation behaviour of a macromolecule in a dilute system. Two relaxation branches have been shown to exist in a system of entangled macromolecules.

4.2.2 Effect of Local Anisotropy

Derived from linear approximation of the equations (3.37), the equilibrium correlation function (4.29), defines two conformation relaxation times τ_α^+ and τ_α^- for every mode. The largest relaxation times have appeared to be unrealistically large for strongly entangled systems, which is connected with absence of effect of local anisotropy of mobility. To improve the situation, one can use the complete set of equations (3.37) with local anisotropy of mobility. It is convenient, first, to obtain asymptotic (for the systems of long macromolecules) estimates of relaxation times, using the reptation-tube model.

Correlation Functions for Pure Reptation

It is not difficult to reproduce an expression for the correlation function $M_\alpha(t)$ and estimate times of relaxation due to the conventional reptation-tube model (see Section 3.5). Indeed, an equation for correlation function follows equation (3.48) and has the form

$$\frac{dM_\alpha}{dt} = -\frac{\pi^2 T \alpha^2}{\zeta Z^3 \xi^2} M_\alpha.$$

The equation has a simple solution

$$M_\alpha(t) = \frac{1}{2\mu\lambda_\alpha} \exp\left(-\frac{t}{2\tau_\alpha^{\text{rep}}}\right), \tag{4.35}$$

$$\tau_\alpha^{\text{rep}} = \frac{\zeta \xi^2 Z^3}{2\pi^2 T} \frac{1}{\alpha^2} = \frac{3\langle R^2 \rangle_0}{\xi^2} \frac{\tau^*}{\alpha^2}, \quad \alpha = 1, 2, \ldots, \ll Z. \tag{4.36}$$

These are exactly the known results (Doi and Edwards 1986, p. 196). The time behaviour of the equilibrium correlation function is described by a formula which is identical to formula for a chain in viscous liquid (equation (4.34)), while the Rouse relaxation times are replaced by the reptation relaxation times. In fact, the chain in the Doi-Edwards theory is considered as a flexible rod, so that the distribution of relaxation times naturally can differ from that given by equation (4.36): the relaxation times can be close to the only disentanglement relaxation time τ_1^{rep}.

One can refer to equations (5.8) to use the other parameter

$$\frac{\langle R^2 \rangle_0}{\xi^2} = \frac{\pi^2}{2\chi},$$

so that the reptation branch of relaxation times can be written as

$$\tau_\alpha^{\text{rep}} = \frac{3}{2} \frac{\pi^2}{\chi} \frac{\tau^*}{\alpha^x}, \quad \alpha = 1, 2, \ldots, \ll Z. \tag{4.37}$$

We have introduced here, instead of index 2, an index x, value of which can be less than 2 according to the results of simulation (see the next subsection, $x \approx 0.5$).

An Estimation of Relaxation Times

The rates of relaxation $\tau_\gamma(t)$ in the moment t, or, in other words, the current relaxation times of the macromolecular coil can be directly calculated as

$$\tau_\gamma(t) = -\frac{1}{2}\left(\frac{d\log(M_\gamma(t)/M_\gamma(0))}{dt}\right)^{-1}, \quad \gamma = 1, 2, \ldots, N. \tag{4.38}$$

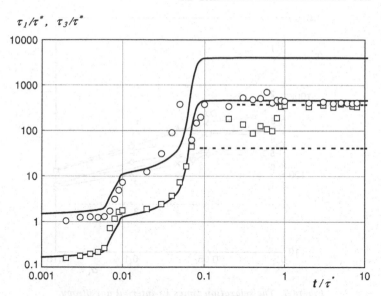

Figure 6. The rate of relaxation of a macromolecule.
The rates of relaxation of the first and the third modes of macromolecule of length
$M = 25M_e$ ($\chi = 0.04$, $B = 429$, $\psi = 8.27$). The results calculated from analytical
correlation function (4.29) are depicted by solid lines. By straight dashed lines, the values
of the relaxation times due to the Doi-Edwards model are presented. The circles (for the
first mode) and squares (for the third mode) depict the results of simulation for above
values of parameters ψ and B and values of parameters of local anisotropy $a_e = 0.3$,
$a_i = 0.06$. Adapted from Pokrovskii (2006).

In Fig. 6, the rates of relaxation $\tau_\gamma(t)$ for two modes are depicted by solid
lines, according to values of the equilibrium correlation function $M_\gamma(t)$ given
by equation (4.29) for linear approximation. The correct result for big times
(slow relaxation) can be found at presence of local anisotropy, which is possible
by numerical integration of non-linear equations (3.52)–(3.54). The calculation
were fulfilled for the chain with $M = 25M_e$ divided into 10 subchains and are
depicted in Fig. 6 by points. The simulation for the small times reproduces the
theoretical dependence of the relaxation rates on the current time (with large
scattering, so as the changes of the correlation functions in this region are
small), while the results shows the existence of the two relaxation branches
as well, in accordance with equation (4.29). The introduction of the local
anisotropy of external resistance alone does not affect the relaxation times, in
contrast to the local anisotropy of the internal resistance. The latter provokes
changes of the largest relaxation times of the macromolecular coil, which is the
bigger, the bigger the coefficient of the local anisotropy of external resistance.
Asymptotic values of the relaxation times are estimated for each case as the
mean values of the rate of relaxation in the interval from $0.7\,\tau^*$ to $10\,\tau^*$.

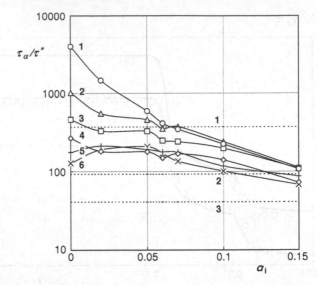

Figure 7. The relaxation times vs internal anisotropy.
Each point is calculated as the asymptotic value of the rate of relaxation for large times (see examples of dependences in Fig. 6) for a macromolecule of length $M = 25M_e$ ($\chi = 0.04$, $B = 429$, $\psi = 8.27$) with the value of the coefficient of external local anisotropy: $a_e = 0.3$. The dashed lines reproduce the values of the relaxation times of the macromolecule due to the reptation-tube model. The labels of the modes are shown at the lines. Adapted from Pokrovskii (2006).

A particular choice of the coefficients $a_e = 0.3$ and $a_i = 0.06$ determines the value $\tau_1 = 417\,\tau^*$ for the relaxation time of the first mode, which is close to the reptation relaxation time $370\,\tau^*$. The calculated relaxation times of the third mode: $\tau_3 = 315\,\tau^*$ is a few times as much as the corresponding reptation relaxation time $41.1\,\tau^*$, which indicates that the dependence of the relaxation times on the mode label is apparently different from the law (4.36). It is clearly seen in Fig. 7, where the dependence of the relaxation times of the first six modes of a macromolecule on the coefficient of internal anisotropy is shown. The relaxation times of different modes are getting closer to each other with increase of the coefficient of internal anisotropy. The values of the largest relaxation time of the first mode for different molecular weights are shown in Fig. 8. The results demonstrate a drastic decrease in values of the largest relaxation times for strongly entangled systems induced by introduction of local anisotropy.

In relaxation processes of the macromolecular coil to equilibrium, the competing mechanisms of mobility of particles are present simultaneously. However, in the region of weakly entangled macromolecules, relaxation occurs due to isotropic mobility of particles of the chain – the diffusive mechanism –

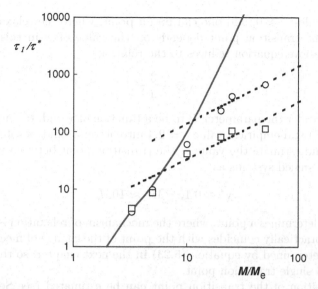

Figure 8. The largest relaxation times of a macromolecule.
Each point is calculated as asymptotic value of the rate of relaxation for large times (see examples of dependences in Fig. 6) for different molecular weights with corresponding values of the parameters B and ψ. The values of the coefficients of local anisotropy are: $a_e = 0.3$, $a_e = 0.06$ for the circles and $a_e = 0.3$, $a_e = 0.15$ for the squares. The solid line depicts analytical results for linear approximation. The dashed lines with the slope 1 reproduce the well-known dependence $\tau_1 \sim M^3$ for the relaxation time of macromolecules in strongly entangled systems. Adapted from Pokrovskii (2006).

and, in the region of strongly entangled systems, the reptation mechanism of relaxation predominates.

4.2.3 Transition Point

To determine the position of transition point, we shall compare the relaxation times (4.27) and (4.37), due to different mechanisms of conformational relaxation, which gives an equation

$$B(1 + \psi) = \frac{3}{2}\frac{\pi^2}{\chi}. \tag{4.39}$$

The transition point can be different, if one uses different modes, but only the transition point for the first mode is considered here. For the strongly entangled systems, according to relation (5.17), $\psi = \pi^2/\chi$, so that, at $\psi \gg 1$, the left-hand side of the above equation is always bigger than the right-hand side: the reptation mechanism of relaxation is realised. However, for short macromolecules $\psi \approx 0$, and one can find a point, where the relaxation times coincide. The transition point depends on the value of ψ in this point. At $\psi = 1$, the above equation reduces to the relation

$$\chi B = \frac{3}{4}\pi^2. \tag{4.40}$$

One can consider the parameter B to be a function of χ and, taking equations (3.17), (3.25) and empirical value $\delta = 2.4$ into account, finds a solution of the equation, and estimate the value of the transition point between weakly and strongly entangled systems as

$$\chi^* \approx 0.1, \quad M^* \approx 10 M_e.$$

This value determines a point, where the mechanism of relaxation is changing. The point practically coincides with the point of the change of mechanisms of diffusion, determined by equation (5.23) in the next chapter, so that one can say about a single transition point.

The position of the transition point can be estimated (see Section 6.4), due to measurements of viscoelastic properties, as $M^* \approx (4.6\text{--}12.0)M_e$. It corresponds to the above value of transition point, though the empirical evaluation of relaxation times could not be done with great accuracy in these investigations.

So, one ought to conclude that large scale (slow) relaxation of the conformation of a macromolecular coil is realised through reptation, instead of the more slow mechanism of rearrangement of all the entangled chains, if the parameter $B > \pi^2/2\chi$.

4.2.4 Conformational Relaxation Times

The dependencies of the relaxation times on the length of the macromolecule are different in two regions. Besides, one has to distinguish the relaxing macromolecule (with molecular weight or length M) and the neighbouring macromolecules (with the length M_0), even if all of them are equal, so that the dependencies can be written as

$$\tau_\alpha^{\text{conf}} = \begin{cases} B\tau_\alpha^{\text{R}} \sim M_0^\delta M^2, & 2\chi B < \pi^2, \text{ weakly entangled system} \\ \frac{3}{2}\frac{\pi^2}{\chi}\frac{\tau^*}{\alpha^x} \sim M_0^0 M^3, & 2\chi B > \pi^2, \text{ strongly entangled system} \end{cases} . \tag{4.41}$$

This relations are valid for small mode numbers, in any case, $\alpha \ll M/M_e$. The index δ in the above formula can be estimated theoretically ($\delta > 2$) and empirically according to the measurements of the characteristics of viscoelasticity ($\delta \approx 2.4$). It remains to be a dream to get a unified formula for relaxation times from the system of dynamic equations (3.37). One can expect that the all discussed relaxation branches will emerge as different limiting cases from one expression for general conformation branch.

The mechanism of small-scale (fast) relaxation of conformation of the macromolecule does not change at the transition from weakly to strongly entangled systems; the times of relaxation are defined by formulae (4.31). However, one has to take into account, that $\psi \ll 1$ for weakly entangled systems, whereas $\psi \gg 1$ for strongly entangled systems, so that one has for the largest of the fast relaxation times

$$\bar{\tau}_{\max} = \begin{cases} \tau^* \chi \sim M, & 2\chi B < \pi^2, \text{ weakly entangled system} \\ \tau^* \chi / \psi \sim M^0, & 2\chi B > \pi^2, \text{ strongly entangled system} \end{cases} . \quad (4.42)$$

One can see that the relaxation times τ_α^- at $\alpha > (\psi/\chi)^{1/2}$ are the Rouse relaxation times of the part of the macromolecule that correspond approximately to the length of the macromolecule between adjacent entanglements M_e. There is an interval between slow and fast relaxation times, which is the bigger the longer the macromolecules.

4.3 Macromolecular Coil in a Flow

4.3.1 Non-Equilibrium Correlation Functions

The non-equilibrium moments of the normal co-ordinates for an entangled system are defined by expression (4.17) with accuracy up to first-order terms in the velocity gradients. It is written down once more with the label of normal co-ordinates

$$\langle \rho_i^\nu \rho_k^\nu \rangle = \langle \rho_i^\nu \rho_k^\nu \rangle_0 + 2 \int_0^\infty \mu_\nu(s) M_\nu(s) \gamma_{ik}(t-s) \mathrm{d}s.$$

The functions $M_\nu(s)$ and $\mu_\nu(s)$ are defined in previous sections. To describe the most slow relaxation, we use expressions (4.32) and (4.33) and find that

$$\langle \rho_i^\nu \rho_k^\nu \rangle = \frac{1}{2\mu\lambda_\nu} + \frac{B^2(1+E/B)}{\mu\lambda_\nu} \left(\frac{\tau_\nu^{\mathrm{R}}}{\tau_\nu}\right)^2 \int_0^\infty \exp\left(-\frac{s}{\tau_\nu}\right) \gamma_{ik}(t-s)\mathrm{d}s, \quad (4.43)$$

where $\gamma_{ik}(x)$ is the velocity gradient as an arbitrary function of time x. The assumption that

$$\gamma_{ik} = \begin{cases} const, & t < 0, \\ 0, & t > 0 \end{cases}$$

determines the law for relaxation of the moments to their equilibrium values

$$\langle \rho_i^\nu \rho_k^\nu \rangle = \frac{1}{2\mu\lambda_\nu} \delta_{ik} + 2 \frac{B^2(1+E/B)(\tau_\nu^{\mathrm{R}})^2}{\tau_\nu} \gamma_{ik} \exp\left(-\frac{t}{\tau_\nu}\right). \quad (4.44)$$

This expression demonstrates that the relaxation time τ_ν, defined by relation (4.27), is the relaxation time of the mean square normal co-ordinate, or the mode labelled ν, which can be separately approximated for the small and big indices

$$\tau_\nu = \begin{cases} \frac{\tau^* B(1+\psi)}{\nu^2}, & \nu^2 \ll \frac{1+\psi}{\chi}, \\ \tau^* B\chi, & \nu^2 \gg \frac{1+\psi}{\chi}. \end{cases} \tag{4.45}$$

At a constant velocity gradient, expression (4.43) takes the form

$$\langle \rho_i^\nu \rho_k^\nu \rangle = \frac{1}{2\mu\lambda_\nu} \left(\delta_{ik} + 2\frac{B^2(1+E/B)(\tau_\nu^{\mathrm{R}})^2}{\tau_\nu}\gamma_{ik} \right). \tag{4.46}$$

For the small indices, this expression can be written as

$$\langle \rho_i^\nu \rho_k^\nu \rangle = \frac{1}{2\mu\lambda_\nu} \left(\delta_{ik} + 2B\tau_\nu^{\mathrm{R}}\gamma_{ik} \right), \quad \nu^2 \ll \frac{1+\psi}{\chi}. \tag{4.47}$$

There are no major difficulties in calculating the mean square normal co-ordinate when more general formulae (4.28) and (4.29) for the functions $M_\nu(s)$ and $\mu_\nu(s)$ are used. In this case three sets (branches) of relaxation times

$$\tau_\nu^+, \quad \tau_\nu^-, \quad \tau_\alpha^0 = \frac{2\tau_\alpha^+\tau_\alpha^-}{\tau_\alpha^+ + \tau_\alpha^-}$$

appear as the relaxation times of the macromolecular coil. One of the branches contains large relaxation times τ_ν^+, the other two small. This is a characteristic feature of polymer melts, as is revealed in experiments.

4.3.2 Size and Form of the Macromolecular Coil

A macromolecular coil at equilibrium has a spherical form (Section 1.4). Under deformation of the system, the macromolecular coil change its form that is characterised in this case by the tensor of gyration

$$\langle S_i S_k \rangle = \frac{1}{N+1} \sum_{\nu=0}^{N} \langle (r_i^\nu - q_i)(r_k^\nu - q_k) \rangle$$

where $N + 1$ is the number of Brownian particles in the chain which represents a macromolecule, q_i is the co-ordinate of the centre of mass of the macromolecular coil, and r_i^α is the co-ordinate of the particle labelled α.

In normal co-ordinates, to which we transform in accordance with the rule defined by (1.13), the expression for the tensor of gyration acquires the form

$$\langle S_i S_k \rangle = \frac{1}{N+1} \sum_{\nu=1}^{N} \langle \rho_i^\nu \rho_k^\nu \rangle. \tag{4.48}$$

This relation is valid both for the macromolecular coil in a viscous liquid and for the macromolecular coil in an entangled system.

At constant velocity gradient γ_{ik}, the moments $\langle \rho_i^\alpha \rho_k^\alpha \rangle$ are given by relations (4.46) for the diffusive mechanism of relaxation, and by similar formula (there is a difference in definition of the relaxation times only) for the reptation mechanism, so that we can evaluate the expression for the tensor of gyration of the macromolecular coil, taking into account alternative mechanisms of relaxation

$$\langle S_i S_k \rangle = \frac{1}{3} \langle S^2 \rangle_o \times \begin{cases} \delta_{ik} + \frac{2\pi^2}{15} B \tau^* \gamma_{ik}, & \chi > \chi^*, \text{ non-reptation} \\ \delta_{ik} + \frac{2\pi^2}{15} \frac{\pi^2}{\chi} \tau^* \gamma_{ik}, & \chi > \chi^*, \text{ reptation} \end{cases} \tag{4.49}$$

The sizes of macromolecular coils in flows can be estimated in experiments with light and neutron scattering. For illustration, we refer to the results of measuring the sizes of coils under flow by small angle neutron scattering (Muller et al., 1990). A blend of hydrogenated and deuterated polystyrene with $M \approx 8 \times 10^5$ was used. For elongational flow ($\nu_{22} = \nu_{33} = \frac{1}{2}\nu_{11}$) at $T = 123°C$, the sizes of the deuterated coils were measured in a both parallel and perpendicular to the stretching direction. The data by Muller et al. (1990) allow one to estimate the mean relaxation time of the macromolecular coil as 400 s. This relaxation time is certainly to be a reptation relaxation time. To confirm the statement, it would be interesting to have similar measurements for samples with different molecular weights to determine the dependence of the relaxation time on the length of the macromolecule. In this case, the relaxation mechanism could be revealed.

The function of density distribution $\rho(r)$ can also be introduced in the non-equilibrium state; it must now satisfy the following conditions

$$\int \rho(r) dr = N, \qquad \frac{1}{N} \int \rho(r) r_i r_k dr = \langle S_i S_k \rangle.$$

In a deformed system, the average form of the macromolecular coil can be approximated by an ellipsoid. The effective volume of the macromolecular coil depends on the velocity gradients. The expansion of the effective volume as a series in powers of the velocity gradients does not contain the first-order term, so $\nu_{ii} = 0$. This means that, at low velocity gradients, the coil does not change its volume (one says: the coil is orientated by flow). At larger velocity gradients, the volume of the coil is increased.

Chapter 5
The Localisation Effect

Abstract In this chapter, peculiarities of thermal motion of a macromolecule in an equilibrium system of entangled macromolecules will be discussed. It will be shown, that the mesoscopic stochastic equation of macromolecule dynamics, considered in the previous chapters, is followed by a localisation effect. This means that the time dependence of the mean square displacements of the centre of mass of the macromolecule and the chain particles are non-linear, so a dynamical internal length (a scale of localisation) can be introduced. This internal length coincides practically with the radius of a tube conventionally used in reptation theories. The macromolecule wobbles around in the tube-like region, remaining near its initial position for some time (a time of localisation), which is the larger, the longer the macromolecule is. A very long macromolecule appears, in fact, to behave exactly as if confined in a tube, though no other restrictions than the fundamental equation exist. Localisation of a macromolecule or tube formation is a linear mesoscopic phenomenon, which can be revealed directly in experiments on neutron scattering, while the reptation of the macromolecule is a non-linear effect.

5.1 Mobility of a Macromolecule

Diffusion of a macromolecule is understood as the diffusion of a co-ordinate of the centre of mass of a chain, which is, according to relation (1.18), proportional to the zeroth normal co-ordinate, that is,

$$q \sim \rho^0.$$

The mean square displacement of the centre of mass of a diffusing macromolecule for a time t is calculated as

$$\Delta(t) = \sum_{i=1}^{3} \langle [q_i(t) - q_i(0)]^2 \rangle.$$

V.N. Pokrovskii, *The Mesoscopic Theory of Polymer Dynamics*,
Springer Series in Chemical Physics 95,
DOI 10.1007/978-90-481-2231-8_5, © Springer Science+Business Media B.V. 2010

One can use the expression

$$q_i(t) - q_i(0) = \int_0^t \dot{q}_i(s)\,\mathrm{d}s,$$

to present the mean square displacement in another form

$$\Delta(t) = \int_0^t \int_0^t \langle \dot{\boldsymbol{q}}(s)\dot{\boldsymbol{q}}(u)\rangle \mathrm{d}s\,\mathrm{d}u. \qquad (5.1)$$

This expression reduces the calculations to the evaluation of the time-dependent velocity correlation function

$$L(u-s) = \frac{1}{3}\langle \dot{\boldsymbol{\rho}}^0(s)\dot{\boldsymbol{\rho}}^0(u)\rangle \sim \langle \dot{\boldsymbol{q}}(s)\dot{\boldsymbol{q}}(u)\rangle$$

which appears to be dependent on the properties of the environment, that is specific for different cases (see Section 4.2.1).

5.1.1 A Macromolecule in a Viscous Liquid

In this case, according to relations (2.43) and (4.15), the correlation function of the zero normal co-ordinate is determined by equation

$$L(x) = \frac{T}{m}\exp\left(-\frac{\zeta}{m}x\right)$$

which allows one, after simple calculations, to find that the macromolecule moves like a Brownian particle in a viscous liquid, and its displacement is given by the standard relation

$$\Delta(t) = 6D_0 t, \qquad (5.2)$$

where the coefficient of the macromolecule diffusion is inversely proportional to the mobility of the macromolecular coil

$$D_0 \sim \frac{T}{\zeta_{\mathrm{M}}}. \qquad (5.3)$$

The dependence of the friction coefficient ζ_{M} of the macromolecule on its length M is affected by exclude-volume effects and effects of draining or non-draining (permeability of macromolecular coils). Taking into account equation (2.14), the coefficient of diffusion can be written as

$$D_0 \sim M^{-(z-2)\nu}.$$

The value of the index in this formula changes from $1/2$ to 1 for different situations. For example, $(z-2)\nu = 1$ for a freely-draining macromolecule without volume effects (Rouse case).

The description of the diffusion of macromolecular coils (5.2)–(5.3) appears to be in good agreement with experimental evidence for dilute solutions (Doi and Edwards 1986; Gennes 1979). It has been also shown (Meerwall et al. 1982; Fleisher and Appel 1995) that short macromolecules in melts at $M < 2M_e$ can be considered to diffuse according to law (5.2) with the coefficient of diffusion

$$D_0 \sim M^{-1}. \tag{5.4}$$

Computer simulations (Paul and Smith 2004) and calculations (Rostiashvili et al. 1999) show, nevertheless, that, due to interactions between chains, the index in the law (5.4) for a macromolecule in a melt can be less than unity.

5.1.2 A Macromolecule in an Entangled System

Diffusive Mobility of a Macromolecule

The mobility of a macromolecule, constrained by other macromolecules, can be also calculated as (5.1). In the linear approximation, the zeroth normal co-ordinates of the macromolecule (equation (4.1), at $\nu_{ij} = 0$) define diffusive mobility of macromolecule. The one-sided Fourier transform velocity correlation function is determined by expression (4.15), so that we can write down the Fourier transform

$$L(\omega) = \frac{T}{-i\omega m + B[\omega]} + \frac{T}{i\omega m + B[-\omega]}.$$

Multiplying this expression by $\frac{1}{2\pi} e^{-i\omega t}$ and integrating with respect to ω from $-\infty$ to ∞, we find

$$L(t) = \frac{T}{\pi} \int_{-\infty}^{\infty} \frac{\cos \omega t}{-i\omega m + B[\omega]} \, d\omega.$$

We use the formula to write down the general expression for the mean square displacement of a particle in an arbitrary viscoelastic liquid

$$\Delta(t) = \frac{6T}{\pi(1+N)} \int_{-\infty}^{\infty} \frac{1 - \cos \omega t}{\omega^2(-i\omega m + B[\omega])} \, d\omega.$$

Turning to the particular memory functions (3.15), one finds for this simple case

$$B[\omega] = \zeta + \frac{\zeta D}{1 - i\omega\tau},$$

and calculates the displacement of the centre of mass of a macromolecule

$$\Delta(t) = \frac{6T\tau}{\zeta NB} \left(\frac{t}{\tau} + 1 - \exp\left(-\frac{t}{\tau}B \right) \right). \tag{5.5}$$

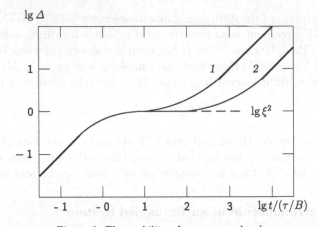

Figure 9. The mobility of a macromolecule.
The mean square displacement of the centre of mass of a macromolecule is measured
in units of the intermediate length ξ. The displacement has been calculated for macromolecules of different lengths according to equation (5.5) at $B = 100$ and $B = 1000$
(curves 1 and 2). The curves demonstrate the existence of the intermediate scale ξ – the
value of the displacement on the plateau which is the longer the longer macromolecules
are.

The time dependence of the displacement of a macromolecule, shown in Fig. 9
as a function of the ratio t/τ, is typical for diffusion of Brownian particle in
viscoelastic fluid (Zanten and Rufener 2000; Zanten et al. 2004). The function (5.5) for big values of B can be approximated as

$$\Delta(t) = \begin{cases} \frac{6T}{N\zeta} t, & t \ll \tau/B, \\ \frac{6T\tau}{N\zeta B}, & \tau/B \ll t \ll \tau, \\ \frac{6T}{N\zeta B} t, & t \gg \tau. \end{cases} \tag{5.6}$$

The mean mobility of the macromolecule changes at $t = \tau/B$, but the
displacement remains constant over a certain time of observation, and is given
by

$$\xi^2 = \frac{6T\tau}{N\zeta B} = \frac{\langle R^2 \rangle \tau}{\pi^2 \tau^* B} = \frac{2}{\pi^2} \langle R^2 \rangle \chi. \tag{5.7}$$

The characteristic time τ/B and the characteristic scale thus appear in the
theory. The non-dimensional quantity χ, defined by equation (3.16), can be
interpreted as the ratio of the square of twice the characteristic scale to the
mean square end-to-end distance of the macromolecule

$$\chi = \frac{\tau}{2B\tau^*} = \frac{\pi^2}{8} \frac{(2\xi)^2}{\langle R^2 \rangle} \approx \frac{(2\xi)^2}{\langle R^2 \rangle}. \tag{5.8}$$

The quantity ξ may be expected to be a characteristic of the system of chains, and is independent of both the length M of the diffusing macromolecules and the lengths M_0 of macromolecules of matrix, so that

$$\chi \sim M_0^0 M^{-1}. \tag{5.9}$$

For short times of observation, $t \ll \tau/B$, the expression for the displacement is identical to (5.2) which was written for the displacement of a macromolecule in a viscous liquid. The situation is similar for long observation times $t \gg \tau$. So, we can see that the coefficient of diffusion of a macromolecule can be defined differently for different displacements: for distances which are less than ξ

$$D = \frac{T}{N\zeta}$$

and for distances which are bigger than ξ

$$D = \frac{T}{N\zeta B}. \tag{5.10}$$

In the last particular situation, the motion of a test chain is also coupled to the motion of neighbouring macromolecules, and the diffusion coefficient is determined both by the length of the test macromolecule M and by the length of the ambient macromolecules M_0

$$D = D_0 B^{-1} \sim M_0^{-\delta} M^{-1}. \tag{5.11}$$

The self-diffusion coefficient for long chains is proportional to $M^{-\delta-1}$ and is small. In this situation, however, a competing mobility mechanism gives a different dependence of the diffusion coefficient on the length of the macromolecule. This will be discussed further in this section.

Localisation of a Macromolecule

One can see that the investigated equations of dynamics even in linear approximation describe anomalous diffusion of the mass centre of macromolecule moving amongst the other macromolecules. The displacement of every particle of the chain is also anomalous in comparison with case of a macromolecule in a viscous liquid. Now we shall consider, following work by Kokorin and Pokrovskii (1990, 1993), the displacement of each internal particle of the chain

$$\Delta_\alpha(t) = \sum_{i=1}^{3} \langle [r_i^\alpha(t) - r_i^\alpha(0)]^2 \rangle, \quad \alpha = 0, 1, \ldots, N. \tag{5.12}$$

It is convenient to transform the expression to normal co-ordinates and to separate the zeroth normal co-ordinate

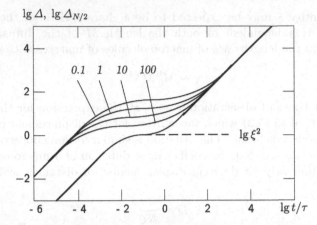

Figure 10. The mobility of a macromolecule and its particles.
The mean square displacement Δ of the centre of mass of a chain (thick solid line) and
the mean square displacement $\Delta_{N/2}$ of the central particle are measured in units of the
intermediate length ξ. The curves are calculated according to formulae (5.5) and (5.13)
for the values of the parameters: $B = 100$; $\chi = 10^{-2}$. The displacement of the centre of
mass does not depend on parameter ψ, but the mean square displacement of the internal
particles does. The values of parameter ψ are shown at the curves for $\Delta_{N/2}$. The picture
demonstrates the existence of the universal intermediate scale ξ. Adapted from the papers
of Pokrovskii and Kokorin (1985) and Kokorin and Pokrovskii (1990).

$$\Delta_\alpha(t) = \frac{1}{N+1}\langle[\boldsymbol{\rho}^0(t)-\boldsymbol{\rho}^0(0)]^2\rangle + 2\sum_{\gamma=1}^{N}Q_{\alpha\gamma}Q_{\alpha\gamma}\Big(\langle\boldsymbol{\rho}^\gamma(t)\boldsymbol{\rho}^\gamma(t)\rangle - \langle\boldsymbol{\rho}^\gamma(t)\boldsymbol{\rho}^\gamma(0)\rangle\Big)$$

where the transform matrix $Q_{\alpha\gamma}$ is defined by (1.16). The first term of the expression present the displacement of the centre of mass of the macromolecule
$\Delta(t)$, which is defined by (5.5). To calculate the second term, we use expression (4.29) for the equilibrium moments and find the displacement

$$\Delta_\alpha(t) = \Delta(t) + 6\sum_{\gamma=1}^{N}Q_{\alpha\gamma}Q_{\alpha\gamma}\frac{1}{2\mu\lambda_\gamma}$$

$$\times\left\{S_\gamma^+\left[1 - \exp\left(-\frac{t}{2\tau_\gamma^+}\right)\right] - S_\gamma^-\left[1 - \exp\left(-\frac{t}{2\tau_\gamma^-}\right)\right]\right\}. \quad (5.13)$$

The quantity $\Delta_\alpha(t) - \Delta(t)$ represents the mean square displacement of the
particle relatively to the displacement of the mass centre of the macromolecule.
The quantity can be evaluated in experiment (Kehr et al. 2007).

As an example, the time dependence of the displacement of the central
particle is shown in Fig. 10 for certain values of the parameters. We can see
that the dependence of any particle of the chain is similar to the dependence of
the entire macromolecule. Both dependencies are characterised by the different
mobility for short and long times of observation.

The asymptotic ($\chi \ll 1, \psi \gg 1$) evaluations for the mean square displacement of a particle were found by Kokorin and Pokrovskii (1990, 1993). For a short time of observation, $t \ll \frac{\tau}{B}$, the mobility of the particle is $N + 1$ times more than the mobility of the macromolecule

$$\Delta_\alpha(t) = \frac{6T}{\zeta}t. \tag{5.14}$$

In the internal interval $\frac{\tau}{B} < t < \tau$, when the change of the displacement is negligible, we can find the expression

$$\Delta_\alpha(t) = \frac{12\pi T \tau^*}{\zeta(N+1)}\left(\frac{B\chi}{B+E}\right)^{1/2}. \tag{5.15}$$

For a long time of observation, $t > \tau^* B$, the displacement of any particle is identical to the displacement of the entire macromolecule.

$$\Delta_\alpha(t) = \frac{6T}{\zeta(N+1)B}t. \tag{5.16}$$

Formula (5.15) defines a certain intermediate scale, which can be compared to the intermediate scale revealed in the consideration of the diffusion of a macromolecule (see expression (5.7)). We ought to believe that the local displacement of any point of the macromolecule should depend neither on the number of subchains nor on the length of the macromolecule, so that we can identify the quantities (5.7) and (5.15) to find the relation between the parameters of the theory at $B \gg 1$

$$\psi = \frac{\pi^2}{\chi}. \tag{5.17}$$

This formula ought to be taken as an asymptotic relation, which requires large values of the parameter ψ for the strongly entangled systems. The above value of ψ ensures that any Brownian particle of the chain does not move more than ξ during the times $t < \tau$. For this time of observation, the large-scale conformation of the macromolecule is frozen, but the small-scale motion of the particles confined to the scale ξ can take place, and the macromolecule, indeed, can be considered to be in a "tube" with radius ξ. Reptation of the macromolecule inside the tube is possible. Localisation of a macromolecule in a tube was assumed by Edwards (1967a) and by Gennes (1971), who have introduced reptation motion for the macromolecule as well. The latter is needed to explain the observed law of diffusion of very long macromolecules in entangled systems. The tube and its diameter are postulated in the earlier theories (Doi and Edwards 1986).

Reptation Mobility of a Macromolecule

It is important to remember now that there is anisotropy of mobility, which can bring the reptation mode of motion and a different law of mobility of macromolecules. To consider these effects, one can refers to the system of equation (3.37) or to the Doi-Edwards model of dynamics of the macromolecule, which was described in Section 3.5. We start with the last model and specify some parameters of the model. As we saw in the previous section, a macromolecule in the system is confined by the scale ξ during the short time τ/B. Up to this scale, a particle moves as in a viscous liquid according the law of diffusion (5.2), so that one can chose the time step as

$$\Delta t = \frac{\xi^2}{6D_0} \tag{5.18}$$

where $D_0 = T/\zeta Z$ is coefficient of diffusion of macromolecule in a viscous liquid. Then, the Doi-Edwards model assumes that the macromolecule moves only along its axis by the specific mechanism, substituting the particles by neighbouring ones, which is possible at $b \geq \xi$. Following Doi and Edwards (1978), we accept that $b = \xi$, so that $\langle R^2 \rangle_0 = Z\xi^2$.

To consider the mobility of the macromolecule, one ought to calculate the displacement of the centre of mass of the chain

$$q(t) = \frac{1}{N+1} \sum_{\nu=0}^{Z} r^\nu(t).$$

After summing equations (3.43), we find

$$\Delta q(t) = q(t + \Delta t) - q(t) = \frac{1}{Z} R(t)\phi(t) + \frac{1}{Z} v(t).$$

We assume that $Z \gg 1$ here, so $Z + 1 \approx Z$, whereas $R = r^Z - r^0$ is the end-to-end distance of the chain. Then, one calculates the correlation function

$$\langle \Delta q(t) \Delta q(u) \rangle = \frac{1}{Z^2} \langle \phi(t)\phi(u)R(t)R(u) \rangle + \frac{1}{Z^2} \langle v(t)v(u) \rangle$$
$$+ \frac{1}{Z^2} [\langle \phi(t)R(t)v(u) \rangle + \langle \phi(u)R(u)v(t) \rangle].$$

Taking into account the properties (3.44) of random quantities, we find

$$\langle \Delta q(t) \Delta q(u) \rangle = \delta_{tu} \left(\frac{\xi^2}{Z} + \frac{\xi^2}{Z^2} \right).$$

Thus, the one-step mean square displacement is determined, at $Z \gg 1$, by relation

$$\Delta(t) = \frac{\xi^2}{Z}.$$

One multiplies the above quantity by the step number $\frac{t}{\Delta t} = \frac{6D_0}{\xi^2} t$, to find the mean square displacement in time t

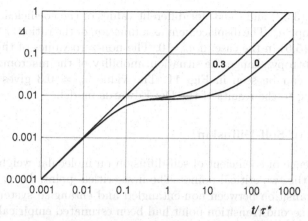

Figure 11. Displacement of macromolecule vs time.
The straight solid line – the consequence of equations (3.52) and (3.53) at $\Phi_i^\alpha = 0$ –
depicts the analytical result for the Rouse dynamics. The solid curves represents the
displacement for a macromolecule of length $M = 25M_e$ ($\chi = 0.04$) with corresponding
(according to relations (3.25) and (3.29)) values of parameters $B = 429$ and $\psi = 8.27$
and the values of parameters of local anisotropy $a_e = 0$ and $a_e = 0.3$. For the isotropic
situation ($a_e = 0$), the curve can be calculated analytically according to equation (5.5),
but for the parameter of local anisotropy $a_e = 0.3$, the displacement ought to be calculated numerically. Internal resistance (parameters E and a_i) does not affect mobility of
macromolecular coil. Adapted from Pokrovskii (2006).

$$\Delta(t) = \frac{\xi^2}{Z}\frac{t}{\Delta t} = \frac{6D_0}{Z}t.$$

The written relation defines the diffusion coefficient of the chain as

$$D = D_0\frac{\xi^2}{\langle R^2 \rangle_0}.$$

Referring to the definition of the intermediate scale (5.7), we write down the
diffusion coefficient in another form

$$D = \frac{2}{\pi^2}D_0\chi. \qquad (5.19)$$

One can see that the diffusion coefficient of the macromolecule due to reptation
does not depend on the length of the ambient macromolecules

$$D \sim M_0^0 M^{-2}. \qquad (5.20)$$

The reptation diffusion is connected with the local anisotropy of mobility
of particles, which can be confirmed by investigation of equations (3.37). As an
example, Fig. 11 contains the results for displacement of a macromolecule of
length $M = 25M_e$ (value of parameter $\chi = 0.04$) due to numerical integration

of equations (3.52) and (3.53) for different values of the coefficient of external local anisotropy a_e. The displacement as a function of the ratio t/τ^* follows the dependence (5.5) in the case, if $a_e = 0$. The non-zero values of the coefficient of local anisotropy change the situation: mobility of the macromolecular coil increases as can be seen in Fig. 11. The value $a_e = 0.3$ gives the results corresponding to the results of the Doi-Edwards model.

Coefficient of Self-Diffusion

The dependence of coefficient of self-diffusion on molecular weight shows the existence of the two critical points. The first critical point, $M_c \approx 2M_e$, determines the transition between non-entangled and entangled systems. The position of the second transition point had been estimated empirically by Wang (2003), who analysed data of various scholars on diffusion of macromolecules and showed, that for both melts and solutions of linear polymers (with a few exceptions, among them hydrogenated polybutadiene – hPB) there is a point $M^* = 10M_e$ dividing regions of different dependences of self-diffusion coefficient on molecular weight, while in the region of higher molecular weights the reptation law of diffusion (5.20) is firmly valid. It was also noted earlier (Tao et al. 2000), that 'if there is a *non-universal* crossover to an exponent of -2.0, for hPB it occurs at or beyond $M/M_e \approx 10^2$, whereas for PS and PDMS it might occur near $M/M_e \approx 10$.' So, for the linear polymers, there is a point about $M^* = 10M_e$, as a rule, where the mechanisms of mobility change, while the reptation mechanism of mobility dominates above the transition point $M^* = 10M_e$ (or in the formulation, which has no exceptions: below the value $\chi^* = 0.1$),[1] and one can write for empirical dependence

$$D \sim \begin{cases} M^{-3}, & M < M^*, \text{ weakly entangled systems} \\ M^{-2}, & M > M^*, \text{ strongly entangled systems} \end{cases}.$$

Relation (5.4) is used to describe the dependence in the region of lengths below $2M_e$, whereas in the region above $2M_e$, the two mechanisms of the displacement of the centre of mass of the macromolecule are optional, so that the resulting coefficient of self-diffusion has to be defined as

$$D = D_{\mathrm{dif}} + D_{\mathrm{rep}}.$$

---

[1] Some other scholars (Lodge 1999; Tao et al. 2000) consider that the transition point coincides with the entanglement point $2M_e$ and find, considering the data for the whole region above $2M_e$, the empirical law of molecular-weight dependence of self-diffusion coefficient with the index about -2.3 instead of non-amended reptation law (5.20). However, to estimate a real empirical value of index in the reptation law of diffusion, one needs much longer macromolecules and, in any case, one has to exclude the transition interval below $10M_e$. Note also that the measurements of diffusion of labelled chains in a melt matrix of significantly higher molecular weight (tracer diffusion) show the index -2 (Wang 2003).

The first and the second diffusion coefficients, D_{dif} and D_{rep} are defined by relations (5.11) and (5.20), respectively. The two competing mechanisms have a different length dependence of the self-diffusion coefficient

$$D \sim \begin{cases} c^{-2\delta}M^{-1-\delta}, & \text{non-reptation} \\ c^{-2}M^{-2}, & \text{reptation} \end{cases} . \tag{5.21}$$

One can see from the comparison of equations (5.10) and (5.19) that the reptation motion of the macromolecules is revealed at the condition

$$\chi B \gg \frac{1}{2}\pi^2. \tag{5.22}$$

This relation determines a molecular weight M^* at which the mechanism of diffusion changes. To determine the dynamic transition point M^*, which separates the strongly entangled ($M > M^*$) and weakly entangled ($M < M^*$) systems, one considers the parameter B to be dependent on χ and, taking equations (3.17) and (3.25) into account, finds a solution of the equation

$$\chi B(\chi) = \frac{1}{2}\pi^2. \tag{5.23}$$

It is not difficult to solve this equation, taking empirical value $\delta = 2.4$ into account, and estimate the number value of the transition point between weakly and strongly entangled systems as

$$\chi^* \approx 0.1, \quad M^* \approx 10 M_e.$$

This is the point, above which the reptation mechanism of diffusion predominates.

The results of estimation of coefficient of self-diffusion due to simulation for macromolecules with different lengths are shown in Fig. 12. The introduction of local anisotropy practically does not affect the coefficient of diffusion below the transition point M^*, the position of which depends on the coefficient of local anisotropy. For strongly entangled systems ($M > M^*$), the value of the index -2 in the reptation law is connected only with the fact of confinement of macromolecule, and does not depend on the value of the coefficient of local anisotropy. At the particular value $a_e = 0.3$, the simulation reproduces the results of the conventional reptation-tube model (see equation (5.21)) and corresponds to the typical empirical situation ($M^* = 10 M_e$).

5.2 Quasi-Elastic Neutron Scattering

The convincing confirmation of existence of intermediate length is given by experimental results on the neutron scattering. To observe the scattering on a single macromolecule in a system of entangled chains, the investigators have

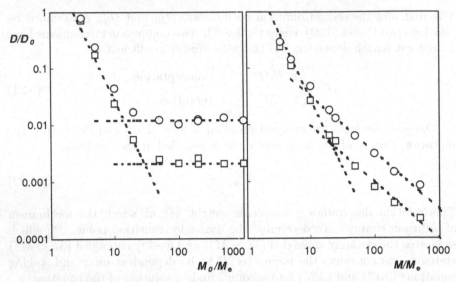

Figure 12. The coefficient of diffusion of a macromolecule.
Diffusion of a macromolecule of length $25M_e$ among macromolecules with various lengths
M_0 (left plot), and among macromolecules of the same length – self-diffusion (right plot) is
depicted. Each point is calculated as the ratio of asymptotic values of the displacement of
a macromolecule for large times (see the dependence in Fig. 11) to values of displacement
for the Rouse chain at the corresponding values of the parameter B and the values of the
parameter $\psi = 0$. The values of the parameters of local anisotropy are $a_e = 0.3$, $a_e = 0$
for the circles and $a_e = 0.1$, $a_e = 0$ for the squares. The slopes of the dashed lines on the
left plot are -2.4 for short macromolecules and 0 for long ones, so that the simulation
determines the point of transition between diffusive and reptation modes of motion. The
slopes of the dashed lines on the right plot are -2.4 for short macromolecules and -1
for long ones, so that the simulation gives the well-known dependence $D \sim M^{-2}$ for
coefficient of self-diffusion of macromolecules above the point of transition. The results
do not depend on the arbitrary number N of subchains, whereas it is taken $N = 10$ for
the sample calculations. All calculations are fulfilled at the step of integration $h = 0.001$,
and the number of realisations is 100 in each case. Adapted from Pokrovskii (2008).

taken blends of chemically identical deuterated and non-deuterated macromolecules (Daoud et al. 1975; Higgins and Roots 1985; Richter et al. 1990).
A small quantity of non-deuterated macromolecules among deuterated macromolecules determines scattering which can be considered to be scattering on
a single macromolecules.

5.2.1 The Scattering Function

An introduction to the theory of neutron scattering can be found, for instance,
in a books by Hansen and Donald (1986) and Higgins and Benoit (1994). The
scattering function for a single macromolecule is known for some models of

polymer chain in dilute solutions (Gennes 1967; Dubuis-Violette and Gennes 1967; Akcasu and Gurol 1976) and in entangled systems (Gennes 1981; Ronca 1983; Kokorin and Pokrovskii 1990; Des Cloizeaux 1993; Wishnewski et al. 2002). We shall consider the entangled systems, using the mesoscopic model of dynamics of macromolecule, which allows us to avoid some assumptions, which were included in the previous calculations, and present a generalised scattering function.

The neutron scattering on a single macromolecule is determined by the dynamic structure function or scattering function

$$S(\boldsymbol{k},t) = \frac{1}{N+1} \sum_{\alpha,\gamma} \langle \exp[i\boldsymbol{k}(\boldsymbol{r}^{\alpha}(t) - \boldsymbol{r}^{\gamma}(0))] \rangle. \tag{5.24}$$

The double sum is evaluated over all the Brownian particles of the macromolecule. In equation (5.24), $\boldsymbol{k}$ is the vector in the direction of the scattering, having the length

$$k = \frac{4\pi}{\lambda} \sin \frac{\theta}{2}$$

where λ is the wave-length of the initial beam and θ is the scattering angle. To investigate the motion of the internal parts of a macromolecular coil, the relation $k\langle R^2 \rangle^{1/2} \gg 1$, or $\lambda \ll \langle R^2 \rangle^{1/2}$ must be fulfilled. For typical macromolecules, it gives for the wave-length the value $\lambda \ll 10^{-6}$ cm.

We are considering the scattering function (5.24) and can see that the expansion of the expression

$$\langle \exp[i\boldsymbol{k}(\boldsymbol{r}^{\alpha}(t) - \boldsymbol{r}^{\gamma}(0))] \rangle = 1 - \frac{1}{2} \sum_{i=1}^{3} k_i^2 \langle (r_i^{\alpha}(t) - r_i^{\gamma}(0))^2 \rangle - \cdots \tag{5.25}$$

has real components only. Since the averaged values of the quantity $(r_i^{\alpha}(t) - r_i^{\gamma}(0))^2$ do not depend on the label of co-ordinates i, expression (5.25) can be written as

$$\langle \exp[i\boldsymbol{k}(\boldsymbol{r}^{\alpha}(t) - \boldsymbol{r}^{\gamma}(0))] \rangle = \exp\left[-\frac{1}{6} k^2 \sum_{i=1}^{3} \langle (r_i^{\alpha}(t) - r_i^{\gamma}(0))^2 \rangle \right].$$

This is an exact relation in the case, if the averaging is fulfilled over a Gaussian distribution function.

Hence, the scattering function (5.24) takes the form

$$S(\boldsymbol{k},t) = \frac{1}{N+1} \sum_{\alpha,\gamma} \exp\left[-\frac{1}{6} k^2 \sum_{i=1}^{3} \langle (r_i^{\alpha}(t) - r_i^{\gamma}(0))^2 \rangle \right]. \tag{5.26}$$

To evaluate the function, we omit the correlation between particles with different labels and consider the non-coherent scattering function

S(k,t)/S(k,0)

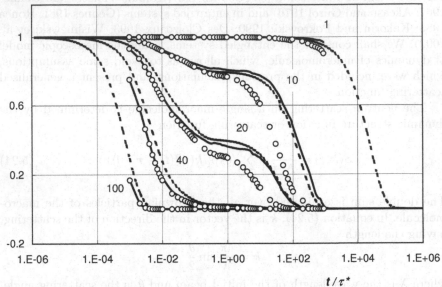

t / τ^*

Figure 13. The neutron scattering function.
The results for a macromolecule of the length $M = 25M_e$ ($\chi = 0.04$, $B = 429$, $\psi = 8.27$).
The analytical results due to equation (5.26), in which $\Delta_\alpha(t)$ is defined by (5.5) and (5.13),
are shown by dashed lines. The results of calculation of function (5.24) due to simulation
are shown for the above values of parameters B and ψ and for values $a_e = 0$, $a_i = 0$ by
solid lines and $a_e = 0.3$, $a_i = 0.1$ by open circles. The values of $k\langle R^2\rangle^{1/2}$ are shown at
the curves. Adapted from the paper of Pokrovskii (2006).

$$S(\boldsymbol{k},t) = \frac{1}{N+1} \sum_{\alpha=0}^{N} \exp\left(-\frac{1}{6}k^2\Delta_\alpha(t)\right), \qquad (5.27)$$

where the mean square displacement $\Delta_\alpha(t) = \langle[\boldsymbol{r}^\alpha(t) - \boldsymbol{r}^\alpha(0)]^2\rangle$ of the particle
α is determined by expression (5.13). We can easily see that all the properties
of function $\Delta_\alpha(t)$ from Section 5.1.2 are reflected in the scattering function,
which is shown in Fig. 13 for some values of parameters. In particular, a
plateau is revealed on the plot of the scattering function.

5.2.2 An Estimation of Intermediate Length

The value of the scattering function in the plateau region is connected directly
with the intermediate length ξ

$$S(k,t) = \exp\left(-\frac{1}{6}k^2\xi^2\right), \qquad \frac{\tau}{B} < t < \tau. \qquad (5.28)$$

Similar results are reported by Ronca (1983) and Des Cloizeaux (1993). Their calculations were based on models, which contained the tube diameter d as an introduced parameter. Though the tube diameter is an arbitrary parameter, it appears, that this quantity is practically the same as the double intermediate length

$$d^2 = \frac{6}{5}(2\xi)^2.$$

Though their theories appear to be not quite consistent, the final results provide an estimate of the intermediate length.

The intermediate length is defined by equation (5.7) and, taking into account formulae (1.4) and (3.30), can be represented as

$$\xi^2 \sim c^{-2}C_\infty^{-2}(T). \tag{5.29}$$

This relation was confirmed by Kholodenko (1996) who started from detailed picture of geometrically confined polymer chain and used more sophisticated methods of calculation.

Results on neutron scattering by specially prepared samples (Higgins and Roots 1985; Richter et al. 1990) reveal the plateau region. The beam of neutrons characterised by k in the range from 0.058 Å^{-1} till 0.135 Å^{-1} was used by Richter et al. (1990). It gives the values of $k\langle R^2\rangle^{1/2}$ from 5.8 to 13 for the typical size of macromolecular coils. We can see that the actual values expose the special time dependence revealing the intermediate length. The time dependence of the observed scattering functions appears to be very similar to the depicted scattering functions.

For an interpretation of the results and evaluation of the intermediate length, Richter et al. (1990) have used the scattering function calculated by Ronca (1983). Richter et al. (1993), Ewen and Richter (1995) and Ewen et al. (1994) have found the typical values of the tube diameter at different temperatures which are 35–50 Å for alternating poly(ethylen-propylene) and 65–70 Å for poly(dimethylsiloxane); the corresponding values of intermediate length ξ are 16–23 Å and 30–32 Å. According to relation (5.29), the quantity $\xi c C_\infty(T)$ does not depend on the temperature. Nevertheless, Richter et al. (1993), at investigating the temperature dependence of the intermediate length in polyethylene-propylene melts, have found that the quantity $\xi c C_\infty(T)$ increases slightly when the temperature increases. One ought to remember that the written relations follow the simplest schematisation of the entangled system.

The results of the investigations of Higgins and Roots (1985) and Richter et al. (1990) confirm the existence of the dynamic intermediate length and this is satisfactory from the point of view of all theories. However, it does not mean that investigators confirm the reptation of long macromolecules. The existence of an intermediate length is the effect of first order in respect of co-ordinates in the equation of macromolecule dynamics, while the reptation of a macromolecule is connected to terms of higher orders. Other arguments are needed to confirm the reptation mobility.

Chapter 6
Linear Viscoelasticity

Abstract In the course-grained approximation, polymer solutions and melts can be considered as a suspension of interacting Brownian particles, which allow us to determine a general expression for the stress tensor, following a method developed in the theory of liquids (Rice and Gray in Statistical mechanics of simple liquids (Wiley, New York), 1965; Gray in Physics of simple liquids, ed by H.N.V Temperley (North Holland, Amsterdam, pp. 507–562), 1968). The general theory is specified to calculate dynamic modulus both for dilute and concentrated polymer systems. The approach allows one correctly to describe the linear viscoelastic behaviour of dilute polymer solutions over a wide range of frequencies, if the effects of excluded volume, hydrodynamic interaction, and internal viscosity are taken into account. As far as the very concentrated solutions and melts of polymers – entangled polymers – are concerned, the results for the linear approximation of macromolecular dynamics are only available now. As one can anticipate, it is not sufficient for complete description of relaxation processes in strongly entangled systems, though some relations for terminal characteristics are obtained for these systems. Remarkably, the mesoscopic theory appears to be self-consistent for entangled systems: the relaxation time of the environment is equal to the relaxation time of the entire system, which is calculated in this chapter. The intermediate scale introduced in Chapter 5 appears here once more as connected with the well-known length of a macromolecule between adjacent entanglements M_e. It casts a new light on the old terms and old theories. The pictures given earlier by different theories appear to be consistent.

6.1 Stresses in the Flow System

6.1.1 The Stress Tensor

As before, we shall consider each macromolecule either in dilute or in concentrated solution to be schematically represented by a chain of $N + 1$ Brownian

V.N. Pokrovskii, *The Mesoscopic Theory of Polymer Dynamics*,
Springer Series in Chemical Physics 95,
DOI 10.1007/978-90-481-2231-8_6, © Springer Science+Business Media B.V. 2010

particles, so that a set of the equations for motion for the macromolecule can be written as a set of coupled stochastic equations

$$m\frac{d^2 r^\alpha}{dt^2} = F^\alpha + G^\alpha + K^\alpha + \phi^\alpha, \quad \alpha = 0, 1, \dots, N, \tag{6.1}$$

where m is the mass of a Brownian particle associated with a piece of the macromolecule of length $M/(N+1)$, r^α are the co-ordinates of the Brownian particles. The dissipative forces F^α and G^α acting on the particles were discussed in Chapter 2 for dilute solutions and in Chapter 3 for entangled systems.

We consider n to be the number density of macromolecular coils in the system, so that the system contains $n(N+1)$ Brownian particles in unit volume. This number is sufficiently large to introduce macroscopic variables for the suspension of Brownian particles, namely, the mean density

$$\rho(x, t) = \sum_{a,\alpha} m\langle \delta(x - r^{a\alpha}) \rangle = m(N+1)n(x, t) \tag{6.2}$$

and the mean density of the momentum

$$\rho v_j(x, t) = \sum_{a,\alpha} m\langle u_j^{a\alpha} \delta(x - r^{a\alpha}) \rangle. \tag{6.3}$$

The angle brackets denote averaging over the ensemble of the realisation of random forces in the equations of motion of the particles. The sum in (6.2) and (6.3) is evaluated over all the Brownian particles. The double index $a\alpha$ consists of the label of a chain a and the label of a particle α in the chain.

The methods developed in the theory of liquids (Rice and Gray 1965, Gray 1968) was used by Pokrovskii and Volkov (1978a) to determine the stress tensor for the set of Brownian particles in this case. One can start with the definition of the momentum density, given by (6.3), which is valid for an arbitrary set of Brownian particles. Differentiating (6.3) with respect to time, one finds

$$\frac{\partial}{\partial t}\rho v_j = -\frac{\partial}{\partial x_i} \sum_{a,\alpha} m\langle u_i^{a\alpha} u_j^{a\alpha} \delta(x - r^{a\alpha}) \rangle + \sum_{a,\alpha} \left\langle m\frac{du_j^{a\alpha}}{dt}\delta(x - r^{a\alpha}) \right\rangle. \tag{6.4}$$

The right-hand side of equation (6.4) has to be reduced to a divergent form. To transform the second term, we use the dynamic equation (6.1), which ought to be multiplied by $\delta(x - r^{a\alpha})$. After summing over all the particles of the macromolecule and averaging, one uses the requirement that there is no mean volume force, that is,

$$\sum_{\alpha=0}^{N} \langle (F^{a\alpha} + \phi^{a\alpha})\delta(x - r^{a\alpha}) \rangle = 0, \quad a = 1, 2, \dots, n. \tag{6.5}$$

So, for each macromolecular coil, one can write

$$m \sum_{\alpha=0}^{N} \left\langle \frac{d\boldsymbol{u}^{a\alpha}}{dt} \delta(\boldsymbol{x} - \boldsymbol{r}^{a\alpha}) \right\rangle = \sum_{\alpha=0}^{N} \langle (\boldsymbol{K}^{a\alpha} + \boldsymbol{G}^{a\alpha}) \delta(\boldsymbol{x} - \boldsymbol{r}^{a\alpha}) \rangle, \quad a = 1, 2, \ldots, n.$$

Next, the formal expansion of the δ-function into a Taylor's series about the centre of mass q^a of the ath macromolecule can be used, retaining only the first two terms of the expansion

$$\delta(\boldsymbol{x} - \boldsymbol{r}^{a\alpha}) = \delta(\boldsymbol{x} - \boldsymbol{q}^a) - (r_k^{a\alpha} - q_k^a) \frac{\partial}{\partial x_k} \delta(\boldsymbol{x} - \boldsymbol{q}^a).$$

So, the above formula is transformed into

$$-\frac{\partial}{\partial x_k} \sum_{\alpha=0}^{N} \langle (K_j^{a\alpha} r_k^{a\alpha} + G_j^{a\alpha} r_k^{a\alpha}) \delta(\boldsymbol{x} - \boldsymbol{q}^a) \rangle, \quad a = 1, 2, \ldots, n.$$

Here, the sum is conducted over all the particles in a given macromolecule. Assuming that all the macromolecules are identical, and neglecting the statistical dependence of the position of the centres of mass of the macromolecules on the other co-ordinates, one obtains an expression for the second term on the right-hand side of equation (6.4) in the divergent form

$$\sum_{a,\alpha} \left\langle m \frac{du_j^{a\alpha}}{dt} \delta(\boldsymbol{x} - \boldsymbol{r}^{a\alpha}) \right\rangle = -\frac{\partial}{\partial x_k} n \sum_{\alpha=0}^{N} \langle K_j^{a\alpha} r_k^{a\alpha} + G_j^{a\alpha} r_k^{a\alpha} \rangle.$$

The first term on the right-hand side of (6.4) can also be rewritten in a more convenient form. One uses the definition of the mean velocity v_i and, taking only the first term of the expansion of the δ-function into account, one finds that

$$m \sum_{a,\alpha} \langle u_j^{a\alpha} u_i^{a\alpha} \delta(\boldsymbol{x} - \boldsymbol{q}^a) \rangle = nm \sum_{\alpha=0}^{N} \langle (u_j^\alpha - v_j)(u_i^\alpha - v_i) \rangle + \rho v_i v_j.$$

Thus, an equation, which has the sense of a law of conservation of momentum has been obtained. There is an expression for the momentum flux $\rho v_i v_j - \sigma_{ij}$ under the derivation symbol, which allows one to write down the expression for the stress tensor

$$\sigma_{kj} = -n \sum_{\alpha=0}^{N} \left[m \langle (u_j^\alpha - v_j)(u_k^\alpha - v_k) \rangle + \langle K_k^\alpha r_j^\alpha + G_k^\alpha r_j^\alpha \rangle \right]. \quad (6.6)$$

The assumption that the particle velocities are described by the local-equilibrium distribution yields

$$\sigma_{kj} = -n(N+1)T\delta_{jk} - n\sum_{\alpha=0}^{N}\langle K_j^\alpha r_k^\alpha + G_j^\alpha r_k^\alpha\rangle. \tag{6.7}$$

As was demonstrated by Pyshnograi (1994), the last term in (6.7) can be written in symmetric form, if the continuum of Brownian particles is considered incompressible. In equation (6.7), the sum is evaluated over the particles in a given macromolecule. The monomolecular approximation ensures that the stress tensor of the system is the sum of the contributions of all the macromolecules. In this form, the expression for the stresses is valid for any dynamics of the chain. One can consider the system to be a dilute polymer solution or a concentrated solution and melt of polymers. In any case the system is considered as a suspension of interacting Brownian particles.

6.1.2 Oscillatory Deformation

Experimentally a variety of quantities are used to characterise linear viscoelasticity (Ferry 1980). There is no need to consider all the characteristics of linear viscoelastic response of polymers which are measured under different regimes of deformation: in linear region, they are connected with each other. The study of the reaction of the system in the simple case, when the velocity gradients are independent of the co-ordinates and vary in accordance with the law

$$\gamma_{ik} \sim e^{-i\omega t}$$

for different deformation frequencies ω, gives a clear picture of the phenomena of linear viscoelasticity and yields important information about the relaxation processes in the system. For this case, the expression for the stress tensor can be written in the form

$$\sigma_{ik}(t) = -p\delta_{ik} + 2\eta(\omega)\gamma_{ik}(t) \tag{6.8}$$

which defines the complex viscosity coefficient – dynamic viscosity

$$\eta(\omega) = \eta'(\omega) + i\eta''(\omega).$$

Since the velocity gradient is related to the displacement gradient by the expression $\nu_{12} = -i\omega\lambda_{12}$, it follows that, instead of the dynamic viscosity, the use may be made of another characteristic – the dynamic modulus

$$G(\omega) = G'(\omega) - iG''(\omega) = -i\omega\eta(\omega). \tag{6.9}$$

The components of the above complex quantities are linked by the relation

$$G' = \omega\eta'', \qquad G'' = \omega\eta'. \tag{6.10}$$

Dynamic modulus is a convenient characteristic of viscoelasticity. To analyse the results, it is convenient also to consider the asymptotic behaviour of the dynamic modulus at high and low frequencies. In the latter case

$$G(\omega) = -i\omega\eta + \omega^2\nu. \tag{6.11}$$

The expansion determines the terminal quantities: the viscosity coefficient η and the elasticity coefficient ν which, in their turn, determine the terminal relaxation time and steady-state compliance, correspondingly,

$$\tau = \frac{\nu}{\eta}, \qquad J_e = \frac{\nu}{\eta^2}. \tag{6.12}$$

Both the dynamic modulus and the terminal quantities are characteristics of viscoelasticity of a system and are subject of interest of experimentalists.

Note that the dynamic modulus is the Fourier-transform of the relaxation modulus $G(t)$

$$G(\omega) = -i\omega \int_0^\infty G(t)e^{i\omega t}\, dt,$$

which is also often used to characterise viscoelastic behaviour on the system.

6.2 Macromolecules in a Viscous Liquid

The dilute polymer solution can be considered as a collection of non-interacting macromolecular coils suspended in a viscous liquid, the stress tensor of which is written as

$$\sigma_{ik}^0 = -p\delta_{ik} + 2\eta_s\gamma_{ik}. \tag{6.13}$$

The dynamics of a separate macromolecular coil in the viscous liquid, discussed in Chapter 2, allows one to determine the problem.

6.2.1 The Stress Tensor

To find the stress tensor, one can use equation (6.7), in which the elastic and internal viscosity forces, according to equations (2.2) and (2.25), have the form

$$K_i^\alpha = -2T\mu A_{\alpha\gamma} r_i^\gamma, \qquad G_j^\alpha = -G_{\alpha\gamma}(\dot{r}_j^\gamma - \omega_{jl} r_l^\gamma).$$

This gives the expression for the stress tensor

$$\sigma_{ik} = -nT(1+N)\delta_{ik}$$
$$+ n\sum_{\nu=0}^N \left[2\mu T A_{\alpha\gamma}\langle r_i^\alpha r_k^\gamma\rangle - T\delta_{ik} + G_{\alpha\gamma}(\langle \dot{r}_k^\gamma r_i^\alpha\rangle - \omega_{il}\langle r_l^\alpha r_k^\gamma\rangle)\right].$$

Furthermore, it is convenient to switch to normal co-ordinates (1.13). We can use the expressions for forces (2.26) to rewrite the expression for the stress tensor in normal co-ordinates

$$\sigma_{ik} = -n(N+1)T\delta_{ik}$$

$$+ n \sum_{\alpha=1}^{N} \left[2\mu T \lambda_\alpha \langle \rho_i^\alpha \rho_k^\alpha \rangle - T\delta_{ik} + \zeta\varphi_\alpha \left(\langle \dot{\rho}_k^\alpha \rho_i^\alpha \rangle - \omega_{kl} \langle \rho_l^\alpha \rho_i^\alpha \rangle \right) \right]. \quad (6.14)$$

Here the linear terms in respect to the coefficient of internal viscosity φ_α have taken into account only. Averaging with respect to the velocity distribution has been assumed here. One ought to add the stresses (6.13) of carrier viscous liquid to stresses (6.14) to determine the stress tensor for the entire system, that is for the dilute solution of the polymer.

Let us note that the extra stresses arise due to the differences in the rate of diffusion $\boldsymbol{w}^\alpha$ of a Brownian particle and the averaged velocity of the medium v^α at the point where the particle is located. It results in the emergence of forces

$$\boldsymbol{F}^\alpha = -\zeta(\boldsymbol{v}^\alpha - \boldsymbol{w}^\alpha).$$

Accordingly, the extra stresses after averaging can be written as

$$-n\zeta \sum_{\alpha=0}^{N} \langle (v_i^\alpha - w_i^\alpha) r_k^\alpha \rangle \quad (6.15)$$

where the angle brackets denote averaging with respect to the distribution function for the particle co-ordinates. One ought to determine the diffusion velocity $\boldsymbol{w}^\alpha$ to arrive at expression (6.14) for the stress tensor. Expression (6.15) was the starting point in the calculations of the extra stresses in dilute solutions of polymer in works by Cerf (1958), Kirkwood and Riseman (1948), Peterlin (1967), and Zimm (1956).

One can use equation (2.37) to obtain the other form of the stress tensor

$$\sigma_{ik} = -nT(N+1)\delta_{ik} + n\zeta \sum_{\nu=1}^{N} \frac{1}{2} \left[\frac{1}{\tau_\nu^\parallel} \left(\langle \rho_i^\nu \rho_k^\nu \rangle - \frac{1}{2\mu\lambda_\nu} \delta_{ik} \right) + 2\varphi_\nu \langle \rho_i^\nu \rho_j^\nu \rangle \gamma_{jk} \right]$$

$$(6.16)$$

where the times of relaxation $\tau_\nu^\perp$ and $\tau_\nu^\parallel = (1+\varphi_\nu)\tau_\nu^\perp$ were defined earlier by expressions (2.30).

Note that the internal viscosity is a residual of internal relaxation process in the case, when the slow deformation is considered. In a more general case, the elastic and internal viscosity forces acting on the chain, according to equations (2.2) and (2.28), can be written as

$$K_i^\alpha = -2T\mu A_{\alpha\gamma} r_i^\gamma, \qquad G_i^\alpha = -\int_0^\infty G_{\alpha\gamma}(s)(u_i^\gamma - \omega_{ij}r_j^\gamma)_{t-s} \mathrm{d}s. \quad (6.17)$$

Then, equation (6.7) gives, instead of (6.14), the expression for the stress tensor in normal co-ordinate

$$\sigma_{ik}(t) = -n(N+1)T\delta_{ik} + n\sum_{\alpha=1}^{N}\left\{2\mu T\lambda_\alpha\langle\rho_i^\alpha\rho_k^\alpha\rangle - T\delta_{ik}\right.$$

$$\left. + \zeta\int_0^\infty \varphi_\alpha(s)\Big(\langle\dot\rho_i^\alpha(t-s)\rho_k^\alpha(t)\rangle - \omega_{il}(t-s)\langle\rho_l^\alpha(t-s)\rho_k^\alpha(t)\rangle\Big)ds\right\}.$$

$$(6.18)$$

The validity of the theory for the non-linear region is restricted by terms of the second power with respect to the velocity gradient for non-steady-state flow and by terms of the third order for steady-state flow, due to approximations described in Chapter 2, when the relaxation modes of a macromolecule were being determined.

6.2.2 Dynamic Characteristics

Let us write down first of all the stress tensor for dilute solution (6.16) as a function of the velocity gradients. We can use expressions (2.41) for moments, in order to determine the stresses with accuracy within the first-order term with respect to velocity gradients

$$\sigma_{ik} = -p\delta_{ik} + 2\eta_s\gamma_{ik}$$
$$+ 2nT\sum_{\nu=1}^{N}\left[\frac{1-\varphi_\nu}{1+\varphi_\nu}\int_0^\infty \exp\left(-\frac{s}{\tau_\nu^{\|}}\right)\gamma_{ik}(t-s)ds + \varphi_\nu\tau_\nu^{\perp}\gamma_{ik}\right]. \quad (6.19)$$

This equation contains two sets of relaxation times, which are defined by equations (2.30), that is,

$$\tau_\alpha^{\|} = \tau_\alpha^{\perp}(1+\varphi_\alpha), \qquad \tau_\alpha^{\perp} = \tau_1\alpha^{-z\nu}, \qquad \varphi_\alpha = \varphi_1\alpha^\theta, \quad \alpha = 1, 2, \ldots \ll N.$$

The exponents in the above expressions can be estimated beforehand from the dependence of the limiting values of the characteristic viscosity at low and high frequencies on the length of the macromolecule.

In the case of the oscillatory motion, equation (6.19) defines, in accordance with equation (6.8), the complex shear viscosity $\eta(\omega) = \eta' + i\eta''$ with components

$$\eta'(\omega) = \eta_s + nT\sum_{\nu=1}^{N}\tau_\nu^{\perp}\left[\varphi_\nu + \frac{1-\varphi_\nu}{1+(\tau_\nu^{\|}\omega)^2}\right],$$

$$(6.20)$$

$$\eta''(\omega) = nT\sum_{\nu=1}^{N}\frac{\omega(\tau_\nu^{\perp})^2}{1+(\tau_\nu^{\|}\omega)^2}.$$

Figure 14 illustrates the dependence of the characteristic viscosity

$$[\eta] = \lim_{n\to0}\frac{\eta-\eta_s}{nT}$$

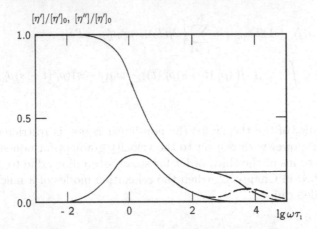

$[\eta']/[\eta']_0,\ [\eta'']/[\eta']_0$

Figure 14. The components of characteristic viscosity.
The real and imaginary components of characteristic viscosity have been calculated according to equations (6.20) for $z\nu = 2$, $\varphi_1 = 0.5$, $\theta = 0.5$. The dashed curves depicts the alternation of the dependencies in the case when an internal relaxation process is taking into account, whereas equations (6.28) are used at $\tau/2\tau_1 = 10^{-5}$.

as defined by equations (6.20) on the non-dimensional frequency $\tau_1\omega$ for some values of the parameters $z\nu$, φ_1 and θ which appeared in the formulae for relaxation times, introduced previously.

Equations (6.20) are followed by the expression for the characteristic dynamic modulus, components of which are

$$G' = nT \sum_{\nu=1}^{N} \frac{(\tau_\nu^\perp \omega)^2}{1 + (\tau_\nu^\| \omega)^2},$$

$$G'' = \eta_s\omega + nT \sum_{\nu=1}^{N} \tau_\nu^\perp \omega \left[\varphi_\nu + \frac{(1 - \varphi_\nu)}{1 + (\tau_\nu^\| \omega)^2}\right]. \tag{6.21}$$

Figure 15 demonstrates a comparison of the characteristic modulus

$$[G] = \lim_{n \to 0} \frac{G - i\eta_s\omega}{nT},$$

calculated according to equation (6.21), with the corresponding experimental values. One can note, that for certain values of the maximum relaxation time τ_1 and certain values of the exponents $z\nu$ and θ (whereas, in virtue of equation (6.27), $\theta = 2\nu - 1$), the theory satisfactorily reproduces the experimental relations for polymer solutions at infinite dilution. We may note yet again that the identifying constants are unambiguously determined by the limiting values of the characteristic viscosity and can be estimated independently.

The results (6.20) and (6.21), which are valid in the first order with respect to the coefficient of internal viscosity φ_1, were found by Peterlin (1967).

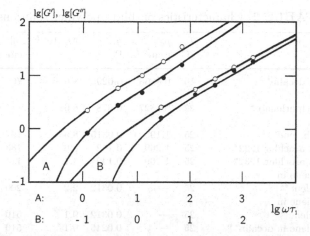

Figure 15. The components of characteristic dynamic modulus.
Frequency dependence of characteristic dynamic modulus for polystyrene solutions in
decalin (A) and in toluene (B). Experimental values due to Rossers et al. (1978) (see
also the last lines of Table 2) are shown by filled points (for the real part) and empty
points (for the imaginary part). The theoretical curves have been plotted for $z\nu = 1.788$,
$\theta = 0.788$, $\tau_1 = 2.5 \times 10^{-3}$ s for case A and for $z\nu = 1.5$, $\theta = 0.5$, $\tau_1 = 8.35 \times 10^{-4}$ s for
case B. Adapted from the paper of Pokrovskii and Tonkikh (1988).

A generalisation of the theory for the case of arbitrary values of internal viscosity was done by Pokrovskii and Tonkikh (1988). We may note that the case
when $\varphi_1 = 0$ and $z\nu = 2$, corresponds to an ideally flexible freely-draining
macromolecule, and reproduces the relations indicated by Rouse (1953).

Thus, one may conclude that, in the region of comparatively low frequencies, the schematic representation of the macromolecule by a subchain, taking
into account intramolecular friction, the volume effects, and the hydrodynamic
interaction, make it possible to explain the dependence of the viscoelastic behaviour of dilute polymer solutions on the molecular weight, temperature,
and frequency. At low frequencies, the description becomes universal. In order to describe the frequency dependence of the dynamic modulus at higher
frequencies, internal relaxation process has to be considered as was shown in
Section 6.2.4.

As an illustration, certain data characterising dilute polymer solutions are
presented in Table 2.

6.2.3 Initial Intrinsic Viscosity

In the study of the linear response, it is convenient to consider quantity independent of concentration and viscosity – the characteristic (intrinsic) viscosity

$$[\eta] = \lim_{c \to 0} \frac{\eta - \eta_s}{c\eta_s} \qquad (6.22)$$

TABLE 2. Characteristics of dilute polymer solutions

System	T °C	ρ_s g cm^{-3}	η_s P	$M \cdot 10^{-5}$	$[\eta]$ cm^3 g^{-1}	$\tau_1 \cdot 10^4$ s
Polystyrene in decalin*	16	0.8868	0.0295	8.6	76	0.35
Polystyrene in di-2-ethyl-hexylphthalate*	22	0.9827	0.678	8.6	—	7.59
Polystyrene in α-chloronaphthalene*	25	1.195	0.0315	8.6	197	1.26
Polystyrene in arochlor 1232*	25	1.269	0.142	8.6	183	3.98
Polystyrene in arochlor 1232*	25	1.269	0.142	4.1	111	1.2
1.4-Polybutadiene in chloronaphthalene**	25	—	0.0312	2.2	200	0.26
1.4-Polybutadiene in chloronaphthalene**	25	—	0.0312	9.1	510	2.75
1.4-Polybutadiene in decalin**	25	—	0.0245	9.1	510	2.14
Poly-α-methylstyrene in α-chloronaphthalene***	25	—	0.0315	14.3	252	2.0
Poly-α-methylstyrene in decalin***	25	—	0.0245	14.3	135	0.79
Polystyrene in decalin****	15	0.887	0.0287	180	300	23
Polystyrene in toluene****	20	0.867	0.0059	180	3100	69

* Johnson et al. (1970); ** Osaki et al. (1972a); *** Osaki et al. (1972b); **** Rossers et al. (1978).

where η_s is the viscosity of the solvent and $c = n M N_A^{-1}$ is the weight concentration of the polymer (N_A = Avogadro number).

The limit of the characteristic viscosity at low frequencies, according to (6.20), is defined as

$$[\eta']_0 = \frac{nT}{c\eta_s} \sum_{\alpha=1}^{N} \tau_\alpha^\perp = \frac{nT}{c\eta_s} \zeta(z\nu) \tau_1 \qquad (6.23)$$

where $\zeta(x)$ is Riemann's zeta-function. This quantity makes it possible to estimate the role of the volume effects and of the hydrodynamic interaction in the dynamics of the macromolecule, which influence the dependence of the quantity under discussion on the molecular weight (the length of the macromolecule)

$$[\eta']_0 = K M^{z\nu-1}. \qquad (6.24)$$

Theoretical estimates of the quantity $z\nu - 1$ are in the range from 0.5 (non-draining Gaussian coil), to 1.11 (draining coil with excluded-volume interaction). A compilation of empirical values of K and of the power exponents for different polymers and different solvents may be found in the literature (Flory 1969, Tsvetkov et al. 1964). The empirical values of the exponent $z\nu-1$ do not exceed 0.9, which indicates significant impermeability of the macromolecular coil in a flow. We may note that once a relation of type (6.24) has been established for a certain polymer, it can be used to determine the molecular weight of the polymer from the characteristic viscosity (Flory 1969, Tsvetkov et al. 1964). If the value of the index $z\nu$ is known, equation (6.23) allows us to estimate the value of the largest relaxation time τ_1.

For a non-draining coil, the characteristic viscosity defined by equation (6.23) can be expressed in the form

$$[\eta']_0 = \Phi \frac{\langle S^2 \rangle^{3/2}}{M}$$
(6.25)

where $\langle S^2 \rangle$ is the average square of the radius of inertia of the coil, while the experimental value of the constant Φ (called the Flory constant) according to Flory (1969)

$$\Phi = (2.66 \pm 0.1) \times 10^{23} \text{ mol}^{-1}.$$

Equation (6.25) makes it possible in this case to interpret a dilute solution of macromolecules as a suspension of solid non-deformable spheres with a radius close to the mean square radius of inertia.

The initial characteristic viscosity defined by equation (6.23) is seen to be independent of the characteristics of intramolecular friction, but this is a consequence of the simplifying assumptions. It has been shown for a dumbbell (Altukhov 1986) that, when account of the internal viscosity and the anisotropy of the hydrodynamic interaction is taken simultaneously, the characteristics of these quantities enter into the expression for a viscosity of type (6.23). This result must be revealed also by the subchain model when account is taken of the anisotropy of the hydrodynamic interaction.

6.2.4 On the Effect of Internal Viscosity

The characteristic viscosity (6.22) is of special interest in the study of the influence of intramolecular friction on the dynamics of a macromolecule in a viscous liquid. At $\omega \to \infty$, characteristic viscosity can be written as

$$[\eta']_\infty = \frac{nT}{c\eta_s} \sum_{\alpha=1}^{N} \tau_\alpha^\perp \varphi_\alpha = \frac{nT}{c\eta_s} \zeta(z\nu - \theta)\tau_1\varphi_1$$
(6.26)

where $\zeta(x)$ is Riemann's zeta-function.

Experimental studies indicate (Cooke and Matheson 1976, Noordermeer et al. 1975) that the limiting characteristic viscosity for a given polymer-homologous series is independent of the length of macromolecule and the type of solvent. Taking into account that $\tau_1 \sim M^{z\nu}, n \sim M^{-1}$ and $\varphi_1 \sim M^{-\theta}$, one can find the relation

$$\theta - z\nu + 1 = 0$$
(6.27)

which follows from equation (6.26) and from the fact that the limiting characteristic viscosity is independent of the length of macromolecule.

The independence of the limiting characteristic viscosity on the type of solvent means that φ_1 is independent of the viscosity of the solvent, that is the dimensional characteristic of the 'internal' friction of the macromolecule $\zeta\varphi_1$ is proportional to the viscosity of the solvent and the "internal" friction is not solely internal. The conclusion that the solvent contributes significantly to the intramolecular viscosity was reported by Schrag (1991), and was dubbed as the "solvent modification effect".

The fact that the value of the characteristic viscosity at high frequencies is not zero indicates the existence of intramolecular (taking into account the solvent molecules) relaxation processes with relaxation times smaller than the reciprocal of the frequency of the measurement. The true limiting value is naturally zero and experiments sometimes reveal a step at a frequency ω which indicates the occurrence of a relaxation process with a relaxation time $\tau \sim \omega^{-1}$ which is compatible to the times of the deformation of a system. This phenomenon may be described by including the relaxing intramolecular viscosity, as it was done by Volkov and Pokrovskii (1978).

One uses expression (6.18) for the stress tensor in which the memory function can be chosen in the simplest way

$$\varphi_\alpha(s) = \frac{\varphi_\alpha}{\tau} \exp\left(-\frac{s}{\tau}\right),$$

where φ_α is a coefficient of the intramolecular viscosity which can be defined by relation (2.27), for example. Then, we use the results of Chapter 4 for the correlation functions to write down the stresses for oscillatory deformation and to find an expression for the coefficient of dynamic viscosity

$$\eta(\omega) = \eta_s + nT \sum_{\alpha=1}^{N} \left(\frac{\tau_\alpha^\perp - \tau_\alpha^-}{\tau_\alpha^+ - \tau_\alpha^-}\right)^2 \left[\frac{\tau_\alpha^+}{\tau_\alpha^\perp} \frac{\tau_\alpha^+[1 - i\omega(\tau_\alpha^+ - \tau_\alpha^\perp)]}{1 - i\omega\tau_\alpha^+}\right.$$
$$- \frac{\tau_\alpha^+ + \tau_\alpha^-}{\tau_\alpha^\perp} \frac{\tau_\alpha^\perp - \tau_\alpha^+}{\tau_\alpha^\perp - \tau_\alpha^-} \frac{\tau_\alpha^0[1 - i\omega(\tau_\alpha^0 - \tau_\alpha^\perp)]}{1 - i\omega\tau_\alpha^0}$$
$$\left. + \frac{\tau_\alpha^-}{\tau_\alpha^\perp}\left(\frac{\tau_\alpha^\perp - \tau_\alpha^+}{\tau_\alpha^\perp - \tau_\alpha^-}\right)^2 \frac{\tau_\alpha^-[1 - i\omega(\tau_\alpha^- - \tau_\alpha^\perp)]}{1 - i\omega\tau_\alpha^-}\right], \tag{6.28}$$

where the relaxation times are defined by

$$2\tau_\alpha^\pm = \tau_\alpha \pm \left(\tau_\alpha^2 - 2\tau\tau_\alpha^\perp\right)^{1/2}, \qquad \tau_\alpha^\perp = \frac{\tau_1}{\alpha^{2\nu}},$$
$$\tau_\alpha = \frac{\tau}{2} + \tau_\alpha^\perp(1 + \varphi_\alpha), \qquad \tau_\alpha^0 = \frac{2\tau_\alpha^- \tau_\alpha^+}{\tau_\alpha^+ + \tau_\alpha^-}. \tag{6.29}$$

In the case, when one neglects the relaxation time of the intramolecular process,

$$\tau_\alpha \to \tau_\alpha^\| = \tau_\alpha^\perp(1 + \varphi_\alpha),$$
$$\tau_\alpha^+ \to \tau_\alpha^\|, \quad \tau_\alpha^- \to 0, \quad \tau_\alpha^0 \to 0$$

and expressions (6.28) reduce to the equation for dynamic viscosity

$$\eta(\omega) = \eta_{\mathrm{s}} + nT \sum_{\alpha=1}^{N} \tau_\alpha^\perp \frac{1 - i\omega(\tau_\alpha^\| - \tau_\alpha^\perp)}{1 - i\omega\tau_\alpha^\|} \qquad (6.30)$$

which has the components (6.20).

Figure 14 illustrates the dependence of the viscosity on the frequency, while taking into account the intramolecular relaxation process with a relaxation time τ according to expression (6.28). It may be hoped that the study of intramolecular relaxation processes from a phenomenological point of view will promote the establishment of the detailed mechanism of the rapid relaxation processes in polymers, although there is no doubt that more detailed models of the macromolecule studied, for example, by Gotlib et al. (1986), Priss and Popov (1971), Priss and Gamlitski (1983) must be used at high frequencies. These models make it possible to describe the small-scale motions of the chain.

6.3 Macromolecules in a Viscoelastic Liquid

One of the first attempts to find a molecular interpretation of viscoelastic behaviour of entangled polymers was connected with investigation of the dynamics of a macromolecule in a form of generalised Rouse dynamics (Pokrovskii and Volkov 1978a; Ronca 1983; Hess 1986). It formally means that, instead of assumption that the environment of the macromolecule is a viscous medium, Brownian particles of the chain are considered moving in a viscoelastic liquid with the stress tensor

$$\sigma_{ij}^0 = -p\delta_{ij} + 2 \int_0^\infty \eta_{\mathrm{s}}(s)\gamma_{ij}(t-s)\mathrm{d}s. \qquad (6.31)$$

The generalised Rouse dynamics is proved to be not sufficient for consistent explanation of viscoelastic behaviour of entangled polymers, but appears to be interesting from methodological point of view.

6.3.1 The Stress Tensor

To obtain the expression for the stress tensor for the set of Brownian particles suspended in a viscoelastic liquid, we use equation (6.7), in which the elastic and internal viscosity forces are specified in Section 3.2

$$K_i^\alpha = -2T\mu A_{a\gamma} r_i^\alpha, \quad G_i^\omega = 0.$$

It is convenient to write the stress tensor (6.7) in terms of normal coordinates:

$$\sigma_{ik}(t) = -n(N+1)T\delta_{ik} + nT \sum_{\alpha=1}^{N} \left(2\mu\lambda_\alpha \langle \rho_i^\alpha \rho_k^\alpha \rangle - \delta_{ik}\right). \qquad (6.32)$$

Expression (4.17) for non-equilibrium moments allows us to determine the stress tensor for a dilute suspension of macromolecular coils in the linear viscoelastic liquid

$$\sigma_{ik} = -p\delta_{ik} + 2\int_0^\infty \left(\eta(s) + nT\sum_{\alpha=1}^{N} 2\mu\lambda_\alpha\mu_\alpha(s)M_\alpha(s)\right)\gamma_{ik}(t-s)\mathrm{d}s. \quad (6.33)$$

Expressions (4.28) and (4.29) allow us to write the functions

$$\mu_\nu(t) = T_\nu^+ \exp\left(-\frac{t}{2\tau_\nu^+}\right) - T_\nu^- \exp\left(-\frac{t}{2\tau_\nu^-}\right) - R(t),$$

$$M_\nu(t) = \frac{1}{2\mu\lambda_\nu}\left[T_\nu^+ \exp\left(-\frac{t}{2\tau_\nu^+}\right) - T_\nu^- \exp\left(-\frac{t}{2\tau_\nu^-}\right)\right],$$

where

$$T_\nu^\pm = \frac{\tau_\nu^R(1+B) - \tau_\nu^\mp}{\tau_\nu^+ - \tau_\nu^-}.$$

In accordance with definitions (4.26) the relaxation times are defined as

$$2\tau_\nu^\pm = \tau_\nu \pm \sqrt{\tau_\nu^2 - 2\tau\tau_\nu^R}, \quad \tau_\nu = \frac{\tau}{2} + \tau_\nu^R(1+B), \quad \tau_\nu^R = \frac{\tau^*}{\alpha^2}.$$

In equation (6.33), the stresses in the moving viscoelastic liquid (6.31) are added to the stresses in the continuum of Brownian particles. When the equations of motion are formulated, we have to take into account the presence of the two interacting and interpenetrating continuous media formed by the viscoelastic liquid carrier and the interacting Brownian particles that model the macromolecules. However, the contribution of the carrier in the case of a concentrated solution is slight, and we shall ignore it henceforth.

6.3.2 Dynamic Characteristics

We are studying the simple case, when the viscoelastic carrier liquid is characterised by the dynamic viscosity

$$\eta_s(\omega) = \eta_s + \frac{\eta_s B}{1 - i\omega\tau} \quad (6.34)$$

where η_s and τ are the coefficient of viscosity and the relaxation time of the carrier liquid. The equation of dynamics of a single macromolecule in a viscoelastic liquid has the form (3.11) in which, for this case, the memory functions are determined by the transforms

$$\beta[\omega] = \zeta + \frac{\zeta B}{1 - i\omega\tau}, \qquad \varphi[\omega] = 0.$$

In this case, expression (6.33) for an oscillatory shear gradient gives the dynamic modulus of the system

$$G(\omega) = G_{\mathrm{s}}(\omega) - nT \sum_{\nu=1}^{N} i\omega \mathcal{L}\{\mu_\nu^2(s)\}$$

where $\mathcal{L}\{\mu_\nu^2(s)\}$ is the Laplace transform of the functions $\mu_\nu^2(s)$.

When calculating the Laplace transform, one finds an enhancement of the dynamic modulus due to the macromolecular coils in the viscoelastic liquid

$$G(\omega) = nT \sum_{\alpha=1}^{N} \left[(T_\alpha^+)^2 \frac{-i\omega\tau_\alpha^+}{1 - i\omega\tau_\alpha^+} - 2T_\alpha^+ T_\alpha^- \frac{-i\omega\tau_\alpha^0}{1 - i\omega\tau_\alpha^0} + (T_\alpha^-)^2 \frac{-i\omega\tau_\alpha^-}{1 - i\omega\tau_\alpha^-} \right].$$

$$(6.35)$$

The dynamic modulus of the suspension of non-interacting macromolecular coils is determined by three sets of relaxation times

$$\tau_\alpha^+, \quad \tau_\alpha^-, \quad \tau_\alpha^0 = \frac{2\tau_\alpha^+ \tau_\alpha^-}{\tau_\alpha^+ + \tau_\alpha^-} \approx 2\tau_\alpha^-. \tag{6.36}$$

Further on we shall consider the case of large values of parameter B, when the first terms in the expansion of the relaxation times in powers of the quantity $1/B$ are

$$\tau_\alpha^+ \approx \tau_\alpha^{\mathrm{R}} B(1 + \chi\alpha^2) \left[1 - \frac{2\chi\alpha^2}{B(1 + \chi\alpha^2)^2} \right], \tag{6.37}$$

$$\tau_\alpha^- \approx \frac{2\tau^* \chi}{1 + \chi\alpha^2}, \quad \chi = \frac{\tau}{2\tau^* B}. \tag{6.38}$$

At large values of B, the whole set of relaxation times can be divided into two sets: large relaxation times τ_α^+ and small relaxation times τ_α^- and τ_α^0, while the times τ_α^+ are B times the largest times from the set τ_α^- and τ_α^0.

One can see that the frequency dependence of the dynamic modulus is determined by two parameters B and χ

$$G(\omega) = nT f(\omega\tau^*, B, \chi).$$

Before we discuss the frequency dependencies of the dynamic modulus, which are shown in Fig. 16 for typical values of parameters, we shall find expressions for the characteristic quantities at $B \gg 1$. The latter assumption allows us to use expressions (6.36)–(6.38) and to define

$$T_\nu^+ \approx \frac{\tau_\nu^{\mathrm{R}} B}{\tau_\nu^+ - \tau_\nu^-}, \qquad T_\nu^+ \approx \frac{\tau_\nu^{\mathrm{R}} B \chi\alpha^2}{\tau_\nu^+ - \tau_\nu^-}.$$

Expressions for viscosity η, elasticity ν and the real value of the dynamic modulus on the intermediate plateau, when $\tau_\alpha^+ \gg \frac{1}{\omega} \gg \tau_\alpha^-$, follow from formula (6.35)

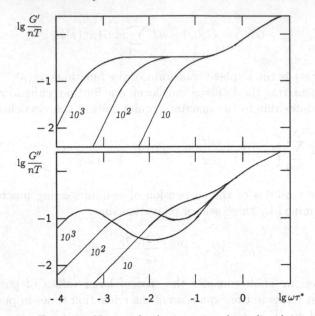

Figure 16. Macromolecules in a viscoelastic liquid.
The real and the imaginary components of dynamical modulus of a dilute suspension of
macromolecules in a viscoelastic liquid are calculated at values of B shown at the curves
and at $\chi = 1$. Adapted from the paper of Pokrovskii and Volkov (1978a).

$$\eta = nT\tau^* B \sum_{\alpha=1}^{N} \frac{1}{\alpha^2(1 + \chi\alpha^2)},$$

$$\nu = nT(\tau^* B)^2 \sum_{\alpha=1}^{N} \frac{1}{\alpha^4},$$

$$G_e = nT \sum_{\alpha=1}^{N} \frac{1}{(1 + \chi\alpha^2)^2}.$$

The replacement of the sums by integrals allows us to estimate the characteristic quantities at $N \to \infty$. One can find that the elasticity does not
depend on the parameter χ

$$\nu = \frac{\pi^4}{90} nT(\tau^* B)^2. \tag{6.39}$$

The viscosity and dynamic modulus value for the plateau can be estimated
at large and small values of χ

$$\eta \approx \frac{\pi^4}{90} nT\tau^* B\chi^{-1}, \qquad G_e \approx \frac{\pi^4}{90} nT\chi^{-2}, \quad \chi \gg 1,$$

$$\eta \approx \left(\frac{\pi^2}{6} - \frac{\pi}{2}\chi^{1/2}\right) nT\tau^* B, \qquad G_e \approx \frac{\pi}{4} nT\chi^{-1/2}, \quad \chi \ll 1. \tag{6.40}$$

Now, we can try to relate the above results to the experimental data on the viscoelasticity of concentrated solutions of polymers. For the systems of long macromolecules, the estimated values of parameter χ are small. Having used expressions (6.40) for this case, one can evaluate the terminal relaxation time of the system

$$\bar{\tau} = \frac{\nu}{\eta} = \frac{\pi^2}{15}\tau^* B, \tag{6.41}$$

which, due to the requirement of assumed self-consistency, ought to coincide with the given relaxation time τ. This requirement, in virtue of definition (6.38), determines the self-consistent value of the parameter

$$\chi = \frac{\tau}{2\tau^* B} = \frac{\pi^2}{30} \approx 0.33. \tag{6.42}$$

In the alternative case of large values of χ one can use the upper line of equation (6.40) to calculate the terminal relaxation time of the system, which coincides with the given relaxation time in order of magnitude

$$\bar{\tau} = \frac{\nu}{\eta} = \frac{\tau}{2}.$$

The suspension of dilute macromolecular coils in a viscoelastic liquid is suitable for the interpretation of results on the viscoelasticity of concentrated systems with macromolecules, which are not long ($M \approx M_e$). This case was carefully investigated by Leonov (1994). He has confirmed the possibility of a self-consistent description for a system of very short macromolecules.

6.4 Entangled Macromolecules

Investigation of viscoelastic behaviour of linear polymer solutions and melts shows that there are universal laws for dependencies of the terminal characteristics on the length of macromolecules, which allows to interpret these phenomena on the base of behaviour of a single macromolecule in the system of entangled macromolecules (Ferry 1980, Doi and Edwards 1986). The validity of the mesoscopic approach itself rests essentially on the fundamental experimental fact that quantities that characterise the behaviour of a polymer system have a well-defined unambiguous dependence on the length of the macromolecule.

The dependence of the characteristics on molecular weight was used for the classification of the systems (Ferry 1980; Graessley 1974; Watanabe 1999).

The law for coefficient of viscosity, which was unambiguously established by Fox and Flory (1948) for polystyrene and polyisibutilene and confirmed for many polymer system investigated later (Berry and Fox 1968, Ferry 1980), determines the first critical point $M_c \approx 2M_e$ separating entangled and non-entangled systems of linear polymers

$$\eta \sim \begin{cases} M, & \text{non-entangled systems,} \quad M < M_c, \\ M^{3.4}, & \text{entangled systems,} \qquad\quad M > M_c. \end{cases} \tag{6.43}$$

While the law with index 3.4 for viscosity is valid in the whole region above M_c, the dependence of terminal relaxation time is different for weakly and strongly entangled systems (Ferry 1980) and determines the second critical point M^*

$$\tau \sim \begin{cases} M^{4.4}, & \text{weakly entangled systems,} \quad M < M^*, \\ M^{3.4}, & \text{strongly entangled systems,} \quad M > M^*. \end{cases} \tag{6.44}$$

The data for melts of different polymers collected by Ferry (1980, p. 379, Table 13-III) allows us to estimate the second critical point[1] M^*. Assuming that $M_c = 2M_e$, one has

$$M^* \approx (4.6\text{--}12.0)M_e.$$

The critical value of molecular weight can be identified with the transition point between weakly and strongly entangled systems, the position of which was estimated in Sections 4.2.3 and 5.1.2 as

$$M^* \approx 10M_e.$$

The difference in the molecular-weight dependence of the terminal relaxation time can be attributed to the change of the mechanisms (diffusive and reptation, correspondingly) of conformational relaxation in these systems. Further on in this section, we shall calculate dynamic modulus and discuss characteristic quantities both for weakly and strongly entangled systems.

6.4.1 The Stress Tensor

To calculate the characteristics of viscoelasticity in the framework of meso-scopic approach, one can start with the system of entangled macromolecules, considered as a dilute suspension of chains with internal viscoelasticity moving in viscoelastic medium, while the elastic and internal viscosity forces, according to equations (3.4)–(3.6) and (3.8), have the form

$$K_i^\alpha = -2T\mu A_{a\gamma} r_i^\alpha, \qquad G_i^\alpha = -G_{\alpha\gamma} \int_0^\infty \varphi(s)(u_i^\gamma - \omega_{ij}r_j^\gamma)_{t-s}\mathrm{d}s.$$

---

[1] To avoid many subscripts, instead of Ferry's symbol M_c' for the second critical point, I use the symbol M^*.

For calculation, it is convenient to write the stress tensor (6.7) in terms of normal co-ordinates in the following form:

$$
\sigma_{ik}(t) = -n(N+1)T\delta_{ik} + nT \sum_{\alpha=1}^{N} \left\{ 2\mu\lambda_\alpha \langle \rho_i^\alpha \rho_k^\alpha \rangle - \delta_{ik} \right.
$$

$$
\left. + \frac{1}{T} \int_0^\infty \varphi(s)\Big(\langle \dot{\rho}_i^\alpha(t-s)\rho_k^\alpha(t) \rangle - \omega_{il}(t-s)\langle \rho_l^\alpha(t-s)\rho_k^\alpha(t) \rangle \Big)\mathrm{d}s \right\}.
$$

$$(6.45)$$

The contribution of the carrier segment liquid in the case of a concentrated solution is slight, and we shall ignore it henceforth. The contribution of separate macromolecules, which is presented by the terms under the sum in the above equation, can be divided into two parts. The first terms describe external frictions due to connectivity of the particles, while the integral terms present stresses due to intramolecular resistance of the coils. The last can be interpreted, remembering the speculation in Section 3.3.3, as stresses emerging at orientation of separate Kuhn segments in dense medium among the other segments. Let us note that expression (6.45) can be considered as a generalisation of the known (Cerf 1958, Peterlin 1967) expressions for stress in dilute solutions of polymers with internal viscosity. Indeed, if $\varphi(s) \sim \delta(s)$, expression (6.45) for the stress tensor reduces to (6.14).

The expression (6.45) for the stress tensor can be applied to both weakly and strongly entangled systems, but, let us note, that the macromolecular dynamics is different in these cases. We use the expression (6.45) to calculate the stress tensor for entangled systems in linear approximation of macromolecular dynamics. Using expressions for moments (4.17), (4.20) and (4.21) one obtains

$$
\sigma_{ik}(t) = -p\delta_{ik} + 2nT \sum_{\alpha=1}^{N} \left\{ \int_0^\infty 2\mu\lambda_\alpha \mu_\alpha(s) M_\alpha(s)\gamma_{ik}(t-s)\mathrm{d}s \right.
$$

$$
+ \frac{1}{2T} \int_0^\infty \varphi(s) \int_0^\infty \big[\mu_\alpha(u+s)\dot{M}_\alpha(u)
$$

$$
\left. + \dot{\mu}_\alpha(u)M_\alpha(u+s)\big]\gamma_{ik}(t-s-u)\mathrm{d}u\,\mathrm{d}s \right\}.
$$

$$(6.46)$$

The mesoscopic analysis, similar to truly phenomenological analysis, includes some mesoscopic parameters in final expressions for the stress tensor and for viscoelastic characteristics and assumes the necessity of investigation on the base of more specified models of the system. Some theories were based on the image of the structure of polymer systems as a network with temporary knots (entanglements) (Ferry et al. 1955; Lodge 1956; Chompff and Duiser 1966; Chompff and Prins 1968). Those attempts helped us to understand some features of polymer dynamics. A recent work by Schieber et al. (2003) gives us an example of a very detailed picture of flowing entangled polymer system.

6.4.2 Dynamic Modulus and Relaxation Branches

There are plenty of measurements of dynamic modulus of nearly monodisperse polymers starting with pioneering works of Onogi et al. (1970) and Vinogradov et al. (1972a). The more recent examples of the similar dependencies can be found in papers by Baumgaertel et al. (1990, 1992) for polybutadiene and for polystyrene and in paper by Pakula et al. (1996) for polyisoprene.

To calculate the dynamic modulus, we turn to the expression for the stress tensor (6.46) and refer to the definition of equilibrium moments in Section 4.1.2, while memory functions are specified by their transforms as

$$\beta[\omega] = \zeta + \frac{\zeta B}{1 - i\omega\tau}, \qquad \varphi[\omega] = \frac{\zeta E}{1 - i\omega\tau}. \qquad (6.47)$$

It means, according to the speculations in Chapter 3 that the environment of the chosen macromolecule is considered a viscoelastic medium, and, in addition, the internal resistance or the internal viscosity is taken into account. The latter was not considered in the previous section.

We are calculating dynamic modulus and characteristic quantities for entangled systems, when the linear approximation of dynamic equation is used.

The Case of Low Frequencies

To begin with, let us consider the simple case, when ζ can be neglected in comparison to ζB in equations (6.47), which can be done, if one considers low-frequency properties of the systems with long macromolecules – the strongly entangled systems. In this case, according to (4.32) and (4.33), we have

$$\mu_\alpha(s) = \frac{B\tau_\alpha^R}{\tau_\alpha} \exp\left(-\frac{s}{2\tau_\alpha}\right) - \frac{B\tau_\alpha^R}{\tau_\alpha} R(s),$$

$$M_\alpha(s) = \frac{1}{2\mu\lambda_\alpha} \left[\frac{(B+E)\tau_\alpha^R}{\tau_\alpha} \exp\left(-\frac{s}{2\tau_\alpha}\right) + \frac{\tau}{2\tau_\alpha} R(s)\right],$$

$$\dot{\mu}_\alpha(s) = -\frac{B\tau_\alpha^R}{\tau_\alpha} \frac{1}{2\tau_\alpha} \exp\left(-\frac{s}{2\tau_\alpha}\right) + \frac{2B\tau_\alpha^R}{\tau_\alpha} \delta(s),$$

$$\dot{M}_\alpha(s) = -\frac{1}{2\mu\lambda_\alpha} \left[\frac{(B+E)\tau_\alpha^R}{\tau_\alpha} \frac{1}{2\tau_\alpha} \exp\left(-\frac{s}{2\tau_\alpha}\right) + \frac{\tau}{\tau_\alpha} \delta(s)\right]$$

where $\tau_\alpha^R = \tau^*/\alpha^2$ are the Rouse relaxation times and

$$\tau_\alpha = \frac{\tau}{2} + \tau_\alpha^R(B+E).$$

Under oscillatory motion, the stress tensor (6.46) gives us an expression for the dynamic modulus

$$G(\omega) = nT \sum_{\alpha} \frac{(\tau_{\alpha}^{R})^2 B(B+E)}{\tau_{\alpha}}$$

$$\times \left[\frac{-i\omega}{1 - i\omega\tau_{\alpha}} + \frac{-i\omega 2E\tau_{\alpha}^{R}}{2\tau_{\alpha} + \tau - 2i\omega\tau\tau_{\alpha}} - \frac{-i\omega 2E\tau_{\alpha}^{R}}{(2\tau_{\alpha} + \tau - 2i\omega\tau\tau_{\alpha})(1 - i\omega\tau_{\alpha})} \right].$$

We can introduce a new set of relaxation times

$$\tau_{\alpha}^{*} = \frac{2\tau\tau_{\alpha}}{2\tau_{\alpha} + \tau} \tag{6.48}$$

and, after some rearrangement, write an expression for the dynamic modulus in the standard form

$$G(\omega) = nT \sum_{\alpha} \left(\frac{\tau_{\alpha}^{R}}{\tau_{\alpha}} \right)^2 B \left(B \frac{-i\omega\tau_{\alpha}}{1 - i\omega\tau_{\alpha}} + E \frac{\tau_{\alpha}}{\tau} \frac{-i\omega\tau_{\alpha}^{*}}{1 - i\omega\tau_{\alpha}^{*}} \right), \tag{6.49}$$

where for small α and large B, we have

$$\tau_{\alpha} \gg \tau_{\alpha}^{*}, \quad \tau_{\alpha}^{*} \approx \tau.$$

One can, thus, see that, at low frequencies, the viscoelastic behaviour of the system is determined by two sets of relaxation times, or, we can say also, by two relaxation branches. The first term in (6.49) is determined by relaxation of conformation of the macromolecule. The second term in (6.49), as will be shown in the next chapter, is connected with orientational relaxation processes.

Note that the first and the second terms in (6.49) at $\omega \to \infty$ have the orders of magnitudes $nT\psi^{-2}$ and $nT\chi^{-1}$, respectively. The ratio of the quantities is very small for systems of long macromolecules, so that the contribution of the first, conformation branch to the linear viscoelasticity is negligibly small at $\chi \ll \chi^{*}$. Note also that, for strongly entangled systems, at $\chi \ll \chi^{*}$ or $M \gg M^{*}$, as it was shown in Section 4.2.3, conformational relaxation cannot be occurred via the diffusive mechanism (considered here), but via the reptation mechanism, so that the first term in equation (6.49) ought to be replaced by other term, for example, in the form

$$nT \sum_{\alpha=1}^{\pi/\chi} \frac{-i\omega \, p_{\alpha}\tau_{\alpha}^{rep}}{1 - i\omega\tau_{\alpha}^{rep}}, \quad \tau_{\alpha}^{rep} = \frac{\pi^2}{\chi} \frac{\tau^{*}}{\alpha^{0.5}}.$$

Though the reptation relaxation times are defined by equation (4.37), the weights p_{α} of the contributions of separate relaxation processes remain unknown, and in fact, the replacement is forbidden, so that we prefer, as an initial approximation, to consider evaluation of dynamic modulus without any modification.

The Case of Higher Frequencies

To extend the theory for higher frequencies, we have to consider the general case, when the micro-viscoelasticity is given by (6.47). Using equations (4.28) and (4.29), after some rearrangement, one can find the dynamic modulus

$$G(\omega) = nT(-i\omega) \sum_{a=1}^{5} \sum_{\alpha} \frac{p_\alpha^{(a)} \tau_\alpha^{(a)}}{1 - i\omega\tau_\alpha^{(a)}} \tag{6.50}$$

where the times of relaxation and the corresponding weights are given by the following expressions

$$\tau_\alpha^{(1)} = \tau_\alpha^+ = \frac{1}{2}\left(\tau_\alpha + \left(\tau_\alpha^2 - 2\tau\tau_\alpha^R\right)^{1/2}\right),$$

$$\tau_\alpha^{(2)} = \frac{2\tau\tau_\alpha^+}{\tau + 2\tau_\alpha^+}, \qquad \tau_\alpha^{(3)} = \frac{2\tau_\alpha^+\tau_\alpha^-}{\tau_\alpha^+ + \tau_\alpha^-}, \qquad \tau_\alpha^{(4)} = \frac{2\tau\tau_\alpha^-}{\tau + 2\tau_\alpha^-},$$

$$\tau_\alpha^{(5)} = \tau_\alpha^- = \frac{1}{2}\left(\tau_\alpha - \left(\tau_\alpha^2 - 2\tau\tau_\alpha^R\right)^{1/2}\right),$$

$$p_\alpha^{(1)} = T_\alpha^+ S_\alpha^+ \left(1 - \frac{2E\tau_\alpha^R}{2\tau_\alpha^+ - \tau}\right),$$

$$p_\alpha^{(2)} = S_\alpha^+ \frac{E\tau_\alpha^R}{\tau} + T_\alpha^+ S_\alpha^+ \frac{2E\tau_\alpha^R}{2\tau_\alpha^+ - \tau} - (T_\alpha^+ S_\alpha^- + T_\alpha^- S_\alpha^+)\frac{2E\tau_\alpha^R}{2\tau_\alpha^- - \tau},$$

$$p_\alpha^{(3)} = (T_\alpha^+ S_\alpha^- + T_\alpha^- S_\alpha^+)\left(\frac{E\tau_\alpha^R}{2\tau_\alpha^+ - \tau} + \frac{E\tau_\alpha^R}{2\tau_\alpha^- - \tau} - 1\right),$$

$$p_\alpha^{(4)} = -S_\alpha^- \frac{E\tau_\alpha^R}{\tau} + T_\alpha^- S_\alpha^- \frac{2E\tau_\alpha^R}{2\tau_\alpha^- - \tau} - (T_\alpha^+ S_\alpha^- + T_\alpha^- S_\alpha^+)\frac{E\tau_\alpha^R}{2\tau_\alpha^+ - \tau},$$

$$p_\alpha^{(5)} = T_\alpha^- S_\alpha^- \left(1 - \frac{2E\tau_\alpha^R}{2\tau_\alpha^- - \tau}\right).$$

Expression (6.50) for the dynamic modulus includes now five relaxation branches and generalises formula (6.49) for higher frequencies.

The situation is illustrated in Fig. 17, which contains experimental values of dynamic shear modulus for polystyrenes with different molecular weights and theoretical dependences calculated according to equation (6.50) and presented by the solid lines. This comparison illustrates insufficiency of linear approximation for macromolecule dynamics to describe the effects of linear viscoelasticity of entangled systems. For polymers with the length $M > 10M_e$ – strongly entangled systems, the most essential contribution is given by the second relaxation branch, that is the orientation relaxation branch with relaxation times close to τ, which determines terminal characteristics (see the next section). The largest conformational relaxation times, contribution of which are shown by the dashed lines, have appeared to be unrealistically large for strongly entangled systems in linear approximation of macromolecular dynamics. It was shown (see Section 4.2.2) that introduction of local anisotropy

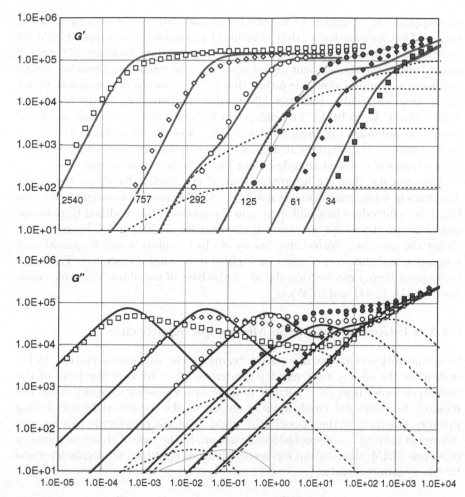

Figure 17. Dynamic modulus of typical polymers.
The experimental points (taken from the review by Watanabe 1999) are due to the measurements of Schausberger et al. (1985) for polystyrenes. The numbers indicate the lengths of macromolecules $10^{-3} \cdot M$. The reference temperature is $T = 180°$ C, $G'_e = 2 \times 10^5$ Pa. The length between entanglement is $M_e = 16000$, so that the theoretical dependences, shown by the solid lines, are calculated for the numbers of entanglements per macromolecule $Z = 2.125, 3.813, 7.813, 18.25, 47.31, 158.75$, which induce, according to relations (3.17), (3.25) and (3.29), the corresponding values of parameters χ, B, and E. The separate contributions from the conformational relaxation branches are shown by dashed lines.

of mobility helps one to improve the situation: the largest relaxation times decrease when the coefficient of local anisotropy increases. However, one can see that the contribution of the conformation reptation branch into dynamic modulus appears to be negligible for the high-molecular-weight polymers in the region of low frequencies, so that, whichever mechanism of conformational relaxation is realised, the second branch gives a good approximation of terminal quantities for the strongly-entangled systems. The remaining branches merge and form a group of slow relaxation times. The absence of non-linear terms in the macromolecular dynamics affects also the behaviour in the transition region about $M \approx 10M_e$. The difference between theoretical and empirical results for polymers with length $M < 10M_e$ – weakly entangled systems, can be also connected, in particular, by polydispersity of polymers, which is larger for low-molecular weight samples, than for high-molecular weight ones.

One can see that the approximation of the theory, based on the linear dynamics of a macromolecule, is not adequate for strongly entangled systems. One has to introduce local anisotropy in the model of the modified Cerf-Rouse modes or use the model of reptating macromolecule (Doi and Edwards 1986) to get the necessary corrections (as we do in Chapters 4 and 5, considering relaxation and diffusion of macromolecules in entangled systems). The more consequent theory can be formulated on the base of non-linear dynamic equations (3.31), (3.34) and (3.35).

6.4.3 Self-Consistency of the Mesoscopic Approach

One can notice that the dissipative terms in the dynamic equation (3.11) (taken for the case of zero velocity gradients, $\nu_{ij} = 0$) have the form of the resistance force (D.3) for a particle moving in a viscoelastic liquid, while the memory functions are (with approximation to the numerical factor) fading memory functions of the viscoelastic liquid. The macromolecule can be considered as moving in a viscoelastic continuum. In the case of choice of memory functions (3.15), the medium has a single relaxation time and is characterised by the dynamic modulus

$$G_s(\omega) = \frac{-i\omega\eta_s}{1 - i\omega\tau}, \qquad G_e = \lim_{\omega\to\infty} G(\omega) = \frac{\eta_s}{\tau}$$

where τ is the correlation time introduced in (3.15), and η_s is a constant. One can say that the written dynamic modulus characterises the micro-viscoelasticity.

On the other hand, the properties of the system as a whole can be calculated and the macroscopic dynamic modulus can be determined. Here the question of the relation between the postulated micro-viscoelasticity and the resulting macro-viscoelasticity appears. The answer requires a properly formulated self-consistency condition. Simple speculations show that equality of the micro- and macro-viscoelasticity cannot be obtained. Nevertheless, it is natural to require the equality of relaxation times of micro- and macro-viscoelasticities. It will be shown in this section that this condition can be satisfied.

First, we shall consider in detail the characteristic quantities: the viscosity coefficient η and the elasticity coefficient ν, defined by expansion (6.11), and the value of the dynamic modulus on the plateau G_e. The latter can be calculated as the limiting value of the modulus at frequencies satisfying the relation

$$\tau_\alpha^- < \omega^{-1} < \tau_\alpha^+ \approx \tau.$$

The estimation of the main terms of expansion of dynamic modulus (6.49) determine the expressions for the terminal quantities

$$G_e = nT \sum_{\alpha=1}^{N} \frac{2\chi\alpha^2 + \psi(\chi\alpha^2 + 1 + \psi)}{2\chi\alpha^2(\chi\alpha^2 + 1 + \psi)^2},$$

$$\eta = nT\tau^* B \sum_{\alpha=1}^{N} \left(\frac{1}{\alpha^2(\chi\alpha^2 + 1 + \psi)} + \frac{\psi}{\alpha^2(2\chi\alpha^2 + 1 + \psi)} \right),$$

$$\nu = nT(\tau^* B)^2 \sum_{\alpha=1}^{N} \frac{2\chi\psi(\chi\alpha^2 + 1 + \psi)}{\alpha^2(2\chi\alpha^2 + 1 + \psi)}.$$

A preliminary estimate of χ which, according to (5.8), can be interpreted as the ratio of the square of the tube diameter $(2\xi)^2$ to the mean square end-to-end distance $\langle R^2 \rangle_0$, shows that $\chi \ll 1$ for strongly entangled systems. For large N, this enables us to replace summation by integration and, according to the rules of Appendix G, to obtain expressions for the characteristic quantities

$$G_e = nT \left[\frac{\pi^2}{12} \frac{\psi}{\chi(1+\psi)} + \frac{\pi}{8} \frac{2-\psi}{(1+\psi)^{3/2}} \chi^{-1/2} \right],$$

$$\eta = nT\tau^* B \left[\frac{\pi^2}{6} - \frac{\pi}{2} \left(\frac{\chi}{1+\psi} \right)^{1/2} \right], \qquad (6.51)$$

$$\nu = nT(\tau^* B)^2 \left[\frac{\pi^2}{3} \frac{\chi\psi}{1+\psi} - \frac{\pi}{2} \left(\frac{2\chi}{1+\psi} \right)^{3/2} 2\psi \right].$$

These expressions are valid for arbitrary ψ and small χ. We can then distinguish between two cases, namely: for systems consisting of very long molecules in the almost complete absence of the solvent (strongly entangled systems) we have $\psi \gg 1$, whereas $\psi \ll 1$ for a concentrated system consisting of not very long macromolecules (weakly entangled systems). In the latter case expressions (6.51) at $\psi = 0$ are identical to expressions (6.39) and (6.40). Here, we shall consider the former case, when $\psi \gg 1$ and find from (6.51) the zeroth-order terms in power of ψ^{-1}.

$$G_e = \frac{\pi^2}{12} nT \chi^{-1}, \qquad \eta = \frac{\pi^2}{6} nT\tau^* B, \qquad \nu = \frac{\pi^2}{3} nT(\tau^* B)^2 \chi. \qquad (6.52)$$

One can note that the relaxation times of the second branch are very close to each other, so that the frequency dependence of the modulus could be approximated by a expression with the single relaxation time determined by the relation

$$\bar{\tau} = \frac{\eta}{G_e} = 2\tau^* B\chi = \tau.$$

The relaxation time that we have determined may be referred to as the terminal viscoelastic relaxation time; it is equal to the relaxation time which was introduced to characterise the medium surrounding the chosen macromolecule. Thus, for $\psi \to \infty$, the theory is self-consistent and this confirms the statement of Section 3.1.1 that chains of Brownian particles are moving independently in a liquid made of interacting Kuhn segments.

The condition of self-consistency, as a requirement of the identity of the times of relaxation of macro- and micro-viscoelasticity, gives the following relation for the first-order terms in power of ψ^{-1} of expansion of (6.51)

$$\psi = \frac{4\pi^2}{9}\frac{1}{\chi}. \tag{6.53}$$

This relation is practically identical to relation (5.17).

Equation (6.52) and the experimental data allow us to estimate the parameters of the theory χ and $\tau^* B$ which can be also estimated directly by other methods discussed in Chapter 5. So, the consistency of the theory can be tested.

6.4.4 Modulus of Elasticity and the Intermediate Length

Initially, the elasticity of concentrated polymer systems was ascribed to the existence of a network in the system formed by long macromolecules with junction sites (Ferry 1980). The sites were assumed to exist for an appreciable time, so that, for observable times which are less than the lifetime of the site, the entangled system appears to be elastic. Equation (1.44) was used to estimate the number density of sites in the system. The number of entanglements for a single macromolecule $Z = M/M_e$ can be calculated according to the modified formula

$$G_e = nT\frac{M}{M_e} \tag{6.54}$$

where n is the density of the number of macromolecules and T is temperature in energy unit.

The length of a macromolecule between adjacent entanglements M_e is used as an individual characteristic of a polymer system. Table 1 contains values of M_e for certain polymer systems. The more complete list of estimates of the quantity M_e can be found in work by Aharoni (1983, 1986). One can compare expressions (6.52) and (6.54) for the value of the modulus on the plateau to see that the length of a macromolecule between adjacent entanglements M_e is closely connected with one of the parameters of the theory

$$\chi = \frac{\pi^2}{12}\frac{M_e}{M}. \tag{6.55}$$

We should note, recalling the interpretation of χ as the ratio of the doubled intermediate length to the size of the coil discussed previously (formula (5.8)), that the length M_e, determined in the usual way, is actually related to the intermediate length ξ. Expression (6.52) can be rewritten in a form which is identical to the relation by Doi and Edwards (1986)

$$G_e = \frac{2}{3} nT \frac{\langle R^2 \rangle}{(2\xi)^2}.$$

Note that the squared diameter of the Doi-Edwards tube relates to our intermediate length as follows

$$d^2 = \frac{6}{5}(2\xi)^2. \tag{6.56}$$

The intermediate length (tube diameter) 2ξ can be estimated from the modulus with the aid of the above equations. Comparison of values of the intermediate length found from dynamic modulus and from neutron-scattering experiments was presented by Ewen and Richter (1995). They found the values to be close to each other, though there is a difference in the temperature dependence of the values of intermediate length found by different methods.

Although a network is not present in a concentrated solution, there exists a characteristic length, which had earlier been assumed the distance between neighbouring network sites. The characteristic length is a dynamic one. There are no temporary knots in a polymer system, though there is a characteristic time, which is the lifetime of the frozen large-scale conformation of a macromolecule in the system. So, the conceptions of intermediate length and characteristic time are based on deeper ideas and are reflected in the theory.

6.4.5 Concentration and Macromolecular Length Dependencies

Thus in the mesoscopic approximation or, in other words, in the mean-field approximation, the dynamic shear modulus of the melt or the concentrated solution of the polymer (strongly entangled systems) is represented by a function of a small number of parameters

$$G(\omega) = nT f(\tau^* \omega, B, \chi). \tag{6.57}$$

In this case, one assumes that $B \gg 1$, and, hence, it follows that $\tau > \tau^*$, which fact imposes certain restrictions on χ, so that $1/B < \chi \ll 1$. For these values of B and χ, the theory is found to be self-consistent for $\psi \gg 1$, so that once again, as was shown in Section 6.4.3, the formulae for the dynamic modulus lead to expressions for the characteristic quantities

$$\eta = \frac{\pi^2}{6} nT\tau^* B, \qquad \tau = 2\tau^* B \chi,$$

$$\nu = \frac{\pi^2}{3} nT(\tau^* B)^2 \chi, \qquad G_e = \frac{\pi^2}{12} nT\chi^{-1}. \tag{6.58}$$

Experiments reveal that the dynamic modulus and the characteristic quantities (6.58) depend on the polymer concentration c and length M of the macromolecule (Ferry 1980), and these dependencies are implied through the parameters of the theory.

In accordance with equations (1.33) and (4.25) we can write

$$n \sim \frac{c}{M}, \qquad \tau^* \sim \frac{\zeta_0 M^2 C_\infty}{T} \qquad (6.59)$$

where ζ_0 is the monomer friction coefficient and C_∞ is a quantity connected with the temperature dependence of the size of a macromolecular coil (see Section 1.1). The values of parameter C_∞, which reflects the thermodynamic rigidity of the macromolecule, are given for different polymers in tables of the monographs by Flory (1969) and by Tsvetkov et al. (1964).

The mesoscopic parameters χ and B, as was shown earlier in Section 3.3.4, can be written as functions of a single argument, which can now be rewritten as

$$n \langle R^2 \rangle^{3/2} \sim C_\infty^{3/2} c M^{1/2}. \qquad (6.60)$$

This allows one to write the dependencies of the characteristic quantities on the concentration of polymer and on the thermodynamic rigidity, if the dependence on molecular weight of the macromolecule, for example, is known. With help of the result of Section 3.3.4 (see formulae (3.30)), one can obtain for the strongly entangled systems

$$\eta \sim \zeta_0 C_\infty^{3\delta+1} c^{2\delta+1} M^{\delta+1}, \qquad \nu \sim \zeta_0^2 C_\infty^{3\delta-1} c^{4\delta-1} M^{2\delta+2},$$

$$G_e \sim T C_\infty^3 c^3 M^0, \qquad \tau \sim \frac{1}{T} \zeta_0 C_\infty^{3\delta-2} c^{2\delta-2} M^{\delta+1}. \qquad (6.61)$$

These equations allow one to establish various relations between the characteristic quantities, while the only index δ ought to be evaluated empirically. The data obtained for almost monodisperse samples of polymer melts of different molecular weight allows one to evaluate for high molecular weights $\delta = 2.4$ (Berry and Fox 1968, Ferry 1980). Empirical estimate corresponds to the coarse theoretical estimation in Section 3.3.2, according to which $\delta = 2$ or $\delta = 3$. The molecular-weight dependencies of other quantities in (6.61) are typical for high-molecular-weight polymers: $G_e \sim M^0$, the dependence of η and of τ on the length of a macromolecule is the same (Ferry 1980).[2].

---

[2] The reptation-tube model, being used for interpretation of viscoelastic behaviour of the system, has allowed to obtain (Doi and Edwards 1986) the relation for terminal characteristics

$$\eta \sim M^3, \qquad \tau \sim M_0^0 M^3.$$

The small deviation of the derived value of the index 3 from the empirical value 3.4 (see equations (6.43) and (6.44)) gave rise to the hopes that some improvements of the model could bring the correct results, at least, for strongly entangled systems. However, it appeared that the results delivered by the model far from empirical results (6.43) and (6.44) more, than one could earlier imagine (Altukhov et al. 2004). To appreciate these results

At the comparison of concentration dependencies of the characteristic quantities (6.61) with experimental determinations, one has to remember that effect of excluded volume was not taken into account in equations (6.61), which allow us to say only about qualitative correspondence. The behaviour of the initial viscosity is the most widely studied (Poh and Ong 1984, Takahashi et al. 1985). The concentration dependence of the viscosity coefficient in the "melt-like" region can be represented by a power law (Phillies 1995). The index can be found to be approximately $2\delta + 1$, in accordance with (6.61). There are some differences in the behaviour of polymer solutions, which are connected with different behaviour of macromolecular coils at dilution.

One should note once again that the above discussion and expressions are valid only for very long macromolecules and in the limit of very high concentrations. For semi-dilute solutions, the analysis should also include another non-dimensional parameter (see Sections 1.5 and 1.6), but then the results would become more complicated.

6.4.6 Frequency–Temperature Superposition

The dependence of the characteristic quantities (6.58) on temperature is mainly determined by the monomer friction coefficient ζ_0, which depends on temperature, concentration, and (for small M) of molecule length (Berry and Fox 1968). The dependencies were recently discussed by Tsenoglou (2001). The monomer friction coefficient ζ_0 is a material characteristic of the system, its value is strongly determined by chemical structure of macromolecule as was shown for polybutadiene by Allal et al. (2002).

The value of the coefficient of friction is connected with relative motion of small portions of the macromolecule, so that its temperature dependence is similar to that found for low-molecular-weight liquids, and can be written in the following form at temperatures much higher than the glass transition point

$$\zeta_0 \sim \exp \frac{U}{T} \qquad (6.62)$$

where U is the activation energy that depends on the molecular weight (for small M), on the concentration, and also on the temperature, if the temperature range in which the viscosity is considered is large. Near the glass transition point T_g, we have

---

properly, one has to consider the terminal relaxation time, distinguishing the probe macromolecule (with molecular weight or length M) and the neighbouring macromolecules (with the length M_0), even if all of them are equal. The reptation relaxation time, derived by Doi and Edwards, does not depend on the length of neighbouring macromolecules, which strongly contradicts to empirical evidence (see Section 6.5.3, equation (6.78)). The numerous attempts to improve the situation were controversial, so that there is a strong conviction that the Doi-Edwards model does not provide the first or even zero approximation to the theory of viscoelasticity of entangled system, though the reptation motion itself exists and influences effects of viscoelasticity as will be discussed later in this chapter and in Chapter 9.

$$\zeta_0 \sim \exp\left[\frac{A}{f_g - \alpha(T - T_g)}\right] \tag{6.63}$$

where A is an individual parameter, f_g is the volume fraction of free volume, and α is the expansion coefficient of the liquid. Quantities A and f_g are practically independent of the concentration and molecular weight, so that the dependence of ζ_0 on c and M is determined by the dependence of T_g on these quantities.

We note that, since the parameters B and χ are practically independent of temperature, the shape of the curves showing G/nT as a function of the non-dimensional frequency $\tau^*\omega$ does not change as the temperature increases, so that we can make a superposition using a reduction coefficient obtained from the temperature dependence of the viscosity.

To determine the procedure for the reduction, we shall write down the dynamic modulus at two different temperatures, one of which is a reference temperature T_{ref} and the other is an arbitrary temperature T,

$$G(\omega, T_{\text{ref}}) = nT_0 f(\tau^*_{T_{\text{ref}}} \omega, B, \chi),$$
$$G(\omega, T) = nT f(\tau^*_T \omega, B, \chi).$$

One can consider the parameters B and χ to be independent of the temperature and change the argument in the first line in such a way as to exclude the non-dimensional function. Then we write down the rule for reduction as

$$G(a_T\omega, T_{\text{ref}}) = \frac{\rho_{T_{\text{ref}}} T_{\text{ref}}}{\rho_T T} G(\omega, T), \tag{6.64}$$

where the shift coefficient is given by

$$a_T = \frac{\tau^*_T}{\tau^*_{T_{\text{ref}}}} = \frac{T_{\text{ref}} (C^{3\delta}_\infty \rho^{2\delta+1})_{T_{\text{ref}}}}{T (C^{3\delta}_\infty \rho^{2\delta+1})_T} \frac{\eta_T}{\eta_{T_{\text{ref}}}}. \tag{6.65}$$

The above expressions confirm the known (Ferry 1980) method of reducing the dynamic modulus measured at different temperatures to an arbitrarily chosen standard temperature T_{ref}, while offering a relatively insignificant improvement on the usual shift coefficient

$$a_T = \frac{T_{\text{ref}} \, \rho_{T_{\text{ref}}} \, \eta_T}{T \, \rho_T \, \eta_{T_{\text{ref}}}}.$$

6.5 Dilute Blends of Linear Polymers

The change in the stress produced by the small amount of macromolecules of another kind is, clearly, determined by the dynamics of the non-interacting impurity macromolecules among the macromolecules of another length, so that this case is of particular interest from the standpoint of the theory of the viscoelasticity of linear polymers. By studying a mixture of two polymers, one of which is present in much smaller amounts, – a dilute blend, one has a unique opportunity to obtain direct information about the dynamics of a chosen single macromolecule among the neighbouring macromolecules (Pokrovskii and Kokorin 1984).

6.5.1 Relaxation of Probe Macromolecule

Consider a system consisting of linear polymer with molecular weight M_0 and a small additive of a similar polymer with another molecular weight M. We shall assume that the amount of the additive is so small that its molecules do not interact with each other. The matrix is characterised by two characteristic length: M_e – the length of macromolecule between adjacent entanglements and $M^* \approx 10 M_e$ – the critical length dividing weakly (macromolecules of the matrix do not reptate) and strongly (macromolecules of the matrix do reptate) entangled systems. To uncover which mechanism of diffusion and relaxation of a probe macromolecules of the additive is realised, one can consider, following the speculations in Sections 4.2.3 and 5.1.2, the competition between the diffusive and reptation mechanisms of motion of a macromolecule of the additive to obtain the condition for realisation of reptation mechanism

$$2\chi(Z)B(Z_0) > \pi^2, \tag{6.66}$$

where Z_0 and Z are the lengths of macromolecules of the matrix and the additive, respectively, in units of M_e. The function $\chi(Z_0, Z)$ and $B(Z_0)$ are given by equations (3.17) and (3.25). Taking these equation into account, one can find from equation (6.66) that the lengths of the macromolecules of the matrix and the macromolecule of the additive in the point, where the mechanism of relaxation of macromolecules of the additive changes, are connected by relation

$$\frac{M}{M_e} = \frac{1}{3 \cdot 2^{1+\delta}} \left(\frac{M_0}{M_e} \right)^\delta. \tag{6.67}$$

If $\delta = 2.5$, this relation reduces to equation

$$\frac{M}{M_e} = 0.03 \left(\frac{M_0}{M_e} \right)^{2.5}, \tag{6.68}$$

which is identity at $M = M_0 \approx 10 M_e$, in accordance with the results of Section 5.1.2.

Equation (6.68) determines a critical length M^*, above which macromolecules of the additive do not reptate. The dependence of M^*/M_e on M_0/M_e, according to the above equation at $\delta = 2.5$, is depicted in Fig. 18 by solid line. For the matrix of short macromolecules, when $M_0 < 10 M_e$, the transition point is situated in the short-length region, so that the macromolecules of the additive, which are shorter than M_0 but longer than M^*,

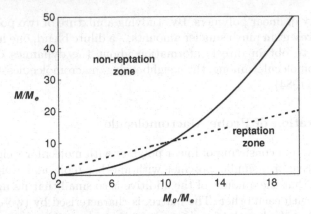

Figure 18. Alternative modes of motion of a macromolecule.
The realisation of a certain mode of motion of a macromolecule among other macromolecules depends on the lengths of both diffusing macromolecule and macromolecules
of the environment. The solid line M^* divides the dilute blends into those, in which
macromolecules of the additive can reptate, and those, where no reptation occurs. The
dashed line marks the systems with macromolecules of equal lengths.

do not reptate. However, if the matrix consists of macromolecules, for which
$M_0 > 10M_e$, there is a region between $10M_e$ and M^* in which a probe macromolecules of the additive reptate. However, the macromolecules of additive
longer that M^* do not reptate in the matrix of shorter macromolecules with
$M_0 > 10M_e$. One has to discuss two cases: non-reptating and reptating macromolecules.

6.5.2 Characteristic Quantities

The considered system contains n_0 macromolecules of the matrix and n macromolecules of the additive per unit volume and can be characterised by dynamic
modulus $G(\omega)$. The medium, in which the macromolecules of the additive
move, is a system consisting of a linear polymer of molecular weight M_0,
which is characterised by the modulus $G_0(\omega) = -i\omega\eta_0(\omega)$. The change of dynamic modulus, taking into account the fact that some of the macromolecules
of the matrix have been replaced by impurity macromolecules, can be written
as

$$G(\omega) - G_0(\omega) = n\left(g(\omega) - \frac{M}{M_0}g_0(\omega)\right) \qquad (6.69)$$

where $g(\omega)$ and $g_0(\omega)$ are the contributions to the dynamic modulus, respectively, from a single macromolecule of the impurity and the matrix, which can
be easily found from the derived expressions. We shall consider the case of
low frequencies, for which the dynamic modulus can be written in the form
of the expansion given by (6.11), and introduce the characteristic quantities

$$[\eta] = \lim_{c \to 0} \frac{\eta - \eta_0}{c\eta_0}, \qquad [\nu] = \lim_{c \to 0} \frac{\nu - \nu_0}{c\nu_0}, \qquad (6.70)$$

which are apparently functions of the length (or molecular weight) of the macromolecules of the matrix and the additive. The index 0 refers to the matrix and c is the impurity concentration.

To calculate the characteristic quantities both for the matrix and for the additive, we use equation (6.39), if $\psi \ll 1$, or (6.52), if $\psi \gg 1$. We shall assume that the macromolecules of the matrix are long enough, so that one can write, taking also relations (6.69) into account, for coefficients of viscosity and elasticity

$$\eta_0 = \frac{\pi^2}{6} n_0 T \tau_0^* B, \qquad \nu_0 = \begin{cases} \frac{\pi^4}{90} n_0 T (\tau_0^* B)^2, & M_0 < 10 M_e, \\ \frac{\pi^2}{3} n_0 T (\tau_0^* B)^2 \chi_0, & M_0 > 10 M_e. \end{cases} \qquad (6.71)$$

To choose a formulae for calculation the contributions of macromolecules of the additive, one have to estimate value of ψ, which, according to equation (3.29) depends on both macromolecules of the matrix and macromolecules of the additive. One can consider that the conditions of reptation correspond also to the big values of ψ, which is realised at $M < M^*$, and the case $M > M^*$ corresponds to the small values of ψ, so that one can write expressions for coefficients of viscosity and elasticity of the system of independent macromolecules of the additive suspended in the matrix as

$$\eta = \frac{\pi^2}{6} n T \tau^* B, \qquad \nu = \begin{cases} \frac{\pi^4}{90} n T (\tau^* B)^2, & M > M^*, \\ \frac{\pi^2}{3} n T (\tau^* B)^2 \chi, & M < M^*. \end{cases} \qquad (6.72)$$

In equations (6.71) and (6.72), the quantities B and τ_0^* are considered as functions of M_0, and the characteristic relaxation time of the macromolecules of the additive τ^* as a function of M.

Taking all this into account, one can find increments of viscosity and elasticity in the form

$$\eta - \eta_0 = \frac{\pi^2}{6} n T \tau^* B \left(1 - \frac{M_0}{M}\right),$$

$$\nu_b - \nu_0 = \begin{cases} \frac{\pi^4}{90} n T (\tau^* B)^2 \left(1 - \frac{M_0^3}{M^3}\right), & M_0 < 10 M_e, \ M > M_0, \\ \frac{\pi^2}{3} n T (\tau^* B)^2 \frac{M_e}{M} \left(1 - \frac{M_0^2}{M^2}\right), & M_0 > 10 M_e, \ M < M^*, \\ \frac{\pi^4}{90} n T (\tau^* B)^2 \left(1 - \frac{30}{\pi^2} \frac{M_e M_0^2}{M^3}\right), & M_0 > 10 M_e, \ M > M^*. \end{cases} \qquad (6.73)$$

Using the above relations and equations (6.58), one finds that for $M \gg M_0$

$$[\eta] \sim M_0^{-1} M, \qquad [\nu] \sim \begin{cases} M_0^{-3} M^3, & M_0 < 10 M_e, \\ M_0^{-2} M^2, & M_0 > 10 M_e, \ M < M^*, \\ M_0^{-2} M^3, & M_0 > 10 M_e, \ M > M^*. \end{cases} \qquad (6.74)$$

On the other hand, when $M \ll M_0$ (this condition excludes the case $M_0 <$ $10 M_e$) the characteristic quantities are negative and are independent of the length of the matrix and of the impurity macromolecules

$$[\eta] \sim M_0^0 M^0, \qquad [\nu] \sim \begin{cases} M_0^0 M^0, & M_0 < 10 M_e, \\ M_0^{-1} M^0, & M_0 > 10 M_e. \end{cases} \tag{6.75}$$

Results (6.74) and (6.75) do not depend upon any choice of the dependence of B on the length (molecular weight) of the macromolecule.

The viscoelastic behaviour of dilute blends of polymers of different length and narrow molecular weight distributions was investigated experimentally for polybutadiene by Yanovski et al. (1982) and by Jackson and Winter (1995) and for polystyrene by Watanabe and Kotaka (1984) and Watanabe et al. (1985) (the results can be found in the work by Jackson and Winter (1995)). The results for polybutadiene were approximated by Pokrovskii and Kokorin (1984) by the dependencies

$$[\eta] \sim M_0^{-0.8} M^{0.5}, \qquad [\nu] \sim M_0^{-(1.8 \rightarrow 2.2)} M^{1.3 \rightarrow 3.0}. \tag{6.76}$$

The comparison of the theoretical formulas (6.74) with the experimental ones (6.76) shows the consistency of the results, though the absolute values of indexes in formula for characteristic viscosity has appeared to be less that theoretical value 1. Unfortunately, the accuracy of original empirical data (in fact, the required linear dependence of quantities on concentration had never been reached in the work by Yanovski et al. 1982) does not allow one to say whether there are any certain deviations from relations (6.74) or not. If relations (6.76) are confirmed, it could mean that there are some unaccounted issues (intra-chain hydrodynamic interaction, for example), which would decrease in values of the index. Apparently, one needs in extra experimental data for different polymer systems in both weakly and strongly entangled states to analyse the situation in more details. Nevertheless, the above results confirm that the contribution of the orientational relaxation branch of a macromolecule in an entangled system dominates over the contribution of the reptation relaxation branch in phenomena of linear viscoelasticity. Otherwise, by considering the competing mechanism of relaxation – the reptation of the macromolecules, one would apparently have, following Daoud and Gennes (1979), instead of relation (6.74), the other expression for characteristic viscosity of blends for $M \gg M_0$

$$[\eta] \sim M_0^{-3} M^3 \tag{6.77}$$

which deviates from empirical evidence (6.76) more than relations (6.74).

6.5.3 Terminal Relaxation Time

It was assumed that the quantity B is a function of M_0, but, luckily, one does not need in expression for explicit dependence to obtain the final results (6.74) and (6.75) for characteristic quantities for dilute blends of linear polymers. However, the dependence of the quantity B on M_0 can be recovered due to empirical data. To estimate this dependence, one can consider terminal relaxation time

$$\tau = \frac{\nu - \nu_0}{\eta - \eta_0}$$

and use equations (6.73) to obtain for $M > M^*$

$$\tau \sim \begin{cases} B(M_0)M^2, & M_0 < 10M_e, \\ B(M_0)M^2, & M_0 > 10M_e. \end{cases} \tag{6.78}$$

The first line is valid for the case when matrix is a weakly entangled matrix, the second line – a strongly entangled matrix.

Watanabe (1999, p. 1354) has deducted that, according to experimental data for polystyrene/polystyrene blends, when the matrix is a weakly entangled system, terminal time of relaxation depends on the lengths of macromolecules as

$$\tau \sim M_0^3 M^2, \tag{6.79}$$

while also for polystyrene/polystyrene blends, Montfort et al. (1984) found different values of indexes (2.3 instead of 3 and 1.9 instead of 2); the difference is discussed by Watanabe (1999, p. 1356). No empirical relation, similar to relation (6.79), is available for strongly entangled matrices, but, as it can be seen in plots of the paper (Watanabe 1999), that the value of the first index are less that 3 in this case. It is possible that situation is different for weakly and strongly entangled matrices, so that values of the index in formula (6.79) could be different for these two types of systems.

The comparing formulae (6.78) and (6.79) allows one to estimate the dependence of coefficient of enhancement on the lengths of macromolecules as

$$B \sim M_0^3, \tag{6.80}$$

that is $\delta = 3$, in contrast with previous estimate of index as 2.4. The last value of the index, as discussed in the end of the previous subsection, is followed the suggestion that hydrodynamic interaction inside macromolecular coils is ignored. One cannot exclude that this index could be greater, but, in this case, value of the second index in equation (6.79) must be less.

The empirical result (6.80) does not correspond to the reliable results for monodisperse ($M_0 = M$) system well. Indeed, taking result (6.80) into account, the terminal relaxation time (6.58) can be written as

$$\tau \sim \begin{cases} M^5, & M < 10M_e, \\ M^4, & M > 10M_e. \end{cases} \tag{6.81}$$

To provide the validity of empirical dependencies of viscosity and terminal relaxation time on the molecular length (relations (6.43) and (6.44)), the sum of the two indexes in equations (6.81) must have value 4.4 in the case, when the matrix is a weakly entangled system, and value 3.4, when the matrix is a strongly entangled system with macromolecular length M between $10M_e$ and M^*.

6.5.4 A Final Remark

The investigation of viscoelasticity of dilute blends confirms that the reptation dynamics does not determine correctly the terminal quantities characterising viscoelasticity of linear polymers. The reason for this, as has already been noted, that the reptation effect is an effect due to terms of order higher than the first in the equation of motion of the macromolecule, and it is actually the first-order terms that dominate the relaxation phenomena. Attempts to describe viscoelasticity without the leading linear terms lead to a distorted picture, so that one begins to understand the lack of success of the reptation model in the description of the viscoelasticity of polymers. Reptation is important and have to be included when one considers the non-linear effects in viscoelasticity.

Chapter 7
Equations of Relaxation

Abstract The discussion of relaxation and diffusion of macromolecules in very concentrated solutions and melts of polymers showed that the basic equations of macromolecular dynamics reflect the linear behaviour of a macromolecule among the other macromolecules, so that one can proceed further. Considering the non-linear effects of viscoelasticity, one have to take into account the local anisotropy of mobility of every particle of the chains, introduced in the basic dynamic equations of a macromolecule in Chapter 3, and induced anisotropy of the surrounding, which will be introduced in this chapter. In the spirit of mesoscopic theory we assume that the anisotropy is connected with the averaged orientation of segments of macromolecules, so that the equation of dynamics of the macromolecule retains its form. Eventually, the non-linear relaxation equations for two sets of internal variables are formulated. The first set of variables describes the form of the macromolecular coil – the conformational variables, the second one describes the internal stresses connected mainly with the orientation of segments.

7.1 Normal-Modes Form of Dynamic Equation

7.1.1 Transition to the Normal Modes

According to speculations in Chapter 3 (see Section 3.2), the standard equation of macromolecular dynamics can be written in the form

$$m\frac{d^2 r_i^\alpha}{dt^2} = -\zeta(\dot{r}_i^\alpha - \nu_{ij}r_j^\alpha) + F_i^\alpha + G_i^\alpha - 2\mu T A_{\alpha\gamma}r_i^\gamma + \phi_i^\gamma(t). \qquad (7.1)$$

The external resistance force of a particle in equation (7.1) is split into two terms, the first of which is equal to $\zeta(u_j^\gamma - \nu_{jl}r_l^\gamma)$ – the resistance in a corresponding 'monomer' liquid, and the second one, F_i^α, is connected with the neighbouring macromolecules and satisfies the equation, which can be written in the simplest covariant form (see Section 8.4 and Appendix D).

V.N. Pokrovskii, *The Mesoscopic Theory of Polymer Dynamics*,
Springer Series in Chemical Physics 95,
DOI 10.1007/978-90-481-2231-8_7, © Springer Science+Business Media B.V. 2010

$$\tau\left(\frac{dF_i^\alpha}{dt} - \omega_{il}F_l^\alpha\right) + F_i^\alpha = -\zeta BH_{ij}^{\alpha\gamma}(u_j^\gamma - \nu_{jl}r_l^\gamma). \tag{7.2}$$

Similarly, the internal resistance force G_i^α satisfies the covariant-form equation

$$\tau\left(\frac{dG_i^\alpha}{dt} - \omega_{il}G_l^\alpha\right) + G_i^\alpha = -\zeta EG_{ij}^{\alpha\gamma}(u_j^\gamma - \omega_{jl}r_l^\gamma). \tag{7.3}$$

The matrices $H_{ij}^{\alpha\gamma}$ and $G_{ij}^{\alpha\gamma}$ describe the mutual influences of the particles of the chain and depend not only on the direction of particle motion in comparison with direction of chain in the vicinity of the particle labelled α (the local anisotropy) (see equation (3.13)), but also on the mean anisotropy of the medium (the global anisotropy).

Equations (7.2) and (7.3) determine the covariant expressions for the external and internal resistance forces, which, in linear approximation, can be written as expressions (3.6) and (3.7), respectively. We may notice that to obtain a more general linear form of equations for forces, the terms $\gamma_{il}F_l^\alpha$ and $\gamma_{il}G_l^\alpha$ multiplied by arbitrary constants which, nevertheless, could depend on the mode number, should be added to the left-hand side of equations (7.2) and (7.3) respectively. Then, after having calculated the results, the arbitrary quantities can be estimated on the basis of certain requirements (Pyshnograi 1997). However, further on we shall proceed with expressions (7.2) and (7.3), for simplicity's sake.

It is convenient to introduce the normal co-ordinates, using the transformation (1.13). In terms of the new variables

$$r_i^\alpha = Q_{\alpha\gamma}\rho_i^\gamma, \quad \psi_i^\alpha = Q_{\alpha\gamma}\psi_i^\gamma, \quad F_i^\alpha = Q_{\alpha\gamma}\Gamma_i^\gamma, \quad G_i^\alpha = Q_{\alpha\gamma}T_i^\gamma$$

and one can rewrite the set of equations (7.1), (7.2), and (7.3) in the form

$$\frac{d}{dt}\rho_i^\alpha = \psi_i^\alpha,$$

$$m\frac{d}{dt}\psi_i^\alpha = -\zeta(\psi_i^\alpha - \nu_{ij}\rho_j^\alpha) + \Gamma_i^\alpha + T_i^\alpha - 2\mu T\lambda_\alpha\rho_i^\alpha + \xi_i^\alpha,$$

$$\tau\left(\frac{d}{dt}\Gamma_i^\alpha - \omega_{il}\Gamma_l^\alpha\right) + \Gamma_i^\alpha = -\zeta B\mathcal{H}_{ij}^{\alpha\nu}(\psi_j^\nu - \nu_{jl}\rho_l^\nu),$$

$$\tau\left(\frac{d}{dt}T_i^\alpha - \omega_{il}T_l^\alpha\right) + T_i^\alpha = -\zeta E\mathcal{G}_{ij}^{\alpha\nu}(\psi_j^\nu - \omega_{jl}\rho_l^\nu),$$

$$\tag{7.4}$$

where

$$\mathcal{H}_{ij}^{\alpha\nu} = Q_{\mu\alpha}H_{ij}^{\mu\gamma}Q_{\gamma\nu}, \quad \mathcal{G}_{ij}^{\alpha\nu} = Q_{\mu\alpha}G_{ij}^{\mu\gamma}Q_{\gamma\nu}. \tag{7.5}$$

We reproduce the procedure used in Section 3.4 and consider the stochastic forces ξ_i^α in the above system of equations as the sum of two independent processes

$$\xi_i^\alpha(t) = \bar{\xi}_i^\alpha(t) + \tilde{\xi}_i^\alpha(t).$$

It is convenient to introduce a variable $\varphi_i^\alpha = \Gamma_i^\alpha + T_i^\alpha + \tilde{\xi}_i^\alpha(t)$, to write the dynamic equations (7.4) in more compact form

$$\frac{\mathrm{d}}{\mathrm{d}t}\rho_i^\alpha = \psi_i^\alpha,$$

$$m\frac{\mathrm{d}}{\mathrm{d}t}\psi_i^\alpha = -\zeta(\psi_i^\alpha - \nu_{ij}\rho_j^\alpha) + \varphi_i^\alpha - 2\mu T\lambda_\alpha\rho_i^\alpha + \bar{\xi}_i^\alpha,$$

$$\tau\left(\frac{\mathrm{d}\varphi_i^\alpha}{\mathrm{d}t} - \omega_{il}\varphi_l^\alpha\right) = -\varphi_i^\alpha - \zeta B\mathcal{H}_{ij}^{\alpha\nu}(\psi_j^\nu - \nu_{jl}\rho_l^\nu)$$

$$\qquad\qquad - \zeta E\mathcal{G}_{ij}^{\alpha\nu}(\psi_j^\nu - \omega_{jl}\rho_l^\nu) + \sigma_i^\alpha(t). \tag{7.6}$$

The stochastic process $\sigma_i^\alpha(t)$ is related to the random force $\tilde{\xi}_i^\gamma$ (see details in Section 3.4). Both $\bar{\xi}_i^\alpha$ and $\sigma_i^\alpha(t)$ are assumed to be independent Gaussian processes.

The set of stochastic equations given by (7.6) is equivalent (in the linear case) to (3.4)–(3.7) with the memory functions defined in Section 3.3.1 but, in contrast to the latter case, set (7.6) is written as a set of Markov stochastic equations. This enables us to determine the variables that describe the collective motion of the set of macromolecules. In this particular approximation, the interaction between neighbouring macromolecules ensures that the phase variables of the elementary motion are the position co-ordinate, the velocity, and the external random force. The set of elementary modes describes the dynamics of the entire set of entangled macromolecules.

We note once more that the Markovian representation of the equation of macromolecular dynamics cannot be made for any arbitrary case, but only for some simple approximations of the memory functions. The above system describes the situation when the medium is characterised by the only relaxation time, but generalisation for few relaxation times is possible.

7.1.2 Anisotropy of Particle Mobility

The quantities (7.5) can be determined in some simple cases. In the simplest case, when no hydrodynamic interaction is assumed, one uses equation (3.8) with matrix (3.10) and, omitting the diffusive normal mode with the label 0, has

$$\mathcal{H}_{ij}^{\alpha\nu} = \delta_{\alpha\nu}\delta_{ij}, \qquad \mathcal{G}_{ij}^{\alpha\nu} = \delta_{\alpha\nu}\delta_{ij}. \tag{7.7}$$

A more complicated cases take into account global and local anisotropy.

Global Anisotropy

The system of entangled macromolecules becomes anisotropic when velocity gradients are applied, and one can assume that each Brownian particle of the chain moves in the anisotropic medium. The expressions for the discussed quantities (7.5) for case, when one can neglect the hydrodynamic interaction and consider the particles of the chain moving in anisotropic medium[1] can be written as

$$\mathcal{H}_{ij}^{\alpha\nu} = \delta_{\alpha\nu}\beta_{ij}, \qquad \mathcal{G}_{ij}^{\alpha\nu} = \delta_{\alpha\nu}\epsilon_{ij}, \tag{7.8}$$

where β_{ij} and ϵ_{ij} are tensor functions of the anisotropy tensor a_{ij} which is discussed later. In linear approximation, the functions have the simple form

$$\begin{aligned}
\beta_{ik} &= \delta_{ik} - 3\beta a'_{ik} - \kappa a_{ll}\delta_{ik}, \\
\epsilon_{ik} &= \delta_{ik} - 3\epsilon a'_{ik} - \nu a_{ll}\delta_{ik}
\end{aligned} \tag{7.9}$$

where the notation for the deviator of the tensor of anisotropy is used

$$a'_{ik} = a_{ik} - \frac{1}{3}a_{ll}\delta_{ik}.$$

The linear approximation (7.9) is insufficient to describe the variation of the friction coefficient at large velocity gradients. In this case, approximation (7.9) can be generalised (Pokrovskii and Pyshnograi 1990, 1991) to become

$$\begin{aligned}
\beta_{ik} &= (\delta_{ik} + 3\beta a'_{ik} + \kappa a_{ll}\delta_{ik})^{-1}, \\
\epsilon_{ik} &= (\delta_{ik} + 3\epsilon a'_{ik} + \nu a_{ll}\delta_{ik})^{-1}.
\end{aligned} \tag{7.10}$$

With accuracy up to the first-order terms in respect of the tensor of anisotropy, expressions (7.9) and (7.10) coincide. Of course, one can use any other approximation that is consistent with (7.9).

The tensor of local anisotropy a_{ik} is assumed to be determined by orientation of the segments of macromolecules which, according to the stress optical law (see Chapter 10), is proportional to the stress tensor, so that

$$a_{ij} \sim \sigma_{ij}. \tag{7.11}$$

In the simplest cases it can be reduced to the tensor of deformation of macromolecular coils

$$a_{ij} = \left(\sum_{\nu}\langle\rho_i^{\nu}\rho_j^{\nu}\rangle \Big/ \sum_{\nu}\langle\rho_k^{\nu}\rho_k^{\nu}\rangle_0\right) - \frac{1}{3}\delta_{ij}.$$

In this case, the friction coefficient of the Brownian particle changes, if the form of the macromolecular coils, described by the terms with parameters β and ϵ, changes or the volume of the macromolecular coil, described by the terms with parameters κ and ν, changes.

---

[1] Expressions for the resistance coefficients of a particles in an anisotropic liquid can be found in papers by Tskhai and Pokrovskii (1985) and by Pokrovskii and Tskhai (1986).

Local Anisotropy

The situation is getting considerably complicated, if one takes local anisotropy into account. The mobility of a particle along the axis of a macromolecule is considered to be bigger than that in the perpendicular direction, so that the entire macromolecule can move easier along its contour. Introduction of the local anisotropy of mobility in Chapter 3 (equations (3.12) and (3.13)) allows us to specify the extra forces of external and internal resistance and define the quantities (7.5) as

$$
\mathcal{H}_{ij}^{\alpha\nu} = Q_{\mu\alpha} H_{ij}^{\mu\gamma} Q_{\gamma\nu}, \qquad \mathcal{G}_{ij}^{\alpha\nu} = Q_{\mu\alpha} G_{ij}^{\mu\gamma} Q_{\gamma\nu},
$$

$$
H_{ij}^{\alpha\gamma} = \delta_{\alpha\gamma} \left(\delta_{ij} - \frac{3}{2} a_{\mathrm{e}} \left(e_i^\alpha e_j^\alpha - \frac{1}{3}\delta_{ij} \right) \right),
$$

$$
G_{ij}^{\alpha\gamma} = \frac{1}{N} \left\{ (N+1)\delta_{\alpha\gamma} \left[\delta_{ij} - \frac{3}{2} a_{\mathrm{i}} \left(e_i^\alpha e_j^\alpha - \frac{1}{3}\delta_{ij} \right) \right] \right. \tag{7.12}
$$

$$
\left. - \left[\delta_{ij} - \frac{3}{2} a_{\mathrm{i}} \left(e_i^\gamma e_j^\gamma - \frac{1}{3}\delta_{ij} \right) \right] \right\},
$$

where a_{e} and a_{i} are measures of local anisotropy. Every internal particle of the chain can be ascribed by the direction vector

$$
e^\alpha = \frac{r^{\alpha+1} - r^{\alpha-1}}{|r^{\alpha+1} - r^{\alpha-1}|}, \quad \alpha = 1, \ldots, N-1,
$$

while the zeroth and the last particles having no direction, so that there are $N - 1$ vectors for a chain. It is convenient formally to consider the product of components of vectors e^0 and e^N to be defined as

$$
e_i^0 e_j^0 = e_i^N e_j^N = \frac{1}{3}\delta_{ij}.
$$

While introducing of the global anisotropy, the equation for the macromolecular dynamics remains linear in co-ordinates and velocities, the introduction of the local anisotropy makes it non-linear in co-ordinates. Both global and local anisotropy are needed to describe the non-linear effects of the relaxation phenomena in the mesoscopic approximation.

7.2 Equations for the Non-Equilibrium Moments

The dynamics of the macromolecule in the form of a set of differential equations of the first order is convenient for derivation of relaxation equations (Volkov and Vinogradov 1984, 1985; Volkov 1990). As a starting point, we use equations (7.6) and consider $m = 0$ in this system, so that the second equation allows us to define

$$\varphi_k^\alpha = \zeta(\psi_k^\alpha - \nu_{kj}\rho_j^\alpha) + 2\mu T \lambda_\alpha \rho_k^\alpha + \bar{\xi}_k^\alpha \qquad (7.13)$$

and to rewrite the system of equations (7.6) as

$$\frac{\mathrm{d}}{\mathrm{d}t}\rho_i^\alpha = \psi_i^\alpha,$$

$$\frac{\mathrm{d}}{\mathrm{d}t}\psi_i^\alpha = 0,$$

$$\tau\frac{\mathrm{d}}{\mathrm{d}t}\varphi_i^\alpha = \tau\omega_{il}\varphi_l^\alpha - \zeta(\psi_i^\alpha - \nu_{ij}\rho_j^\alpha) - 2\mu T\lambda_\alpha \rho_i^\alpha$$
$$- \zeta\mathcal{B}_{ij}^{\alpha\nu}(\psi_j^\nu - \nu_{jl}\rho_l^\nu) - \zeta\mathcal{G}_{ij}^{\alpha\nu}(\psi_j^\nu - \omega_{jl}\rho_l^\nu) + \bar{\xi}_i^\alpha + \sigma_i^\alpha.$$

These equations allow one to find relaxation equation for different moments, if the quantities $\mathcal{B}_{ij}^{\alpha\nu}$ and $\mathcal{G}_{ij}^{\alpha\nu}$ are given. We consider here that the quantities are independent on the co-ordinates; the more complicated case, when these quantities depend on the co-ordinates of particles, is left for other researchers. It is convenient in this section to omit the mode label and write down the above equations in the form

$$\frac{\mathrm{d}\rho_i}{\mathrm{d}t} = \psi_i,$$

$$\frac{\mathrm{d}}{\mathrm{d}t}\psi_i = 0, \qquad (7.14)$$

$$\frac{\mathrm{d}\varphi_i}{\mathrm{d}t} = \kappa_{ij}\rho_j + \lambda_{ij}\psi_j + \omega_{ij}\varphi_j + \frac{1}{\tau}(\bar{\xi}_i^\alpha + \sigma_i^\alpha),$$

where

$$\kappa_{ij} = \frac{\zeta}{\tau}\left(-\frac{1}{2\tau^{\mathrm{R}}}\delta_{ij} + \nu_{ij} + B_{il}\nu_{lj} + E_{il}\omega_{lj}\right),$$

$$\lambda_{ij} = -\frac{\zeta}{\tau}\left(\delta_{ij} + B_{ij} + E_{ij}\right), \qquad (7.15)$$

$$B_{ij} = B\beta_{ij}, \quad E_{ij} = E\epsilon_{ij}, \quad \tau^{\mathrm{R}} = \frac{\zeta}{4\mu T\lambda}.$$

Further on we shall use the following symbols for the moments of the considered variables

$$r_{ik} = \langle\rho_i\rho_k\rangle, \qquad n_{ik} = \langle\rho_i\varphi_k\rangle,$$
$$y_{ik} = \langle\rho_i\psi_k\rangle, \qquad m_{ik} = \langle\psi_i\varphi_k\rangle,$$
$$z_{ik} = \langle\psi_i\psi_k\rangle, \qquad l_{ik} = \langle\varphi_i\varphi_k\rangle.$$

The moments can be found as solutions of the set of equations, which are followed by set (7.14)

$$\frac{\mathrm{d}r_{ik}}{\mathrm{d}t} = y_{ik} + y_{ki},$$

$$\frac{\mathrm{d}y_{ik}}{\mathrm{d}t} = z_{ik},$$

$$\frac{\mathrm{d}z_{ik}}{\mathrm{d}t} = 0,$$

$$\frac{\mathrm{d}n_{ik}}{\mathrm{d}t} = m_{ik} + \kappa_{kj}r_{ji} + \lambda_{kj}y_{ij} + \omega_{kj}n_{ij} + \frac{1}{\tau}\langle\rho_i(\bar{\xi}_i^\alpha + \sigma_i^\alpha)\rangle,$$

$$\frac{\mathrm{d}m_{ik}}{\mathrm{d}t} = \kappa_{kj}y_{ji} + \lambda_{kj}z_{ji} + \omega_{kj}m_{ij} + \frac{1}{\tau}\langle\psi_i(\bar{\xi}_i^\alpha + \sigma_i^\alpha)\rangle,$$

$$\frac{\mathrm{d}l_{ik}}{\mathrm{d}t} = \kappa_{ij}n_{jk} + \kappa_{kj}n_{ji} + \lambda_{ij}m_{jk} + \lambda_{kj}m_{ji} + \omega_{ij}l_{jk} + \omega_{kj}l_{ji}$$

$$+ \frac{1}{\tau}\langle\varphi_i(\bar{\xi}_i^\alpha + \sigma_i^\alpha(t)) + \varphi_k(\bar{\xi}_i^\alpha + \sigma_i^\alpha)\rangle.$$

To determine the average quantities, which contain the random forces in the above-written equations, one can consider the equilibrium situation. The unknown terms can be evaluated through the equilibrium values of moments, which can be used to rewrite the equations for the moments. Finally, the set of relaxation equations has the form of the above-written equations, where the terms with random forces and the terms containing the product of velocity gradient and an equilibrium moment are omitted. Instead of moments, the differences in the moments and their equilibrium values, such as $r_{ik} - r_{ik}^0$ instead of r_{ik}, for instance, ought to be written, so that one has

$$\frac{\mathrm{d}r_{ik}}{\mathrm{d}t} = y_{ik} + y_{ki},$$

$$\frac{\mathrm{d}y_{ik}}{\mathrm{d}t} = z_{ik},$$

$$\frac{\mathrm{d}z_{ik}}{\mathrm{d}t} = 0,$$

$$\frac{\mathrm{d}n_{ik}}{\mathrm{d}t} = m_{ik} - m_{ik}^0 + \kappa_{kj}r_{ji} - \kappa_{kj}^0 r_{ji}^0 + \lambda_{kj}y_{ij} + \omega_{kj}n_{ij}, \tag{7.16}$$

$$\frac{\mathrm{d}m_{ik}}{\mathrm{d}t} = \kappa_{kj}y_{ji} + \omega_{kj}m_{ij},$$

$$\frac{\mathrm{d}l_{ik}}{\mathrm{d}t} = \kappa_{ij}n_{jk} - \kappa_{ij}^0 n_{jk}^0 + \kappa_{kj}n_{ji} - \kappa_{kj}^0 n_{ji}^0 + \lambda_{ij}m_{jk} + \lambda_{kj}m_{ji}$$

$$+ \omega_{ij}l_{jk} + \omega_{kj}l_{ji}$$

where $\kappa_{ij}^0 = -\frac{\zeta}{\tau}\frac{1}{2\tau^R}\delta_{ij}$ is the value of κ_{ij} given by (7.15) at zero velocity gradients. The fact that some of the equilibrium moments are equal to zero, as shown below, has already been taken into account in equations (7.16).

The system of equations (7.16) determines the unknown quantities as functions of time and the parameters of the problem: ζ, B, E, and the parameters of anisotropy. To find a solution, one uses the equilibrium values of moments, three of which are found in Section 4.1.2

$$r_{ik}^0 = \frac{1}{2\mu\lambda}\delta_{ik}, \qquad z_{ik}^0 = \frac{T}{m}\delta_{ik}, \qquad y_{ik}^0 = 0. \tag{7.17}$$

The others are contained in the following relations, which are consequences of relation (7.13)

$$n_{ik} = n_{ik}^0 + \zeta(y_{ik} - \nu_{kj}r_{ji}) + \frac{\zeta}{2\tau^R}(r_{ik} - r_{ik}^0),$$

$$m_{ik} = m_{ik}^0 + \zeta(z_{ik} - \nu_{kj}y_{ji}) + \frac{\zeta}{2\tau^R}y_{ki}, \tag{7.18}$$

$$l_{ik} = l_{ik}^0 + \zeta(m_{ki} - \nu_{kj}n_{ji}) + \frac{\zeta}{2\tau^R}(n_{ki} - n_{ki}^0).$$

One can note that, with help of the one of the above relations, the fourth equation in the set (7.16) can be written as

$$\frac{dn_{ik}}{dt} = \zeta z_{ik} - \zeta\nu_{kj}y_{ji} + \frac{\zeta}{2\tau^R}y_{ki} + \kappa_{kj}r_{ji} - \kappa_{kj}^0 r_{ji}^0 + \lambda_{kj}y_{ij} + \omega_{kj}n_{ij}.$$

From the other side, after having differentiated the first of equations (7.18), one has

$$\frac{dn_{ik}}{dt} = \zeta(z_{ik} - \nu_{kj}(y_{ji} + y_{ij})) + \frac{\zeta}{2\tau^R}(y_{ik} + y_{ki}).$$

These equations are followed by the relation

$$\left(\frac{\tau}{2\tau^R}\delta_{kj} + \delta_{kj} + B_{kj} + E_{kj}\right)y_{ij}$$

$$= -\frac{1}{2\tau^R}(r_{ik} - r_{ik}^0) + \left(\frac{\tau}{2\tau^R}\delta_{kl} + \delta_{kl} + B_{kl} + E_{kl}\right)\omega_{lj}r_{ji} + B_{kl}\gamma_{lj}r_{ji}$$

$$+ \tau\nu_{kj}y_{ij} + \tau\omega_{kj}y_{ji} + \frac{\tau}{\zeta}\omega_{kj}n_{ji}^0 + \frac{\tau}{2\tau^R}\omega_{kj}r_{ji}^0, \tag{7.19}$$

where terms containing velocity gradients in the second power are already excluded.

The equation (7.19) has to be considered as an equation for the quantity y_{ik}. When the velocity gradients are absent,

$$y_{ik} = -\frac{1}{2\tau_\diamond}(r_{ik} - r_{ik}^0), \qquad \tau_\diamond = \frac{\tau}{2} + \tau^R(1 + B + E). \tag{7.20}$$

The symbol $\diamond$ is used here to show the place of label of relaxation times which are identical to relaxation times (4.26). Then, one can see that the last four terms in (7.19) can be neglected and the last equation allows us to find the relation

$$y_{ik} = -\frac{1}{2\tau_\diamond}(r_{ij} - r_{ij}^0)b_{jk} + \omega_{kj}r_{ji} + \frac{B\tau^R}{\tau_\diamond}r_{ij}\gamma_{jl}c_{lk}, \tag{7.21}$$

where the following notations are used

$$b_{ik} = \tau_\diamond \left(\frac{\tau}{2} \delta_{ik} + B\tau^{\mathrm{R}} \left(\beta_{ik} + \frac{E}{B} \epsilon_{ik} \right) \right)^{-1},$$

$$c_{ik} = \beta_{ij} b_{jk}, \qquad\qquad (7.22)$$

$$\beta_{ik} = (\delta_{jk} + \kappa a_{ll} \delta_{jk} + 3\beta a'_{jk})^{-1},$$

$$\epsilon_{ik} = (\delta_{jk} + \nu a_{ll} \delta_{jk} + 3\epsilon a'_{jk})^{-1}.$$

Now one can find the quantity $y_{ik} + y_{ki}$ and then, according to the first equation from (7.16), write down the relaxation equation for the moment r_{ij}

$$\frac{\mathrm{d}r_{ik}}{\mathrm{d}t} - \omega_{ij} r_{jk} - \omega_{kj} r_{ji} - \frac{B\tau^{\mathrm{R}}}{\tau_\diamond} (r_{ij} \gamma_{jl} c_{lk} + r_{kj} \gamma_{jl} c_{li})$$

$$= -\frac{1}{2\tau_\diamond} \left(\left(r_{ij} - r^0_{ij} \right) b_{jk} + \left(r_{kj} - r^0_{ij} \right) b_{ji} \right). \qquad (7.23)$$

We remind the reader that the equations are valid for the case when $m \to 0$, $\zeta \to 0$, $\zeta B \neq 0$, $\zeta E \neq 0$.

7.3 Relaxation of the Macromolecular Conformation

The mean size and form of the macromolecular coil is characterised by the set of the tensors $\langle \rho^\alpha_i \rho^\alpha_j \rangle$, $\alpha = 1, 2, \dots, N$. To describe relaxation of macromolecular coil, it is convenient to use the non-dimensional tensor variables

$$x^\alpha_{ij} = \langle \rho^\alpha_i \rho^\alpha_j \rangle / \langle \rho^\alpha_l \rho^\alpha_l \rangle_0.$$

We have to take into account that there are two competing mechanisms of conformational relaxation, whereas it is clear that only one of them is realised for a given system with a certain values of parameters B, χ and ψ. After having compared the relaxation times of the two competing mechanisms (Section 4.2.3), we ought to conclude that there is a certain value of the parameter $\chi^* \approx 0.1$ at which the mechanism of relaxation changes. So, we have to consider two cases.

7.3.1 Diffusive Relaxation

The relaxation equation for conformation of macromolecule follows directly from the relation (7.23) of the previous section

$$\frac{\mathrm{d}x^\alpha_{ik}}{\mathrm{d}t} - \nu_{ij} x^\alpha_{jk} - \nu_{kj} x^\alpha_{ji} + \frac{E\tau^{\mathrm{R}}_\alpha}{\tau_\alpha} (x^\alpha_{ij} \gamma_{jl} c_{lk} + x^\alpha_{kj} \gamma_{jl} c_{li})$$

$$= -\frac{1}{2\tau_\alpha} \left[\left(x^\alpha_{ij} - \frac{1}{3} \delta_{ij} \right) b_{jk} + \left(x^\alpha_{kj} - \frac{1}{3} \delta_{kj} \right) b_{ji} \right], \qquad (7.24)$$

where the notations of the previous chapters are used for relaxation times

$$\tau_\alpha = \frac{\tau}{2} + B\tau_\alpha^{\mathrm{R}}(1 + E/B), \quad \tau_\alpha^{\mathrm{R}} = \tau^*/\alpha^2.$$

We consider the situations when the values of parameter ψ are small, in other terms

$$\chi^* < \chi < 0.5, \quad \psi \ll 1,$$

so that the first term in the expression for the relaxation time τ_α in the above-written formula can be neglected

$$\tau_\alpha = B\tau_\alpha^{\mathrm{R}}, \quad \alpha = 1, 2, \ldots \ll (1/\chi)^{\frac{1}{2}}.$$

This relation allows us to write relations (7.22) as

$$b_{ik} \approx \beta_{ik}^{-1}, \quad c_{ik} \approx \delta_{ik}$$

and simplify the relaxation equation (7.24) as

$$\frac{\mathrm{d}x_{ik}^\alpha}{\mathrm{d}t} - \nu_{ij}x_{jk}^\alpha - \nu_{kj}x_{ji}^\alpha = -\frac{1}{2\tau_\alpha}\left[\left(x_{ij}^\alpha - \frac{1}{3}\delta_{ij}\right)b_{jk} + \left(x_{kj}^\alpha - \frac{1}{3}\delta_{kj}\right)b_{ji}\right].$$
(7.25)

Let us note that one can neglect the effect of anisotropic environment and have obtained the simpler linear form of relaxation equation

$$\frac{\mathrm{d}x_{ik}^\alpha}{\mathrm{d}t} - \nu_{ij}x_{jk}^\alpha - \nu_{kj}x_{ji}^\alpha = -\frac{1}{\tau_\alpha}\left(x_{ik}^\alpha - \frac{1}{3}\delta_{ik}\right).$$
(7.26)

A solution of equation (7.25) for a steady case at small velocity gradients can be easily found as an expansion in series in powers of velocity gradient. Up to the second-order terms with respect to velocity gradients, equation (7.26) immediately gives

$$x_{ik}^\alpha = \frac{1}{3}\delta_{ik} + \frac{2}{3}B\tau_\alpha^{\mathrm{R}}\gamma_{ik} + \frac{2}{3}(B\tau_\alpha^{\mathrm{R}})^2(\omega_{il}\gamma_{lk} + \omega_{kl}\gamma_{li}) + \frac{4}{3}(B\tau_\alpha^{\mathrm{R}})^2\gamma_{il}\gamma_{lk}. \quad (7.27)$$

Now it is not difficult to calculate the amendment to formula (7.27) due to effect of anisotropy. At small velocity gradients, the tensor of anisotropy a_{ik} is small, so that according to formulae (7.10) and (7.22), in linear approximation

$$b_{ik} \approx \delta_{ik} + 3\beta a_{ik}.$$

Then, the solution of equation (7.25) with approximation up to the terms of the second order in velocity gradients has the form

$$x_{ik}^\alpha = \frac{1}{3}\delta_{ik} + \frac{2}{3}B\tau_\alpha^{\mathrm{R}}\gamma_{ik} + \frac{2}{3}(B\tau_\alpha^{\mathrm{R}})^2(\omega_{il}\gamma_{lk} + \omega_{kl}\gamma_{li})$$

$$+ \frac{4}{3}(B\tau_\alpha^{\mathrm{R}})^2\gamma_{il}\gamma_{lk} - B\beta\tau_\alpha^{\mathrm{R}}(a_{il}\gamma_{lk} + a_{kl}\gamma_{li}). \quad (7.28)$$

One can see that relations (7.27) and (7.28) up to the first term coincides with relation (4.47) which is valid for small values of α, namely, $\alpha^2 \ll 1/\chi$. We have to consider this relation to be a condition of applicability of the equation (7.24).

7.3.2 Reptation Relaxation

Equations of the previous subsection describe relaxation of large-scale conformation of the macromolecule due to diffusive motion of particles through the sea of segments, which is valid, as we considered in Chapters 4 and 5, for weakly entangled systems. For highly entangled system, when

$$\chi < \chi^* \ll 1,$$

relaxation of the macromolecular coil is realised through reptation instead of the more slow mechanism of rearrangement of all the entangled chains.

One has no results for this case derived consequently from the basic equations (7.6) with local anisotropy. Instead, to find conformational relaxation equation, we shall use the Doi-Edwards model, which approximate the large-scale conformational changes of the macromolecule due to reptation. The mechanism of relaxation in the Doi-Edwards model was studied thoroughly (Doi and Edwards 1986; Öttinger and Beris 1999), which allows us to write down the simplest equation for the conformational relaxation for the strongly entangled systems

$$\frac{dx_{ik}^{\alpha}}{dt} - \nu_{ij}x_{jk}^{\alpha} - \nu_{kj}x_{ji}^{\alpha} = -\frac{1}{2\tau_{\alpha}^{\text{rep}}} \left[\left(x_{ij}^{\alpha} - \frac{1}{3}\delta_{ij} \right) b_{jk} + \left(x_{kj}^{\alpha} - \frac{1}{3}\delta_{kj} \right) b_{ji} \right].$$
$$(7.29)$$

We assume in the above equation, that anisotropy of environment is possible. If one neglect the latter, the relaxation equation takes the simpler form

$$\frac{dx_{ik}^{\alpha}}{dt} - \nu_{ij}x_{jk}^{\alpha} - \nu_{kj}x_{ji}^{\alpha} = -\frac{1}{\tau_{\alpha}^{\text{rep}}} \left(x_{ij}^{\alpha} - \frac{1}{3}\delta_{ij} \right). \qquad (7.30)$$

One can compare equations (7.29) and (7.30) with equations (7.25) and (7.26) to see that the only difference between this and previous case is the difference in relaxation times, which for the strongly entangled systems, according to formula (4.37), are

$$\tau_{\alpha}^{\text{rep}} = \frac{\pi^2}{\chi} \tau_{\alpha}^{\text{R}}, \quad \tau_{\alpha}^{\text{R}} = \tau^*/\alpha^2. \qquad (7.31)$$

However, the effect of anisotropy of the environment is expressed differently. One can see from formula (5.17), that the parameter ψ is big in the case of strongly entangled system $(\chi < \chi^*)$, so that, according to equation (7.22),

$$b_{ik} \approx \epsilon_{ik}^{-1}.$$

For a steady-state case, equations (7.29) and (7.30) allow one at small velocity gradients to obtain a solution as an expansion in series in powers of velocity gradient. Up to the first-order terms with respect to velocity gradients, equation (7.30) immediately gives

$$x_{ik}^{\alpha} = \frac{1}{3}\delta_{ik} + \frac{2}{3}\frac{\pi^2}{\chi}\tau_{\alpha}^{R}\gamma_{ik}, \qquad b_{ik} \approx \delta_{ik} + 3\epsilon a_{ik}.$$

Then, the solution of equation (7.29) with approximation up to the terms of the second order in velocity gradients has the form

$$x_{ik}^{\alpha} = \frac{1}{3}\delta_{ik} + \frac{2}{3}\frac{\pi^2}{\chi}\tau_{\alpha}^{R}\gamma_{ik} + \frac{2}{3}\left(\frac{\pi^2}{\chi}\tau_{\alpha}^{R}\right)^2(\omega_{ij}\gamma_{jk} + \omega_{kj}\gamma_{ji})$$

$$+ \frac{4}{3}\left(\frac{\pi^2}{\chi}\tau_{\alpha}^{R}\right)^2\gamma_{il}\gamma_{lk} - \frac{\pi^2}{\chi}\epsilon\tau_{\alpha}^{R}(a_{il}\gamma_{lk} + a_{kl}\gamma_{li}), \qquad (7.32)$$

where one can use the relation

$$\frac{\pi^2}{\chi} = \frac{E}{B}$$

in the case of the strongly entangled system at $\chi \ll \chi^*$.

7.4 Relaxation of Orientational Variables

We can notice that, apart from the deformation of the coil, the stresses (6.7) are determined by the forces of internal viscosity which satisfy equation (7.3) or, in normal form, it is the last equation from set (7.4). It is convenient to consider quantities

$$u_{ik}^{\nu} = -\frac{1}{3T}\langle \rho_k^{\nu}T_i^{\nu}\rangle, \qquad w_{ik}^{\nu} = -\frac{1}{3T}\langle \psi_k^{\nu}T_i^{\nu}\rangle, \qquad v_{ik}^{\nu} = -\frac{1}{3T}\langle \varphi_k^{\nu}T_i^{\nu}\rangle$$

as variables that describe the situation. In fact, one needs in the equations for the first set of variables only, but to get them, the two other sets of variables has to be also included into consideration.

We use the last equation from (7.4) and equations (7.14) to obtain the equation of relaxation for above-defined quantities. After the procedure, which is quite similar to that used in Section 7.2, we write down

$$\frac{d}{dt}u_{ik}^{\alpha} - \omega_{il}u_{lk}^{\alpha} + \frac{1}{\tau}u_{ik}^{\alpha} = w_{ik}^{\alpha} + \frac{1}{3T}\frac{\zeta}{\tau}E_{ij}(y_{kj}^{\alpha} - \omega_{jl}r_{kl}^{\alpha}),$$

$$\frac{d}{dt}w_{ik}^{\alpha} - \omega_{il}w_{lk}^{\alpha} + \frac{1}{\tau}w_{ik}^{\alpha} = \frac{1}{3T}\frac{\zeta}{\tau}E_{ij}\omega_{jl}y_{lk}^{\alpha},$$

$$\frac{dv_{ik}^{\alpha}}{dt} - \omega_{il}v_{lk}^{\alpha} + \frac{1}{\tau}v_{ik}^{\alpha} = \kappa_{kj}u_{ij}^{\alpha} + \lambda_{kj}w_{ij}^{\alpha} + \omega_{kj}v_{ij}^{\alpha}$$

$$+ \frac{1}{3T}\frac{\zeta}{\tau}E_{ij}(m_{jk}^{\alpha} - m_{jk}^{\alpha}(0) - \omega_{jl}n_{lk}^{\alpha}). \qquad (7.33)$$

One can use the relations, which follow equation (7.13), that is

$$v_{ik}^\alpha = \zeta(w_{ik}^\alpha - \nu_{kj}u_{ij}^\alpha) + 2\mu T\lambda_\alpha u_{ik}^\alpha,$$

$$\frac{dv_{ik}^\alpha}{dt} = \zeta\left(\frac{dw_{ik}^\alpha}{dt} - \nu_{kj}\frac{du_{ij}^\alpha}{dt}\right) + 2\mu T\lambda_\alpha\frac{du_{ik}^\alpha}{dt}$$

to exclude the variable v_{ik}^ν and to obtain the relation

$$-\frac{1}{2\tau^R}u_{ik}^\alpha + \left(\frac{\tau}{2\tau^R}\omega_{kj} + \nu_{kj} + B_{kl}\nu_{lj} + E_{kl}\omega_{lj}\right)u_{ij}^\alpha$$

$$-\left(\frac{\tau}{2\tau^R}\delta_{kj} + \delta_{kj} + B_{kj} + E_{kj}\right)w_{ij}^\alpha + \tau\omega_{kj}w_{ij}^\alpha + \tau\nu_{kj}w_{ij}^\alpha$$

$$= \frac{\zeta}{3T}\left(-\frac{1}{\zeta}E_{ij}(m_{jk}^\alpha - m_{jk}^\alpha(0) + \omega_{jl}n_{lk}^\alpha) + E_{ij}\omega_{jl}y_{lk}^\alpha\right.$$

$$\left. - \nu_{kj}E_{kl}y_{il}^\alpha + \frac{1}{2\tau^R}E_{ij}(y_{kj}^\alpha - \omega_{jl}r_{kl}^\alpha)\right).$$

Here terms containing velocity gradients in the second power are already excluded.

One can see that in zeroth approximation

$$w_{ik}^\alpha = -\frac{1}{2\tau_\alpha}u_{ki}^\alpha, \quad \tau_\alpha = \frac{\tau}{2} + \tau_\alpha^R(1 + B + E). \tag{7.34}$$

In the second iteration, some of the terms in the above relation can be neglected, so that this relation is followed

$$w_{ik}^\alpha = -\frac{1}{2\tau_\alpha}b_{kj}u_{ji}^\alpha + \omega_{kj}\,u_{ji}^\alpha + \frac{B\tau_\alpha^R}{\tau_\alpha}e_{kj}\gamma_{jl}u_{li}^\alpha. \tag{7.35}$$

Now one can return to the first equation from the set (7.33) and, also using equation (7.21), obtain the equation of relaxation of the orientational variables

$$\frac{du_{ik}^\alpha}{dt} - \omega_{ij}u_{jk}^\alpha - \omega_{kj}u_{ji}^\alpha$$

$$= -\left(\frac{1}{\tau}\delta_{ij} + \frac{1}{2\tau_\alpha}b_{ij}\right)u_{jk}^\alpha - \frac{E\tau_\alpha^R}{\tau\tau_\alpha}\left[\left(x_{ij}^\alpha - \frac{1}{3}\delta_{ij}\right)d_{jk} - 2B\tau_\alpha^R x_{il}^\alpha\gamma_{lj}f_{jk}\right]$$

$$+ \frac{B\tau_\alpha^R}{\tau_\alpha}e_{ij}\gamma_{jl}u_{lk}^\alpha. \tag{7.36}$$

In equation (7.35) and (7.36), in line with the previously introduced auxiliary quantities (7.22), we use the notation

$$e_{ik} = b_{ij}\beta_{jk}, \quad d_{ik} = b_{ij}\epsilon_{kj}, \quad f_{ik} = c_{ij}\epsilon_{kj}. \tag{7.37}$$

Equation (7.36) contains terms with velocity gradients, which cause deviation of the internal parameter from its equilibrium value and terms which determine the approach to the equilibrium. We can see that the relaxation of quantity u_{ik}^α depends on the quantity x_{ik}^α which is in turn determined by the relaxation equation (7.25) for weakly entangled systems or equation (7.29) for strongly entangled systems. Equation (7.36) can be simplified for these two limiting cases for which, according to equation (4.45), one can write asymptotic expressions for relaxation time

$$\tau_\nu = \begin{cases} \tau_\alpha^R B, & \psi \ll 1, \quad \nu^2 \ll \frac{1}{\chi}, \\ \tau_\alpha^R E, & \psi \gg 1, \quad \nu^2 \ll \frac{\psi}{\chi}. \end{cases}$$

7.4.1 Weakly Entangled Systems

For the weakly entangled systems ($\psi \ll 1$), coefficients (7.22) and (7.37) can be approximated as

$$b_{ik} \approx \beta_{ik}^{-1}, \quad c_{ik} \approx \delta_{ik}, \quad e_{ik} \approx \delta_{ik}, \quad d_{ik} \approx \beta_{ij}^{-1}\epsilon_{kj}, \quad f_{ik} \approx \epsilon_{ik}$$

and the equation of relaxation (7.36) reduces to

$$\frac{du_{ik}^\alpha}{dt} - \omega_{ij}u_{jk}^\alpha - \omega_{kj}u_{ji}^\alpha$$
$$= -\left(\frac{1}{\tau}\delta_{ij} + \frac{1}{2\tau_\alpha}b_{ij}\right)u_{jk}^\alpha - \frac{1}{\tau}\frac{E}{B}\left[\left(x_{ij}^\alpha - \frac{1}{3}\delta_{ij}\right)d_{jk} - 2B\tau_\alpha^R x_{il}^\alpha \gamma_{lj}f_{jk}\right]$$
$$+ \gamma_{il}u_{lk}^\alpha. \tag{7.38}$$

In steady state, the quantity u_{ik}^α can be calculated as an expansion in powers of velocity gradients. Up to the accuracy to the second order, equation (7.38) is followed by the relation

$$u_{ik}^\alpha = -\frac{E}{B}\frac{\tau_\alpha^*}{\tau}\left[\left(x_{ij}^\alpha - \frac{1}{3}\delta_{ij}\right)d_{jk} - 2B\tau_\alpha^R x_{il}^\alpha \gamma_{lj}f_{jk}\right]$$
$$+ \tau_\alpha^*\left(\omega_{ij}u_{jk}^\alpha + \omega_{kj}u_{ji}^\alpha + \gamma_{il}u_{lk}^\alpha\right),$$

where

$$d_{ik} \approx \delta_{ik} - 3(\epsilon - \beta)a_{ik}, \qquad f_{ik} \approx \delta_{ik} - 3\epsilon a_{ik}, \qquad \tau_\alpha^* = \frac{2\tau\tau_\alpha}{2\tau_\alpha + \tau}.$$

One uses equation (7.28) for x_{ij}^α to be convinced that the expansion begins with the second-order terms

$$u_{ik}^\alpha = -\frac{2}{3}EB(\tau_\alpha^R)^2(\omega_{ij}\gamma_{jk} + \omega_{kj}\gamma_{ji}) - E(\beta + \epsilon)\tau_\alpha^R(a_{ij}\gamma_{jk} + a_{kj}\gamma_{ji}). \tag{7.39}$$

7.4.2 Strongly Entangled Systems

In the opposite case ($\psi \gg 1$), coefficients (7.22) and (7.37) can be approximated as

$$b_{ik} \approx \epsilon_{ik}^{-1}, \quad c_{ik} \approx \beta_{il}\epsilon_{lk}^{-1}, \quad e_{ik} \approx \epsilon_{ij}^{-1}\beta_{jk}, \quad d_{ik} \approx \delta_{ik}, \quad f_{ik} \approx \epsilon_{il}^{-1}\beta_{lk}\epsilon_{kj},$$

so that we can obtain from equation (7.36), at $\tau_\alpha = E\tau_\alpha^R$, the simpler form of the relaxation equation

$$\frac{du_{ik}^\alpha}{dt} - \omega_{ij}u_{jk}^\alpha - \omega_{kj}u_{ji}^\alpha$$

$$= -\frac{1}{\tau}u_{ik}^\alpha - \frac{1}{\tau}\left(x_{ik}^\alpha - \frac{1}{3}\delta_{ik} - 2B\tau_\alpha^R x_{il}^\alpha \gamma_{lj} f_{jk}\right) + \frac{B}{E}e_{ij}\gamma_{jl}u_{lk}^\alpha. \quad (7.40)$$

To calculate the quantity u_{ik}^α in steady state as an expansion in powers of velocity gradients, one can use the relation that is following from equation (7.40)

$$u_{ik}^\alpha = -x_{ik}^\alpha + \frac{1}{3}\delta_{ik} + 2B\tau_\alpha^R x_{il}^\alpha \gamma_{lj} f_{jk} + \tau(\omega_{ij}u_{jk}^\alpha + \omega_{kj}u_{ji}^\alpha) + \frac{B}{E}\tau e_{ij}\gamma_{jl}u_{lk}^\alpha, \quad (7.41)$$

where

$$f_{ik} \approx \delta_{ik} - 3\beta a_{ik}, \qquad e_{ik} \approx \delta_{ik} + 3(\epsilon - \beta)a_{ik}.$$

Now, one has to rely on the reptation mechanism of changing conformation and use equations (7.32) to find the expansion of the quantity u_{ik}^α. In the first-order approximation, one can obtain

$$u_{ik}^\alpha = \frac{2}{3}\left(B - \frac{\pi^2}{\chi}\right)\tau_\alpha^R\gamma_{ik} \approx \frac{2}{3}B\tau_\alpha^R\gamma_{ik}. \quad (7.42)$$

We believe that for sufficiently long macromolecules, we can neglect the second term π^2/χ as compared with B, so that from equations (7.32) and (7.41) one has

$$u_{ik}^\alpha = \frac{2}{3}B\tau_\alpha^R\gamma_{ik} + \left[\frac{2}{3}B\tau_\alpha^R\tau - \frac{2}{3}\left(\frac{\pi^2}{\chi}\tau_\alpha^R\right)^2\right](\omega_{ij}\gamma_{jk} + \omega_{kj}\gamma_{ji})$$

$$+ \left[\frac{2}{3}\frac{B^2}{E}\tau_\alpha^R\tau + \frac{4}{3}B\frac{\pi^2}{\chi}\tau_\alpha^R\tau_\alpha^R - \frac{4}{3}\left(\frac{\pi^2}{\chi}\tau_\alpha^R\right)^2\right]\gamma_{ij}\gamma_{jk}$$

$$+ \left(\frac{\pi^2}{\chi}\epsilon - B\beta\right)\tau_\alpha^R(a_{ij}\gamma_{jk} + a_{kj}\gamma_{ji}). \quad (7.43)$$

One can see that dependence of the steady-state values of the moments on the anisotropy coefficients appears in terms of the second order, as was assumed previously.

7.5 Relaxation of the Segment Orientation

7.5.1 Rubber Elasticity and Mean Orientation of Segments

It is impossible, apparently, to discuss the phenomena of elasticity, optical anisotropy and dielectric permittivity of polymer without referring to motion of the Kuhn's segments. Indeed, for example, from the time of classical achievement of Werner Kuhn (the 20–30 years of the last century), it is well known that the elastic stresses in polymers are connected with stochastic rotation of segments. With help of the known (Flory 1969) connection of the tensor of mean orientation of segments of a chain with the end-to-end distance R of a chain

$$\langle e_i e_k \rangle - \frac{1}{3}\delta_{ik} = \frac{3}{5(zl)^2}\left(R_i R_k - \frac{1}{3}R^2 \delta_{ik} \right) \tag{7.44}$$

the results of Section 1.7 can be easily reformulated, so that, for a network in the equilibrium situation, the stress tensor can be easily written as

$$\sigma_{ik} = -p\delta_{ik} + 5\nu z T \left(\langle e_i e_k \rangle - \frac{1}{3}\delta_{ik} \right), \tag{7.45}$$

where ν is a number of chains in the volume unit, z is a number of Kuhn segments in each chain. We have assumed for simplicity, that all segments are in a similar situation.

Considering the entangled systems in the coarse-grained approximation, we forget about segments: the theory contains the effective elastic forces between the fictious adjacent particles, and the stress tensor (equation (6.7)) can be expressed through the variables x_{il}^α and u_{ik}^α in the form

$$\sigma_{ik} = -p\delta_{ik} + 3nT \sum_\nu \left(x_{ik}^\nu - \frac{1}{3}\delta_{ik} + u_{ik}^\nu \right). \tag{7.46}$$

However, in this case, the stresses in entangled systems can also be related to the tensor of mean orientation of the segments with a relation similar to equation (7.45), but with other coefficient of proportionality, because we deal with non-equilibrium situation in this case. In this way one can correspond the two expressions for the stress tensor to each other and relate the introduced variables x_{il}^α and u_{ik}^α to the tensor of mean orientation of the segments

$$\langle e_i e_k \rangle - \frac{1}{3}\delta_{ik} = \frac{k}{z} \sum_{\alpha=1}^{N} \left(u_{ik}^\alpha + x_{ik}^\alpha - \frac{1}{3}\delta_{ik} \right), \tag{7.47}$$

where z is number of segments in a macromolecules and k is a numerical parameter.

The relaxation of orientation of segments is intrisinsigly included in the theory from the very beginning. Indeed, in Chapter 3, discussing the main assumption of the theory, we assumed that there is an underlying relaxation process, which is described by relaxation equation (3.14), that is

$$\frac{d\langle e_i e_k \rangle}{dt} = -\frac{1}{\tau}\left(\langle e_i e_k \rangle - \frac{1}{3}\delta_{ik} \right). \tag{7.48}$$

One can guess that this relaxation process describe the relaxation of mean orientation of segments with the relaxation time τ. It is remarkable that for the strongly entangled systems, there are relaxation variables u_{ik}^{α}, each of them relaxing with the rate $1/\tau$, as can be seen from equations (7.40). When the conformational variables are fixed, equations (7.40) and (7.47) are followed in linear approximation by relaxation equation

$$\frac{\mathrm{d}\langle e_i e_k \rangle}{\mathrm{d}t} = -\frac{1}{\tau}\left(\langle e_i e_k \rangle - \frac{1}{3}\delta_{ik}\right) + \frac{\pi^2}{9}\frac{k}{z}B\frac{\tau^*}{\tau}\gamma_{ik}, \qquad (7.49)$$

which is a generalisation of equation (7.48). In this case of large internal resistance, when $\psi \gg 1$, variables u_{ik}^{α} can be directly corresponded to the orientation of Kuhn segments. A value of the parameter k in equation (7.49) can be estimated due to the stress-optical law. It is known that for polymer systems the tensor of dielectric permittivity is proportional to the stress tensor, while coefficient of proportionality (the stress-optical coefficient) is universal for polymer of given chemical structure and can be found independently (see Chapter 10). One can admit optical anisotropy of the system is determined by the mean orientation of the segments. This allows us to determine the unknown parameter as $k = 3/5$.

In the steady state, one can find the mean orientation of segments from equation (7.49)

$$\langle e_i e_k \rangle = \frac{1}{3}\delta_{ik} + \frac{\pi^2}{15}\frac{1}{z}\tau^* B\gamma_{ik}. \qquad (7.50)$$

One can see that the velocity gradients directly affect the mean orientation of segments, while the effect of the disturbed conformation of macromolecules (the end-to-end distance) can be neglected here in comparison with the latter.

One can apply formula (7.44) to every subchain of the macromolecule, assuming that every subchain of the macromolecule is in the situation of equilibrium. Perhaps, it is possible to reach such a division of a macromolecule in subchains that the distribution of orientation of segment is the equilibrium one, though the entangled system is in deformed state, but the problem about distribution of orientation of the interacting, connected in chains, segments apparently is not solved yet.

7.5.2 Elementary Theory of Dielectric Relaxation

The relative permittivity tensor for the system ε_{ik} is defined (see, for example, Born and Wolf 1970; Landau et al. 1987) by the relation

$$\varepsilon_{ik}E_k = E_i + 4\pi P_i \qquad (7.51)$$

where E_k is the average electric field strength acting in the medium and P_i is the polarisation per unit volume of the system expressed in terms of the polarisabilities of the constituent elements of the system. According to conventional opinion (Riande and Siaz 1992; Adachi and Kotaka 1993; Watanabe 2001), polarisibility of a system of macromolecules, each of them is assumed to consist of z Kuhn segments with electric dipoles μ aligned in the direction of the segment axis (type-A dipole), is connected with mean orientation of segments $\langle e_k \rangle$ as

$$P_k = nz\mu\langle e_k \rangle, \tag{7.52}$$

where n is the number density of macromolecules. We consider here that the applied electric field E_i acts directly on separate dipoles, omitting discussion of the relationship between external and internal fields, which can be found elsewhere (Fröhlich 1958; Havriliak 1990).

The segments are connected in chains, but one can consider each segment to be in a similar situation and regard the mean orientation of segments as a mean orientation of a single segment. In linear approximation, one can imagine that each segment is rotating in the medium as in an isotropic liquid, so that one can adjust the result for the motion of an ellipsoid in the electric field E_i (Pokrovskii 1978, p. 80) and write the relaxation equation

$$\frac{\mathrm{d}\langle e_k \rangle}{\mathrm{d}t} = -\frac{1}{3\tau}\langle e_k \rangle + \frac{1}{6\tau}\frac{\mu}{T}(E_k - \langle e_k e_j \rangle E_j). \tag{7.53}$$

We have taken into account here, according to the conventional theory (Pokrovskii 1978), that the relaxation time of the first-order moment is three times bigger than the relaxation time of the moment of the second order in equation (7.49). A solution of equation (7.53) can be written in the form

$$\langle e_k \rangle = \frac{1}{6\tau}\frac{\mu}{T}\int_0^\infty \exp\left(-\frac{1}{3\tau}s\right)(E_k - \langle e_k e_j \rangle E_j)_{t-s}\,\mathrm{d}s. \tag{7.54}$$

It is known that, in equilibrium situation, the segment mean orientation is linked with the end-to-end distance as a characteristic of the whole macromolecule or separate subchain (see relation (7.44)), so that, if deviation from equilibrium is small, the relaxation of mean orientation of segments follows the conformational relaxation as it is described by the coarse-grained co-ordinates. The statement is assumingly valid for macromolecules in dilute solutions. The result for dielectric relaxation can be presented in terms of relaxation of the end-to-end distance or, considering the subchain model, in terms of the coarse-grained conformational co-ordinates (Zimm 1956). The situation appears to be more complicated in a condense system of strongly entangled macromolecules; the relaxation of the mean orientation of segments appears to be independent on conformational relaxation, and one has to consider relaxation equation (7.53) independently to obtain a result for dielectric relaxation of entangled systems.

One can consider the oscillating amplitude of the field $E \sim \exp(-i\omega)$, while assuming that the second-order moment does not depend on the field, to have from equation (7.54)

$$\langle e_k \rangle = \frac{1}{2}\frac{\mu}{T}\frac{E_k - \langle e_k e_j \rangle E_j}{1 - i3\tau\omega} \tag{7.55}$$

and, using the above equations, together with equation (7.50) and (7.51), to write the tensor of dielectric permittivity for a polar system under steady-state flow

$$\varepsilon_{ik} = 1 + 2\pi n z \frac{\mu^2}{T} \frac{1}{1 - i3\tau\omega} \left(\frac{2}{3}\delta_{ik} - \frac{\pi^2}{15}\frac{1}{z}\tau^* B\gamma_{ik} \right). \tag{7.56}$$

When velocity gradient is absent, the above formula looks like any other formula for dielectric permittivity for a system with the only relaxation process, which is used for estimation of dielectric relaxation time.

A frequency dependence of complex dielectric permittivity of polar polymer reveals two sets or two branches of relaxation processes (Adachi and Kotaka 1993), which correspond to the two branches of conformational relaxation, described in Section 4.2.4. The available empirical data on the molecular-weight dependencies are consistent with formulae (4.41) and (4.42). It was revealed for undiluted polyisoprene and poly(d, l-lactic acid) that the terminal (slow) dielectric relaxation time depends strongly on molecular weight of polymers (Adachi and Kotaka 1993; Ren et al. 2003). Two relaxation branches were discovered for cis-polyisoprene melts in experiments by Imanishi et al. (1988) and Fodor and Hill (1994). The fast relaxation times do not depend on the length of the macromolecule, while the slow relaxation times do. For the latter, Imanishi et al. (1988) have found

$$\tau_\alpha \sim \begin{cases} M^2, & M < M_e \\ M^{4.0\pm0.2}, & M > M_e \end{cases}.$$

This is exactly the molecular-weight dependence of conformational relaxation times of polymer in non-entangled state and for the region of diffusive mobility (see equation (4.41), weakly-entangled system).

Comparison of the dielectric and viscoelastic relaxation times, which, according to the above speculations, obey a simple relation $\tau_n = 3\tau$, has attracted special attention of scholars (Watanabe et al. 1996; Ren et al. 2003). According to Watanabe et al. (1996), the ratio of the two longest relaxation times from alternative measurements is 2–3 for dilute solutions of polyisobutilene, while it is close to unity for undiluted ($M \approx 10M_e$) solutions. For undiluted polyisoprene and poly(d, l-lactic acid), it was found (Ren et al. 2003) that the relaxation time for the dielectric normal mode coincides approximately with the terminal viscoelastic relaxation time. This evidence is consistent with the above speculations and confirms that both dielectric and stress relaxation are closely related to motion of separate Kuhn's segments. However, there is a need in a more detailed theory: experiment shows the existence of many relaxation times for both dielectric and viscoelastic relaxation, while the relaxation spectrum for the latter is much broader that for the former.

Chapter 8
Relaxation Processes in the Phenomenological Theory

Abstract This chapter contains an outline of the phenomenological theory of flow and deformation as a consequence of the conservation laws and the principles of non-equilibrium thermodynamics. We exploit the concept of internal thermodynamic variables that describe the deviation of a state of the system from equilibrium. This concept has a long history beginning with the pioneering work of Mandelstam and Leontovich (Zh. Exper. Theor. Fiziki 7:438–449, 1937) and has appeared to be useful in description of a deformable viscoelastic continuum (Coleman and Gurtin in J. Chem. Phys. 47:597–613, 1967; Pokrovskii in Polym. Mech. 6(5):693–702, 1970; Wood in The thermodynamics of fluid systems (Calendron, Oxford), 1975; Maugin in Thermomechanics of nonlinear irreversible processes (World Scientific, Singapore), 1999). The purpose of the chapter is to show how relaxation processes are included in the phenomenological theory of flow. The principles of the formulation of the phenomenological theory of viscoelasticity for any real materials are clear. In this sense, one can postulate a general phenomenological theory of viscoelasticity, which includes all known particular cases, among them those constitutive equations that are formulated on the basis of macromolecular dynamics in the previous and in the subsequent chapters. Principles of the theory, which allows classify the various phenomenological constitutive equations proposed for a viscoelastic medium, are discussed but no attempt is made to review available constitutive equations.

8.1 The Laws of Conservation of Momentum and Angular Momentum

The general form of transfer equations for a medium of arbitrary structure, including melts and solutions of polymers, is established on the basis of conservation laws of mass, momentum, angular momentum and energy (Landau and Lifshitz 1987a, Shliomis 1966).

V.N. Pokrovskii, *The Mesoscopic Theory of Polymer Dynamics*,
Springer Series in Chemical Physics 95,
DOI 10.1007/978-90-481-2231-8_8, © Springer Science+Business Media B.V. 2010

A continuous medium is characterised by its mean density, a function of co-ordinates and time

$$\rho = \rho(\boldsymbol{x}, t).$$

The motion of a continuous medium is described by its velocity vector $\boldsymbol{v}$, which is a certain mean macroscopic velocity and has three components – functions of the co-ordinates and time –

$$v_i = v_i(\boldsymbol{x}, t), \quad i = 1, 2, 3.$$

The law of conservation of mass can be written in the form of the continuity equation

$$\frac{\partial \rho}{\partial t} + \operatorname{div} \rho \boldsymbol{v} = 0 \tag{8.1}$$

where $\rho \boldsymbol{v}$ is the flux of mass density. Here and further on, the density of some quantity means the amount of this quantity in the volume unit of the medium.

The law of conservation of momentum can be written as

$$\frac{\partial(\rho v_i)}{\partial t} + \frac{\partial \Pi_{ik}}{\partial x_k} = \sigma_i$$

where $\Pi_{ik} = \rho v_i v_k - \sigma_{ik}$ is the tensor flux of momentum density, which consists of the convective flux and the stress tensor; σ_i is the density of the external forces that act on the fluid.

We can use the above relations to rewrite the law of conservation of momentum density in the form

$$\rho \left(\frac{\partial v_i}{\partial t} + v_j \frac{\partial v_i}{\partial x_j} \right) = \frac{\partial \sigma_{ik}}{\partial x_k} + \sigma_i. \tag{8.2}$$

The law of conservation of the angular momentum for the medium can be written under an assumption that there is an internal angular momentum, the density of which S_{ij} obeys the law

$$\frac{\partial S_{ij}}{\partial t} + \frac{\partial(v_l S_{ij})}{\partial x_l} + \frac{\partial g_{ijl}}{\partial x_l} = G_{ij} \tag{8.3}$$

where G_{ij} is the density of force torque which acts on the inner elements of the system, and g_{ijl} is the density of the non-convective flux of angular momentum.

No assumption was stated when equation (8.3) was written down. Without any assumption we can also formulate the law of conservation of the total angular momentum

$$\frac{\partial}{\partial t}(J_{ik} + S_{ik}) + \frac{\partial G_{ikl}}{\partial x_l} = N_{ik} + x_i \sigma_k - x_k \sigma_i \tag{8.4}$$

where $J_{ik} = -\rho(x_i v_k - x_k v_i)$ is the density of the external angular momentum, G_{ikl} is the flux of the total angular momentum, and N_{ik} is the torque density from the external volume forces.

The definition of the density of external angular momentum can be used to express, with the help of equations (8.1) and (8.2), the rate of change of the external angular momentum through the stress tensor σ_{ik}

$$\frac{\partial J_{ik}}{\partial t} + \frac{\partial}{\partial x_l}(v_l J_{ik}) = \sigma_{ki} - \sigma_{ik} - \frac{\partial}{\partial x_l}(x_i \sigma_{kl} - x_k \sigma_{il}) - x_i \sigma_k + x_k \sigma_i. \quad (8.5)$$

After summing equations (8.3) and (8.5), we obtain

$$\frac{\partial(J_{ik} + S_{ik})}{\partial t} + \frac{\partial}{\partial x_l}\Big[(J_{ik} + S_{ik})v_l + x_i \sigma_{kl} - x_k \sigma_{il}) + g_{ikl}\Big]$$
$$= G_{ik} - \sigma_{ik} + \sigma_{ki} - x_i \sigma_k + x_k \sigma_i.$$

The last equation can be compared to (8.4), which determined the relations

$$G_{ikl} = (J_{ik} + S_{ik})v_l + (x_i \sigma_{kl} - x_k \sigma_{il}) + g_{ikl},$$
$$G_{ik} = N_{ik} + \sigma_{ik} - \sigma_{ki}.$$

Then, equations (8.3) and (8.4) can be written in the form

$$\frac{\partial(J_{ik} + S_{ik})}{\partial t} + \frac{\partial}{\partial x_l}\Big[(J_{ik} + S_{ik})v_l + x_i \sigma_{kl} - x_k \sigma_{il} + g_{ikl}\Big]$$
$$= N_{ik} - x_i \sigma_k + x_k \sigma_i, \quad (8.6)$$
$$\frac{\partial S_{ik}}{\partial t} + \frac{\partial}{\partial x_l}(v_l S_{ik} + g_{ikl}) = N_{ik} + \sigma_{ik} - \sigma_{ki}. \quad (8.7)$$

The set of motion equations (8.1), (8.2), (8.6) and (8.7) contains the unknown quantities σ_{ik} and g_{ikl}, which will be determined later.

Before we come to further determinations of the unknown quantities, we shall estimate here the effect of the internal angular momentum on the motion of the liquid. Let a be the characteristic size of internal structural elements, then $S_{ik} \approx \rho a v$, $\sigma_{ik} \approx \eta v/a$, where η is the viscosity coefficient. An estimate of the characteristic relaxation time of the balance of the internal and external rotation follows from equation (8.7)

$$\tau \approx \rho \frac{a^2}{\eta}.$$

For a polymer solution, $\eta \approx 10^{-2}$ P s, $\rho \approx 1$ g/cm^3, and the size of macromolecular coil is $a \approx 10^{-5}$ cm, which allow us to estimate the relaxation time $\tau \approx 10^{-10}$ s. Processes with relaxation times so small are not essential when compared to other relaxation processes in polymer solutions.

For times which are much bigger than the relaxation time, the internal and external rotation are balanced, so equation (8.7) is followed by

$$\sigma_{ik} - \sigma_{ki} = -N_{ik}. \quad (8.8)$$

In this case, the stress tensor is non-symmetric, if there is an external force torque. The law of conservation of angular momentum follows from the law of conservation of momentum.

So, we shall further assume, that the internal and external rotation are balanced in polymer solutions and the stress tensor is symmetric, when there is no external force torque.

8.2 The Law of Conservation of Energy and the Balance of Entropy

We assume that there are no internal sources of energy in the liquid, so that the change of the energy density E is connected with fluxes through the surface of the volume. The law of the conservation of energy can be written in the form

$$\frac{\partial E}{\partial t} + \operatorname{div} \boldsymbol{q} = 0 \qquad (8.9)$$

where $\boldsymbol{q}$ is the flux of energy density.

The law of the conservation of energy is also known as the first principle of thermodynamics. To formulate the motion equation of a liquid, it is necessary to use the second principle of thermodynamics also, which can be written as the equation for the change of the entropy s for unit mass.

The balance equation for the entropy density has the form

$$\frac{\partial(\rho s)}{\partial t} + \operatorname{div}\left(\boldsymbol{v}\rho s + \boldsymbol{H}\right) = \Sigma$$

where $\boldsymbol{H}$ is the non-convective flux of entropy density, Σ is the non-negative quantity of emerging of entropy – entropy production. This equation can be rewritten in another form

$$\rho\left(\frac{\partial s}{\partial t} + v_i\frac{\partial s}{\partial x_i}\right) + \operatorname{div} \boldsymbol{H} = \Sigma. \qquad (8.10)$$

For systems, which are in a state of equilibrium, there is only convective transfer of entropy. This is the case of an ideal fluid, for which

$$\frac{\partial s}{\partial t} + v_i\frac{\partial s}{\partial x_i} = 0. \qquad (8.11)$$

The entropy arises in systems, which can be considered as systems that are locally in equilibrium. The increase of entropy can be connected with heat production in units of volume of fluid or, in other words, with the dissipation of energy Φ.

$$\rho\left(\frac{\partial s}{\partial t} + v_i\frac{\partial s}{\partial x_i}\right) = \frac{\Phi}{T}. \qquad (8.12)$$

Non-equality can be written for the case when we cannot consider the system as to be locally in equilibrium.

$$\rho\left(\frac{\partial s}{\partial t} + v_i \frac{\partial s}{\partial x_i}\right) > \frac{\Phi}{T}.$$

In this general case, equation (8.10) is valid.

So as there is a thermodynamic relation between entropy and internal energy, the unknown quantities q, H in equations (8.9) and (8.10) can be connected with each other and also can be determined production of entropy Σ through other quantities. The density of total energy E in equation (8.9) can be represented as a sum of the kinetic energy and the thermodynamic total energy of the resting volume

$$E = \frac{1}{2}\rho v^2 + E_0. \tag{8.13}$$

In equilibrium situations, the quantity E_0 is internal thermic energy $E_0 = \rho\varepsilon$, which is directly connected with entropy s per unit of mass by relation

$$dE_0 = \rho T\,ds + w\,d\rho \tag{8.14}$$

where ε is internal energy per unit of mass, $w = \varepsilon + p/\rho$ is the enthalpy for unit mass. This relation (8.14) is followed directly from known (Landau and Lifshitz, 1969) thermodynamic relation, which connects change of internal energy ε for unit mass with specific volume v and entropy s

$$d\varepsilon = T\,ds - p\,dv. \tag{8.15}$$

In non-equilibrium situations, local states of the deformed system are described by some internal thermodynamic variables ξ^α, where the label α is used for the number of a variable and its tensor indices. All the equilibrium values of the internal variables are functions of two thermodynamic variables: for example, density and entropy

$$\xi_e^\alpha = \xi_e^\alpha(s, \rho).$$

The deviation of the thermodynamic system from the equilibrium state is described by the differences $\xi^\alpha - \xi_e^\alpha$ which are noted as ξ^α henceforth.

In non-equilibrium situations, the quantity E_0 includes also potential of internal variables (Wood 1975, Maugin 1999, Pokrovskii 2005), so that the differential of this function has the form

$$dE_0 = \rho T\,ds + w\,d\rho + \Xi_\alpha\,d\xi^\alpha \tag{8.16}$$

where the thermodynamic force has appeared:

$$\Xi_\alpha = \left(\frac{\partial E_0}{\partial \xi^\alpha}\right)_{s,\rho} = -T\left(\frac{\partial(\rho s)}{\partial \xi^\alpha}\right)_{T,\rho} > 0.$$

The quantities T, w and Ξ_α are functions of the variables s, ρ, ξ^α. At equilibrium, when there is no external fields, all the $\xi^\alpha = 0$, while the quantities T and w take their equilibrium values. The external field affects the internal variables, which determine the state of the system.

Now, taking relations (8.13) and (8.16) into account, we are ready to write down the rate of change of the density of the total energy of the moving fluid

$$\frac{\partial E}{\partial t} = \rho v_i \frac{\partial v_i}{\partial t} + \rho T \frac{\partial s}{\partial t} + \left(w + \frac{v^2}{2} \right) \frac{\partial \rho}{\partial t} + \Xi_\alpha \frac{\partial \xi^\alpha}{\partial t}.$$

We can use equations (8.1), (8.2) and (8.10) to transform the above expression to the equation which has the form of the law of the conservation of energy

$$\frac{\partial E}{\partial t} + \frac{\partial}{\partial x_k} \left[\rho v_k \left(w + \frac{v^2}{2} \right) - v_i (\sigma_{ik} + p \delta_{ik}) + T H_k \right]$$

$$= T \Sigma - (\sigma_{ik} + p \delta_{ik}) \nu_{ik} + H_i \nabla_i T + \frac{\mathrm{d} \xi^\alpha}{\mathrm{d} t} \Xi_\alpha \qquad (8.17)$$

where, as before, $\nu_{ik} = \frac{\partial v_i}{\partial x_k}$ is a tensor of the velocity gradient.

Comparison of equations (8.9) and (8.17) determines

$$q_k = \rho v_k \left(w + \frac{v^2}{2} \right) - v_i (\sigma_{ik} + p \delta_{ik}) + T H_k,$$

$$\Sigma = \frac{1}{T} \left((\sigma_{ik} + p \delta_{ik}) \nu_{ik} - H_i \nabla_i T - \frac{\mathrm{d} \xi^\alpha}{\mathrm{d} t} \Xi_\alpha \right). \qquad (8.18)$$

Internal variables ξ^α are introduced in relation (8.16) formally. However, the success of the theory depends on the proper choice of the internal variables for the considered case. Consideration of models usually helps to recognise which quantities describe the deviation of the system from its equilibrium state and which can be used as internal variables. A set of internal variables were identified in Chapter 2 for dilute polymer solutions and in Chapter 7 for polymer melts.[1]

8.3 Thermodynamic Fluxes and Relaxation Processes

The laws of conservation determine the equations of fluid motion which, however, contain a few unknown quantities discussed below.

---

[1] Note, that a set of internal variables with labels, which take a continuous set of values, can be considered. Grmela (1985) and Jongschaap (1991) have generalised the above-written relations for this case. They showed that the values of the distribution function itself $W(\rho, t)$ in the problem of dynamics of dumbbells (see Appendix F), for example, can be considered as a set of internal variables, whereas the arguments of the function play the role of the label α with a continuous set of values ρ.

Expression for production of entropy (8.18) can be now compared with the general results of non-equilibrium thermodynamics, which are known for both non-stationary and stationary cases. It is obvious, that last term in the right-hand side of relation (8.18) corresponds to a non-stationary case and includes the equation of change of internal variables that is relaxation equation. The first two terms in formula (8.18) correspond to a stationary case and should be considered as the products of thermodynamic fluxes and thermodynamic forces (it is possible with any multipliers). When the internal variables are absent, we should write a relation between the fluxes and forces in the form

$$\sigma_{ij} + p\delta_{ij} = f_{ij}(\nu_{js}, \nabla_l T),$$
$$-H_i = H_i(\nu_{js}, \nabla_l T).$$

At small gradients, the right parts of these relations can be expanded in a power series. In linear approximation of a parity for the anisotropic environment one gets

$$\sigma_{ik} + p\delta_{ik} = \eta_{ikjs}\nu_{js} + L_{ikj}\nabla_j T,$$
$$-H_i = \bar{L}_{ijs}\gamma_{js} + A_{ij}\nabla_j T.$$

Here one can take advantage of the Onsager principle, that is equate factors at cross members.

In situations when internal relaxation processes cannot be neglected, it is necessary to include in consideration relaxation equation for internal variables, and we write down

$$\sigma_{ij} + p\delta_{ij} = f_{ij}(\nu_{js}, \nabla_l T, \xi^\gamma),$$
$$-H_i = H_i(\nu_{js}, \nabla_l T, \xi^\gamma), \qquad (8.19)$$
$$-\frac{d\xi^\alpha}{dt} = g^\alpha(\nu_{js}, \nabla_l T, \xi^\gamma).$$

One can note that the diffusion of the internal variables, i.e. the diffusion of structural elements at non-homogeneous distribution of the values of internal variables, is neglected here. Otherwise, the quantities $\frac{\partial^2 \xi^\alpha}{\partial x_i \partial x_l}$ must be added to the set of arguments of the right-hand side functions in (8.19). We shall not discuss this situation henceforth.

It is known, that thermodynamic forces are functions of internal variables (not speaking about other thermodynamic variables)

$$\Xi_\alpha = \Xi_\alpha(\xi^\gamma), \qquad (8.20)$$

so that relations (8.19) can be understood in such a way, that the quantities

$$\frac{1}{T}(\sigma_{ik} + p\delta_{ik}), \quad -\frac{1}{T}H_i, \quad -\frac{1}{T}\frac{d\xi^\alpha}{dt}$$

are functions of the thermodynamics forces

$$\nu_{ik}, \quad \nabla_i T, \quad \Xi_\alpha.$$

The application of general thermodynamic theory can be considered, first, in linear approximation. In practice, it is sufficient for the most part of applications. We can use our usual notations for symmetric and antisymmetric tensors of the velocity gradients

$$\gamma_{ij} = \frac{1}{2}(\nu_{ij} + \nu_{ji}), \qquad \omega_{ij} = \frac{1}{2}(\nu_{ij} + \nu_{ji})$$

and divide the stress tensor into symmetric and antisymmetric parts, to write the fluxes as quasi-linear function of the forces

$$\frac{1}{2}(\sigma_{ik} + \sigma_{ki} + 2p\delta_{ik}) = \eta_{ikjs}\gamma_{js} + K_{ikjs}\omega_{js} + L_{ikj}\nabla_j T + M_{ik\alpha}\xi_\alpha,$$

$$\frac{1}{2}(\sigma_{ik} - \sigma_{ki}) = \bar{K}_{ikjs}\gamma_{js} + N_{ikjs}\omega_{js} + \bar{C}_{ikj}\nabla_j T + \bar{D}_{ik\alpha}\xi_\alpha,$$

$$-H_i = \bar{L}_{ijs}\gamma_{js} + C_{ijs}\omega_{js} + A_{ij}\nabla_j T + G_{i\alpha}\xi_\alpha,$$ (8.21)

$$-\frac{d\xi^\alpha}{dt} = \bar{M}_{\alpha js}\gamma_{js} + D_{\alpha js}\omega_{js} + \bar{G}_{\alpha i}\nabla_i T + P_{\alpha\gamma}\xi_\gamma.$$

The matrix coefficients in (8.21) depend on the thermodynamic variables, which, in the case under discussion, are pressure p or density ρ (we can chose any of them, so as there exist an equation of state, connecting these variables), temperature T and internal variables ξ^α. The coefficients can be expanded into series near equilibrium values of internal variables. Zero-order terms of expansions of the components of the matrices in a series of powers of the internal variables are connected due to the Onsager principle (Landau and Lifshitz 1969) by some relations

$$\bar{K}^0_{jsik} = K^0_{ikjs}, \qquad \bar{C}^0_{ikj} = -C^0_{jik}, \qquad \bar{L}^0_{jik} = -L^0_{ikj}.$$

The bars over letters denote matrices, which are obtained from the original matrices (without bars) by simple transformation. Note once more that these relations are valid for equilibrium values. Further on we shall be interested in non-linear relations, so we consider all matrices to depend on the internal variables.

In the simple case when all the internal variables are scalar quantities, the state of the system is isotropic, all the matrix coefficients in (8.21) are expressed in unit matrices, and the relations (8.21) take the simpler form, which can be easily written for every given set of internal variables.

In rheological terms, equations (8.20) and (8.21) make up a set of constitutive relations of the system. Together with equations (8.1), (8.2) and (8.10), they determine the equations of motion of the system.

We should pay special attention to the last relation in (8.20), which is a relaxation equation for the variable ξ^α. One can find examples of relaxation equations in Section 2.7 for dilute solutions of polymers and in Chapter 7 for concentrated solutions and melts of polymers. The presence of internal variables and equations for their change are specific features of the liquids we consider in this monograph.

8.4 The Principle of Relativity for Slow Motions

The form of the above-written relations (8.21) can be specified more by applying some restrictions which follow from the assumption that the motion of structural elements of medium does not change very rapidly, so that the following relation is valid

$$\frac{u\rho a}{\eta} \ll 1. \tag{8.22}$$

Here, a is the characteristic size of the structural element, ρ is the density, which is approximately equal to 1 g/cm^3, η is the effective viscosity coefficient of the medium which is 10^{-2}–10 P s, and u is the characteristic velocity of motion of the particle, which is not more than the mean thermal velocity $(T/m)^{1/2}$. It is easy to see that, at room temperature and with the above values of the parameters, condition (8.22) is valid if $a \gg 10^{-7}$–10^{-6} cm.

As is well known, the equations of mechanics are covariant with the Galileo transform. This can be also said about relations (8.19) and (8.21). In the case, when the motions of the internal particles are slow (in the sense discussed above), we can state that a stronger principle is valid. It says that all the processes run in the same way and, consequently, should be described by similar equations in all the co-ordinate frames which are connected to each other by the transform

$$x_i = a_{ik}x'_k + c_i \tag{8.23}$$

where an orthogonal tensor a_{ik} and a vector c_i are arbitrary functions of time. In contrast to the Galileo principle, the above principle, which is also called the principle of material objectivity (Coleman and Nolle 1961), is valid for the cases when the forces of inertia can be neglected.

Let us consider the restriction imposed on the form of the transfer equations by the discussed principle. It is easy to see that, when transformation (8.23) is applied to the co-ordinates, the tensor of velocity gradients transforms as

$$\nu_{ik} = a_{il}a_{kj}\nu'_{lj} + \dot{a}_{il}a_{kl}.$$

The superscript point denotes differentiation with respect to time.

The symmetrical tensor of velocity gradient transforms as a tensor, which does not depend on time

$$\gamma_{ik} = a_{il}a_{kj}\gamma'_{lj}. \tag{8.24}$$

The antisymmetrical tensor transforms in the following way

$$\omega_{ik} = a_{il}a_{kj}\omega'_{lj} + a_{il}a_{kl}. \tag{8.25}$$

Let us now turn to the internal variables. We consider that one of the internal variables is a tensor of arbitrary rank and transforms as the co-ordinates do, that is, contravariantly

$$\xi_{ik\cdots l} = a_{ij}a_{ks}\cdots a_{ln}\xi'_{js\cdots n}.$$

Differentiating the tensor with respect to time, we find that

$$\frac{d\xi_{ik\cdots l}}{dt} = \dot{a}_{ij}a_{pj}\xi_{pk\cdots l} + \dot{a}_{kj}a_{pj}\xi_{ip\cdots l} + \cdots + \dot{a}_{lj}a_{pj}\xi_{ik\cdots p} + a_{ij}a_{ks}\cdots a_{ln}\frac{d\xi'_{js\cdots n}}{dt}.$$

We can define the expression $\dot{a}_{il}a_{kl}$ from (8.25) to rewrite the last expression in the form

$$\frac{d\xi_{ik\cdots l}}{dt} - \omega_{ip}\xi_{pk\cdots l} - \omega_{km}\xi_{im\cdots l} - \cdots - \omega_{ln}\xi_{ik\cdots n}$$

$$= a_{ij}a_{ks}\cdots a_{ln}\left(\frac{d\xi'_{is\cdots n}}{dt} - \omega'_{jq}\xi'_{qs\cdots n} - \omega'_{sq}\xi'_{jq\cdots n} - \cdots - \omega'_{nq}\xi'_{js\cdots q}\right).$$

We can see that the combination

$$\frac{D\xi_{ik\cdots l}}{Dt} = \frac{d\xi_{ik\cdots l}}{dt} - \omega_{ip}\xi_{pk\cdots l} - \omega_{km}\xi_{im\cdots l} - \cdots - \omega_{ln}\xi_{ik\cdots n} \qquad (8.26)$$

transforms as a tensor, which is independent of time. Expression (8.26) is called the Jaumann derivative of tensor $\xi_{ik\cdots l}$ with respect to time.

There are plenty of covariant derivatives of the tensor $\xi_{ik\cdots l}$ among which the Jaumann derivative has the simplest form. Indeed, expressions (8.24) and (8.25) are followed by the relation

$$\dot{a}_{il}a_{kl} = \omega_{ik} + \kappa\gamma_{ik} - a_{is}a_{kj}(\omega'_{sj} + \kappa\gamma'_{sj})$$

where κ is the arbitrary constant. We can use this relation to introduce derivatives, which are generalisations of (8.26).

Covariant tensors can be considered in a similar way.

8.5 Constitutive Relations for Non-Linear Viscoelastic Fluids

One can now return to the set of transfer equations (8.20) and (8.21), to which the discussed principle of covariance can be applied. The new form of the equations which is covariant under transformation (8.23) is written as follows

$$\frac{1}{2}(\sigma_{ik} + \sigma_{ki} + 2p\delta_{ik}) = \eta_{ikjs}\gamma_{js} + L_{ikj}\nabla_j T + M_{ik\alpha}\xi_\alpha,$$

$$\frac{1}{2}(\sigma_{ik} - \sigma_{ki}) = \bar{D}_{ik\alpha}\xi_\alpha,$$

$$-H_i = \bar{L}_{ijs}\gamma_{js} + A_{ij}\nabla_j T + G_{i\alpha}\xi_\alpha, \qquad (8.27)$$

$$\frac{D\xi^\alpha}{Dt} = \bar{M}_{\alpha js}\gamma_{js} + \bar{G}_{\alpha i}\nabla_i T + P_{\alpha\gamma}\xi_\gamma,$$

where the Jaumann derivative is noted as

$$\frac{D\xi^\alpha}{Dt} = \frac{d\xi^\alpha}{dt} + D_{\alpha js}\omega_{js}.$$

For every given tensor ξ^α, this expression can be compared to (8.26) which determines the matrix $D_{\alpha js}$ and, consequently, in linear approximation, matrix $\bar{D}_{ik\alpha}$ in relations (8.27).

The set of relations (8.27) determines the fluxes as quasi-linear functions of forces. The coefficients in (8.27) are unknown functions of the thermodynamic variables and internal variables. We should pay special attention to the fourth relation in (8.27) which is a relaxation equation for variable ξ^α. The viscoelastic behaviour of the system is determined essentially by the relaxation processes. If the relaxation processes are absent (all the $\xi^\alpha = 0$), equations (8.27) turn into constitutive equations for a viscous fluid.

One can see that the equations of motion for a viscoelastic fluid can always be written, when a set of internal relaxation variables is given, however, a set of internal variables cannot be determined in the frame of phenomenological theory and equations (8.27) cannot be specified any more without extra assumptions.

As an example, we shall consider a simpler case of the isothermal motion of a liquid without the external volume forces and without the external volume force torque, so that equations (8.27) acquire the form

$$\sigma_{ik} + p\delta_{ik} = \eta_{ikjs}\gamma_{js} + M_{ik\alpha}\xi_\alpha,$$

$$-\frac{D\xi^\alpha}{Dt} = \bar{M}_{\alpha js}\gamma_{js} + P_{\alpha\gamma}\xi_\gamma. \tag{8.28}$$

The set of internal variables ξ^γ is usually determined when considering a particular system in more detail. For concentrated solutions and melts of polymers, for example, a set of relaxation equation for internal variables were determined in the previous chapter. One can see that all the internal variables for the entangled systems are tensors of the second rank, while, to describe viscoelasticity of weakly entangled systems, one needs in a set of conformational variables x_{ik}^α which characterise the deviations of the form and size of macromolecular coils from the equilibrium values. To describe behaviour of strongly entangled systems, one needs both in the set of conformational variables and in the other set of orientational variables u_{ik}^α which are connected with the mean orientation of the segments of the macromolecules.

To simplify the situation, one can keep only one internal variables with the smallest number from each set, that is x_{ik}^1 and u_{ik}^1. It allows one to specify equations (8.28) for this case and to write a set of constitutive equations for two internal variables – the symmetric tensors of second rank. The particular case of general equations are equations (9.24)–(9.27) – constitutive equations for strongly entangled system of linear polymer. For a weakly entangled system, one can keep a single internal variable to obtain an approximate description of viscoelastic behaviour of the system. To consider this case in more details, we specify equations (8.28) for a single internal variable – the symmetric tensor of the second rank and rewrite relations (8.28) as follows

$$\sigma_{ik} + p\delta_{ik} = \eta_{ikjs}\gamma_{js} + M_{ikjs}\xi_{js},$$
$$-\frac{\mathrm{D}\xi_{ij}}{\mathrm{D}t} = \bar{M}_{ijls}\gamma_{ls} + P_{ijls}\xi_{ls}. \qquad (8.29)$$

In a more general case, we do not know the dependencies of the matrices in (8.29) on the internal variable, so one can rewrite relations (8.29) in the form

$$\sigma_{ik} + p\delta_{ik} = \eta_{ikjs}\gamma_{js} + \bar{\sigma}_{ik}(\xi_{pq}),$$
$$-\frac{\mathrm{D}\xi_{ij}}{\mathrm{D}t} = \bar{M}_{ijls}(\xi_{pq})\gamma_{ls} + \phi_{ij}(\xi_{pq}). \qquad (8.30)$$

The tensor functions in (8.30) can be written in a general form, according to the rules described, for example, for the arbitrary tensor function in the book by Green and Adkins (1960)

$$\bar{\sigma}_{ik} = \sigma_0\delta_{ik} + \sigma_1\xi_{ik} + \sigma_2\xi_{il}\xi_{lj},$$
$$\phi_{ik} = \phi_0\delta_{ik} + \phi_1\xi_{ik} + \phi_2\xi_{il}\xi_{lk} \qquad (8.31)$$

where the coefficients σ_i and ϕ_i ($i = 0, 1, 2$) are functions of the three invariants of the tensor ξ_{il}

$$I_1 = \sum_{i=1}^{3}\xi_{ii}, \qquad I_2 = \frac{1}{2}\sum_{i,j}(\xi_{ij}\xi_{ji} - \xi_{ii}\xi_{jj}), \qquad I_3 = |\xi_{ij}|.$$

The relations (8.30) and (8.31) make up a general form for a non-linear single-mode constitutive relation. To specify the constitutive equation for a given system, one ought to determine the unknown function in (8.31) relying on experimental evidence. A particular form of relation (8.30) and (8.31), called canonical form (Leonov 1992), embraces many empirical constitutive equations (Kwon and Leonov 1995). One can obtain the canonical form of constitutive relation (Leonov 1992), if one neglects the viscosity term in the stress tensor (8.30), which is quite reasonable for polymer melts, and put an additional assumption on matrix $\bar{M}$

$$\bar{M}_{ijls} = -\frac{1}{2}\kappa(\xi_{il}\delta_{js} + \xi_{js}\delta_{il} + \xi_{is}\delta_{jl} + \xi_{jl}\delta_{is})$$

where κ is a numerical parameter, usually taken as ± 1 or 0. One can look at equations (9.48) and (9.49) in the next chapter as particular case of system (8.30) and (8.31) as well.

Let us note that, according to Godunov and Romenskii (1972) and Leonov (1976), the internal variable ξ_{ij} can be considered to be a second-rank tensor of the recoverable strain. This statement changes neither definition (8.16) of the thermodynamic force Ξ_{ls}, nor the form of equations (8.30), but it does specify the form of the unknown functions and matrices in (8.30). In this case, a form of matrix M_{ikjs} can be determined, taking the relation between the stress tensor and the strain tensor (given by formula (B.7) of Appendix B) into account. Some simplification can be also achieved, because one has for an incompressible continuum an extra condition

$$|\xi_{ij}| = 1.$$

In this case, one has only two invariants of the internal tensor, which makes the general relations for the tensor functions simpler. However, it does not mean that the final relations will be simpler. We can see later (see Section 9.3.5) that there is a relation between the recoverable strain and the deformation of macromolecular coil (see formula (9.75)), so a transfer from one formalism to the other can be performed and the results of the two approaches can be compared.

8.6 Different Forms of Constitutive Relation

All the constitutive relation that we have discussed in this chapter include some relaxation equations for the internal tensor variables which ought to be considered to be independent variables in the system of equations for the dynamics of a viscoelastic liquid.

However, in the earlier times, the constitutive relation for a viscoelastic liquid were formulated when the equations for relaxation processes could not be written down in an explicit form. In these cases the constitutive relation was formulated as relation between the stress tensor and the kinetic characteristics of the deformation of the medium (Astarita and Marrucci 1974).

In this section, we shall show that the constitutive relation with internal variables is followed by two types of constitutive relations which do not include internal variables. For the sake of simplicity, we shall consider the simplest set of equations

$$\sigma_{ik} + p\delta_{ik} = 3\frac{\eta}{\tau}\left(\xi_{ik} - \frac{1}{3}\delta_{ik}\right), \tag{8.32}$$

$$\frac{\mathrm{d}\xi_{ik}}{\mathrm{d}t} - \nu_{ij}\xi_{jk} - \nu_{kj}\xi_{ji} = -\frac{1}{\tau}\left(\xi_{ik} - \frac{1}{3}\delta_{ik}\right) \tag{8.33}$$

where the coefficient of viscosity η and the time of relaxation τ are functions of the invariants of the internal tensor variable ξ_{ik}.

Indeed, we can obtain a relation between the stress tensor and the velocity gradient tensor if we exclude tensor ξ_{ij} from the set of equations (8.32)–(8.33). This can be done in two different ways.

Firstly, from equation (8.32), we can define the tensor ξ_{ij} which can be inserted into the second equation of (8.33). As a result, we obtain a differential equation for the extra stresses

$$\frac{\mathrm{d}\tau_{ik}}{\mathrm{d}t} - \nu_{ij}\tau_{jk} - \nu_{kj}\tau_{ji} = -\frac{1}{\tau}(\tau_{ik} - 2\eta\gamma_{ik}), \quad \tau_{ik} = \sigma_{ik} + p\delta_{ik}. \tag{8.34}$$

The quantities τ and η in equation (8.34) depend on the invariants of the tensor τ_{ik} in accordance with equation (8.32). We ought to note that the behaviour of a non-linear viscoelastic liquid in a non-steady state would be different, if a dependence of the material parameters τ and η on the tensor velocity gradients or on the stress tensor is assumed. This is a point which is sometimes ignored. In any case, if τ and η are constant, equation (8.34) belongs to the class of equations introduced and investigated by Oldroyd (1950).

The linear case of relation (8.34) is the Maxwell equation (see, Landau and Lifshitz 1987b, p. 36).

$$\frac{\mathrm{d}(\sigma_{ik} + p\delta_{ik})}{\mathrm{d}t} + \frac{1}{\tau}(\sigma_{ik} + p\delta_{ik}) = 2\frac{\eta}{\tau}\gamma_{ik} \tag{8.35}$$

where, as before, τ is the relaxation time, and η is the coefficient of shear viscosity. There are different generalisations of equations (8.34) and (8.35) (Astarita and Marrucci 1974).

On the other hand, we can imagine that a solution of equation (8.33) can be found. Below, the solution is written for uniform flow with accuracy up to the second-order terms with respect to the velocity gradient

$$\begin{aligned}
\xi_{ik} = {} & \frac{1}{3}\delta_{ik} + \int_0^\infty \exp\left(-\frac{s}{\tau}\right)\gamma_{ij}(t-s)\mathrm{d}s \\
& + \int_0^\infty \exp\left(-\frac{s}{\tau}\right)\int_0^\infty \exp\left(-\frac{u}{\tau}\right) \\
& \times \left[\nu_{ij}(t-s)\gamma_{jk}(t-s-u) - \nu_{kj}(t-s)\gamma_{ji}(t-s-u)\right]\mathrm{d}u\,\mathrm{d}s.
\end{aligned}$$

Then, the solutions should be inserted into equation (8.32), which determines the stress tensor as a function of the tensor of the velocity gradient in the previous moments of time. The linear term has the form

$$\sigma_{ij} = -p\delta_{ij} + 2\frac{\eta}{\tau}\int_0^\infty \exp\left(-\frac{s}{\tau}\right)\gamma_{ik}(t-s)\mathrm{d}s. \tag{8.36}$$

A generalisation of (8.36) for the case of many relaxation processes can easily be found. In the simplest case of uniform motion one has

$$\sigma_{ik} = -p\delta_{ik} + 2\int_0^\infty \eta(s)\gamma_{ik}(t-s)\mathrm{d}s. \tag{8.37}$$

The memory function $\eta(s)$ can be calculated if a set of internal variables are given.

In general case, the stress tensor ought to be written as

$$\sigma_{ik} + p\delta_{ik} = Y_{s=0}^{\infty}[\nu_{jk}(t-s), \gamma_{lm}(t)]. \tag{8.38}$$

Instead of velocity gradients, displacement gradients can be used in relation (8.38). In this form, relations of the kind (8.38) are established on the basis of the phenomenological theory of so-called simple materials (Coleman and Nolle 1961). To put the theory into practice, function (8.38) should be, for example, represented by an expansion into a series of repeated integrals, so that, in the simplest case, one has the first-order constitutive relation (8.37). Let us note that the first person who used functional relations of form (8.38) for the description of the behaviour of viscoelastic materials was Boltzmann (see Ferry 1980).

Another form of the relation for slow motions can be obtained from equation (8.38). We can expand the velocity gradients in (8.38) into series in powers of time near the moment t. The zeroth terms of the expansion determine a viscous liquid. The next terms take viscoelasticity into account. This description is local in time.

One can see that there are several forms for the representation of the constitutive relation of a viscoelastic liquid. Of course, we ought to say that all the types of constitutive relation we discussed in this section are equivalent. We can use any of them to describe the flow of viscoelastic liquids. However, the description of the flow of a liquid in terms of the internal variables allows one to use additional information, if it is available, about microstructure of the material, and, in fact, appears to be the simplest one for derivation and calculation. We believe that the form, which includes the internal variables, reflects a deeper penetration into the mechanisms of the viscoelastic behaviour of materials. From this point of view, all the representations of deformed material can be unified and classified.

Chapter 9
Non-Linear Effects of Viscoelasticity

Abstract Now we are in a position to formulate a system of constitutive equations for polymer systems on the basis of the mesoscopic approach, described in the previous chapters, to investigate non-linear behaviour of polymeric liquids. In the first section, the known results for dilute polymer solutions are described. The other sections contain derivation of constitutive equations for entangled systems, while the weakly ($2M_e < M < M^*$) and strongly ($2M_e < M^* < M$) entangled systems are considered separately. In the latter case, the reptation motion of macromolecules emerges. Though the reptation motion practically does not contributes to terminal properties of linear viscoelasticity of strongly entangled system, it has to be included in the consideration at higher velocity gradients to obtain the correct dependencies of non-linear effects on the length of the macromolecules. One can demonstrate how different non-linearities can be explained in terms of macromolecular dynamics. Simplifications of the many-modes constitutive equations will be considered in Sections 3. The simplest form of constitutive equations appears to be the well-known Vinogradov equation. Despite of essential simplification, the reduced forms of constitutive equation allow one to describe the non-linear effects for simple flows: shear and elongation.

9.1 Dilute Polymer Solutions

Comparison with experimental data demonstrates that the bead-spring model allows one to describe correctly linear viscoelastic behaviour of dilute polymer solutions in wide range of frequencies (see Section 6.2.2), if the effects of excluded volume, hydrodynamic interaction, and internal viscosity are taken into account. The validity of the theory for non-linear region is restricted by the terms of the second power with respect to velocity gradient for non-steady-state flow and by the terms of the third order for steady-state flow due to approximations taken in Chapter 2, when relaxation modes of macromolecule were being determined.

V.N. Pokrovskii, *The Mesoscopic Theory of Polymer Dynamics*,
Springer Series in Chemical Physics 95,
DOI 10.1007/978-90-481-2231-8_9, © Springer Science+Business Media B.V. 2010

9.1.1 Constitutive Relations

Many-Mode Approximation

The set of constitutive equations for the dilute polymer solution consists of the definition of the stress tensor (6.16), which is expressed in terms of the second-order moments of co-ordinates, and the set of relaxation equations (2.39) for the moments. The usage of a special notation for the ratio, namely

$$x_{ik}^{\nu} = \frac{\langle \rho_i^{\nu} \rho_k^{\nu} \rangle}{\langle \rho^{\nu} \rho^{\nu} \rangle_0} = \frac{2}{3} \mu \lambda_{\nu} \langle \rho_i^{\alpha} \rho_k^{\nu} \rangle,$$

allows us to write down these equations in more compact form

$$\sigma_{ik} = -p\delta_{ik} + 2\eta_s \gamma_{ik}$$

$$+ 3nT \sum_{\nu=1}^{N} \left[\frac{1}{1+\varphi_{\nu}} \left(x_{ik}^{\nu} - \frac{1}{3}\delta_{ik} \right) + \tau_{\nu}^{\perp} \varphi_{\nu} (\gamma_{ij} x_{jk}^{\nu} + \gamma_{kj} x_{ji}^{\nu}) \right], \quad (9.1)$$

$$\frac{dx_{ik}^{\nu}}{dt} - \omega_{ij} x_{jk}^{\nu} - \omega_{kj} x_{ji}^{\nu}$$

$$= -\frac{1}{\tau_{\nu}^{\parallel}} \left(x_{ik}^{\nu} - \frac{1}{3}\delta_{ik} \right) + (1-\varphi_{\nu})(\gamma_{ij} x_{jk}^{\nu} + \gamma_{kj} x_{ji}^{\nu}) \quad (9.2)$$

where $\tau_{\nu}^{\parallel} = (1+\varphi_{\nu})\tau_{\nu}^{\perp}$ and $\tau_{\alpha}^{\perp} = \tau_{\alpha}$ is an orientational relaxation time of the mode α of the macromolecular coils.

In some cases, if we consider, for example, the slow motion of a solution of very long macromolecules, the effect of internal viscosity is negligible, so that the set of constitutive equations can be simplified and written as

$$\sigma_{ik} = -p\delta_{ik} + 2\eta_s \gamma_{ik} + 3nT \sum_{\nu=1}^{N} \left(x_{ik}^{\nu} - \frac{1}{3}\delta_{ik} \right), \quad (9.3)$$

$$\frac{dx_{ik}^{\nu}}{dt} = -\frac{1}{\tau_{\nu}} \left(x_{ik}^{\nu} - \frac{1}{3}\delta_{ik} \right) + \nu_{ij} x_{jk}^{\nu} + \nu_{kj} x_{ji}^{\nu}. \quad (9.4)$$

For the steady-state case, both equations (9.1)–(9.2) and (9.3)–(9.4) are followed by the steady-state form of the stress tensor

$$\sigma_{ik} = -p\delta_{ik} + 2\eta_s \gamma_{ik} + 3nT \sum_{\nu=1}^{N} \tau_{\nu} \left(\nu_{ij} x_{jk}^{\nu} + \nu_{kj} x_{ji}^{\nu} \right). \quad (9.5)$$

This equation makes it possible to calculate stresses for low velocity gradients to within third-order terms in the velocity gradient if one knows the moments to within second-order terms in the velocity gradients. Due to the approximations, used earlier in Chapter 2, the results are applicable for small extensions of the macromolecular coil and hence for low velocity gradients: the results for the moments are valid to within second-order terms in the velocity gradients.

Single-Mode Approximation

We can see that a set of constitutive equations for dilute polymer solutions contains a large number of relaxation equations. It is clear that the relaxation processes with the largest relaxation times are essential to describe the slowly changing motion of solutions. In the simplest approximation, we can use the only relaxation variable, which can be the gyration tensor $\langle S_i S_j \rangle$, defined by (4.48), or we can assume the macromolecule to be schematised by a subchain model with two particles. The last case, which is considered in Appendix F in more detail, is a particular case of equations (9.3) and (9.4), which is followed at $N = 1, \lambda_1 = 2$,

$$\sigma_{ik} = -p\delta_{ik} + 2\eta_s \gamma_{ik} + 3\frac{\eta - \eta_s}{\tau}\left(\xi_{ik} - \frac{1}{3}\delta_{ik}\right), \tag{9.6}$$

$$\frac{d\xi_{ik}}{dt} = -\frac{1}{\tau}\left(\xi_{ik} - \frac{1}{3}\delta_{ik}\right) + \nu_{ij}\xi_{jk} + \nu_{kj}\xi_{ji}. \tag{9.7}$$

The following notations are used in the equations written above

$$\xi_{ik} = x_{ik}^1, \qquad \eta = \eta_s + \frac{3}{2}n\zeta.$$

Equations (9.6) and (9.7) make up the simplest set of constitutive equations for dilute polymer solutions, which, after excluding the internal variables ξ_{ij}, can be written in the form of a differential equation that has the form of the two-constant contra-variant equation investigated by Oldroyd (1950) (Section 8.6).

Note once again that equations (9.6) and (9.7) determines the stresses for the completely idealised macromolecules (without internal viscosity, hydrodynamic interaction and volume effects) in dilute solutions. To remedy the unrealistic behaviour of constitutive equations (9.6) and (9.7), some modifications were proposed (Rallison and Hinch 1988; Hinch 1994).

The expressions for the stress tensor together with the equations for the moments considered as additional variables, the continuity equation, and the equation of motion constitute the basis of the dynamics of dilute polymer solutions. This system of equations may be used to investigate the flow of dilute solutions in various experimental situations. Certain simple cases were examined in order to demonstrate applicability of the expressions obtained to dilute solutions, to indicate the range of their applicability, and to specify the expressions for quantities φ_ν, which were introduced previously as phenomenological constants.

9.1.2 Non-Linear Effects in Simple Shear Flow

We shall consider the case of shear stress when one of the components of the velocity gradient tensor has been specified and is constant, namely $\nu_{12} \neq 0$.

In order to achieve such a flow, it is necessary that the stresses applied to the system should be not only the shear stress σ_{12}, as in the case of a linear viscous liquid, but also normal stresses, so that the stress tensor is

$$\left\| \begin{array}{ccc} \sigma_{11} & \sigma_{12} & 0 \\ \sigma_{21} & \sigma_{22} & 0 \\ 0 & 0 & \sigma_{33} \end{array} \right\|.$$

The shear stress σ_{12} and the differences between the normal stresses $\sigma_{11}-\sigma_{22}$ and $\sigma_{22}-\sigma_{33}$ are usually measured in the experiment.

For the specified in this way motion, equation (9.2) defines, as was shown in Section 2.7.2, the non-zero components of the second-order moments

$$x_{11}^{\nu} = \frac{1}{3}\left[1 + (2 + \varphi_{\nu})(\tau_{\nu}\nu_{12})^2\right],$$

$$x_{22}^{\nu} = \frac{1}{3}\left[1 - \varphi_{\nu}(\tau_{\nu}\nu_{12})^2\right],$$

$$x_{33}^{\nu} = \frac{1}{3},$$
(9.8)

$$x_{12}^{\nu} = \frac{1}{3}\tau_{\nu}\nu_{12},$$

where, in accordance with (2.27) and (2.30), for high molecular weights

$$\varphi_{\alpha} = \varphi_1 \alpha^{\theta}, \quad \varphi_1 \sim M^{-\theta}, \quad 0 < \theta < 1,$$

$$\tau_{\alpha} = \tau_1 \alpha^{-z\nu}, \quad \tau_1 \sim M^{z\nu}, \quad 1.5 < z < 2.1.$$

According to the theoretical estimate of the exponent, $z\nu$ varies from 1.5 (non-draining Gaussian coil) to 2.11 (draining coil with volume interactions).

Then, equation (9.5) defines the non-zero components of the stress tensor, which makes it possible to formulate expressions for the shear viscosity and the differences between the normal stresses:

$$\eta = nT \sum_{\nu=1}^{N} \tau_{\nu} \left[1 - \varphi_{\nu}(\tau_{\nu}\nu_{12})^2\right],$$
(9.9)

$$\sigma_{11} - \sigma_{22} = nT \sum_{\nu=1}^{N} (\tau_{\nu}\nu_{12})^2,$$
(9.10)

$$\sigma_{22} - \sigma_{33} = 0.$$
(9.11)

It follows from equations (9.9) that the viscosity (or, what amounts to the same thing, the characteristic viscosity) is independent of the velocity gradient for flexible chains ($\varphi_1 = 0$). For chains with an internal viscosity, the viscosity diminishes with increase in the velocity gradient; the nature of the variation may be estimated. Using the known dependences of the relaxation times and coefficient of internal viscosity on molecular weight and mode label, one can obtain

$$\eta - \eta_0 \sim M^{3z\nu - \theta - 1}\nu_{12}^2.$$

From empirical equation (6.27), according to which $\theta = z\nu - 1$, the dependence of the viscosity on the molecular weight can be estimated as follows

$$\eta - \eta_0 \sim M^{2z\nu}\nu_{12}^2. \tag{9.12}$$

The dependence of the first difference of normal stresses on the molecular weight follows from equation (9.10)

$$\sigma_{11} - \sigma_{22} \sim M^{2z\nu - 1}. \tag{9.13}$$

In another way, this expression was obtained by Öttinger (1989b).

Experimental data and analysis of the shear-dependent viscosity for dilute solutions of polyethelene oxide in water can be found in work by Kalashnikov (1994). These data show that the deviations in reduced viscosity (9.12) at constant shear rate from initial (at $\nu_{12} \to 0$) values are the more, the more is the molecular weight of the polymer. Other empirical estimates of the exponent $z\nu$ in equation (9.12) for solutions in which the coils are nearly unperturbed yield the exponent $2z\nu \approx 3$ (Lohmander 1964; Tsvetkov et al. 1964).

We may note that it has been shown for the dumbbell (Altukhov 1986) (see Appendix F) that the combined allowance for the internal viscosity and the anisotropy of the hydrodynamic interaction leads to the appearance of a non-zero second difference between the normal stresses $\sigma_{22}-\sigma_{33}$. Since the internal viscosity may be estimated, for example, from dynamic measurements, this effect may serve to estimate the anisotropy of the hydrodynamic interaction in a molecular coil.

9.1.3 Non-Steady-State Shear Flow

In this section we shall continue to investigate shear motion, while, in contrast to the previous section, we shall assume that the velocity gradient depends on the time but, as before, does not depend on the space coordinate. We shall consider a simple case of ideally flexible chains, for which the stress tensor and relaxation equations are defined by equations (9.3) and (9.4).

For simple shear, equation (9.4) is followed by the set of equations for the components of the second-order moment

$$\frac{dx_{11}}{dt} = -\frac{1}{\tau}\left(x_{11} - \frac{1}{3}\right) + 2\nu_{12}x_{12},$$

$$\frac{dx_{22}}{dt} = -\frac{1}{\tau}\left(x_{22} - \frac{1}{3}\right),$$

$$\frac{dx_{33}}{dt} = -\frac{1}{\tau}\left(x_{33} - \frac{1}{3}\right), \tag{9.14}$$

$$\frac{dx_{12}}{dt} = -\frac{1}{\tau}x_{12} + \nu_{12}x_{22},$$

$$\frac{dx_{13}}{dt} = -\frac{1}{\tau}x_{13} + \nu_{12}x_{23},$$

$$\frac{dx_{23}}{dt} = -\frac{1}{\tau}x_{23}.$$

Here and henceforth in this section, the label of mode is omitted for simplicity. Consider the case when the motion with a given constant velocity gradient ν_{12} begins at time $t = 0$. Under the given initial conditions, the set of equations (9.14) has the solution

$$x_{11} = \frac{1}{3}\left[1 + 2\tau^2\left(1 - \frac{t}{\tau}\exp\left(-\frac{t}{\tau}\right) - \exp\left(-\frac{t}{\tau}\right)\right)\nu_{12}^2\right],$$

$$x_{12} = \frac{1}{3}\tau\left[1 - \exp\left(-\frac{t}{\tau}\right)\right]\nu_{12},$$

$$x_{22} = x_{33} = \frac{1}{3}; \qquad x_{13} = x_{23} = 0.$$

Now, we can determine, according to equation (9.3), the non-zero components of the stress tensor

$$\sigma_{11}(t) = -p + 2nT\sum_{\alpha=1}^{N}\tau_\alpha^2\left[1 - \frac{t}{\tau_\alpha}\exp\left(-\frac{t}{\tau_\alpha}\right) - \exp\left(-\frac{t}{\tau_\alpha}\right)\right]\nu_{12}^2,$$

$$\sigma_{12}(t) = \eta^0\nu_{12} + nT\sum_{\alpha=1}^{N}\tau_\alpha\left[1 - \exp\left(-\frac{t}{\tau_\alpha}\right)\right]\nu_{12}, \tag{9.15}$$

$$\sigma_{22}(t) = \sigma_{33}(t) = -p,$$

$$\sigma_{13}(t) = \sigma_{23}(t) = 0.$$

These expressions describe the establishment of stresses for given uniform shear motion.

9.1.4 Non-Linear Effects in Oscillatory Shear Motion

From the methodical point of view, it is very interesting to consider the non-linear terms of the stresses under oscillatory shear velocity gradients, which it is convenient to write in the complex form

$$\nu_{12} \sim e^{-i\omega t}.$$

Akers and Williams (1969), calculating non-linear terms, noticed that the stresses are real quantities, which are determined through real quantities. That is why we ought to remember that formulae always contain the real parts of complex quantities, so that one has to bear in mind that ν_{ik} means $\frac{1}{2}(\nu_{ik} + \bar{\nu}_{ik})$, where the operation of complex conjugation is denoted by the bar above the symbol.

Assuming that the flow is described by the set of equations (9.3) and (9.4), one can use equations (9.14) for arbitrary time dependence of velocity gradient, to obtain for oscillatory simple shear the solution in the form

$$x_{11} = \frac{1}{3}\left[1 + \frac{\tau^2|\nu_{12}|^2}{1+\omega^2\tau^2} + \frac{1}{2}\left(\frac{\tau^2\nu_{12}^2}{(1-i\omega t)(1-2i\omega t)} + \frac{\tau^2\bar{\nu}_{12}^2}{(1+i\omega t)(1+2i\omega t)}\right)\right],$$

$$x_{12} = \frac{1}{6}\left(\frac{\tau\nu_{12}}{1-i\omega t} + \frac{\tau\bar{\nu}_{12}}{1+i\omega t}\right),$$

$$x_{22} = x_{33} = \frac{1}{3},$$

$$x_{12} = x_{23} = 0.$$

Since all non-oscillatory terms in the solution are now omitted, we shall determine the non-zero components of the stress tensor according to equation (9.3)

$$\sigma_{12} = \eta_s \frac{1}{2}(\nu_{12} + \bar{\nu}_{12})$$

$$+ nT \sum_\alpha \left[\frac{\tau_\alpha}{1+\omega^2\tau_\alpha^2}\frac{1}{2}(\nu_{12}+\bar{\nu}_{12}) + \frac{\omega\tau_\alpha^2}{1+\omega^2\tau_\alpha^2}\frac{i}{2}(\nu_{12}-\bar{\nu}_{12})\right], \quad (9.16)$$

$$\sigma_{11} = -p + nT \sum_\alpha \left[\frac{\tau_\alpha^2|\nu_{12}|^2}{1+\omega^2\tau_\alpha^2}\right.$$

$$\left. + \frac{\tau_\alpha^2(1-2\omega^2\tau_\alpha^2)\frac{1}{2}(\nu_{12}^2+\bar{\nu}_{12}^2) + 3\omega\tau_\alpha^3\frac{i}{2}(\nu_{12}^2-\bar{\nu}_{12}^2)}{1+5\omega^2\tau_\alpha^2+4\omega^4\tau_\alpha^4}\right], \quad (9.17)$$

$$\sigma_{22} = \sigma_{33} = -p.$$

Expression (9.16) determines the non-linear dynamic viscosity and dynamical modulus. The first difference of the normal stresses $\sigma_{11}-\sigma_{22}$, defined by expressions (9.17), oscillate with a frequency twice that of velocity gradients (Akers and Williams 1969).

9.2 Many-Mode Description of Entangled Systems

9.2.1 Constitutive Relations

Stress Tensor

The expression for the stress tensor (6.7) allows us to investigate the non-linear with respect to velocity gradient effects. We use the normal co-ordinates (1.13) to write equation (6.7) in the form

$$\sigma_{ik} = -n(N+1)T\delta_{ik} + n\sum_{\nu}\left(2\mu T\lambda_\nu x_{ik}^\nu - T\delta_{ik} - \langle\rho_k^\nu T_i^\nu\rangle\right)$$

where $T_k^\nu = Q_{\nu\gamma}G_k^\gamma$ is the transformed force of the internal viscosity determined by equation (7.4). It is convenient to use the macroscopic mean quantities

$$x_{ij}^\alpha = \frac{2}{3}\mu\lambda_\alpha\langle\rho_i^\alpha\rho_j^\alpha\rangle, \qquad u_{ij}^\alpha = -\frac{1}{3T}\langle\rho_j^\alpha T_i^\alpha\rangle \qquad (9.18)$$

to write the stress tensor in the more compact form

$$\sigma_{ik} = -p\delta_{ik} + 3nT\sum_\nu\left(x_{ik}^\nu - \frac{1}{3}\delta_{ik} + u_{ik}^\nu\right). \qquad (9.19)$$

The pressure p includes both the partial pressure of the gas of Brownian particles $n(N+1)T$ and the partial pressure of the carrier "monomer" liquid. We shall assume that the viscosity of the "monomer" liquid can be neglected. The variables x_{ik}^ν in equation (9.19) characterise the mean size and shape of the macromolecular coils in a deformed system. The other variables u_{ik}^ν are associated mainly with orientation of small rigid parts of macromolecules (Kuhn segments). As a consequence of the mesoscopic approach, the stress tensor (9.19) of a system is determined as a sum of the contributions of all the macromolecules, which in this case can be expressed by simple multiplication by the number of macromolecules n. The macroscopic internal variables x_{ik}^ν and u_{ik}^ν can be found as solutions of relaxation equations which have been established in Chapter 7. However, there are two distinctive cases, which have to be considered separately.

Relaxation Equations for Weakly Entangled Systems

In the cases, when concentration of solution is not very high or melt consists of short macromolecules, the values of parameter χ ascend above the critical one $\chi^* \approx 0.1$. It is also implies the small values of the parameter ψ, that is

$$\chi^* < \chi < 0.3, \quad \psi \ll 1.$$

The internal variables are governed by relaxation equations (7.25) and (7.38) which are valid for the small mode numbers $\alpha^2 \ll 1/\chi$ and can be rewritten in the form

$$\frac{\mathrm{d}x_{ik}^\alpha}{\mathrm{d}t} - \nu_{ij}x_{jk}^\alpha - \nu_{kj}x_{ji}^\alpha$$

$$= -\frac{1}{2\tau_\alpha}\left(\left(x_{ij}^\alpha - \frac{1}{3}\delta_{ij}\right)b_{jk} + \left(x_{kj}^\alpha - \frac{1}{3}\delta_{kj}\right)b_{ji}\right), \tag{9.20}$$

$$\frac{\mathrm{d}u_{ik}^\alpha}{\mathrm{d}t} - \omega_{ij}u_{jk}^\alpha - \omega_{kj}u_{ji}^\alpha$$

$$= -\left(\frac{1}{\tau}\delta_{ij} + \frac{1}{2\tau_\alpha}b_{ij}\right)u_{jk}^\alpha - \frac{1}{\tau}\frac{E}{B}\left[\left(x_{ij}^\alpha - \frac{1}{3}\delta_{ij}\right)d_{jk} - 2B\tau_\alpha^\mathrm{R}x_{il}^\alpha\gamma_{lj}f_{jk}\right]$$

$$+ \gamma_{il}u_{lk}^\alpha \tag{9.21}$$

where the set of relaxation times is defined as

$$\tau, \quad \tau_\alpha = B\tau_\alpha^\mathrm{R}, \quad \tau_\alpha^\mathrm{R} = \frac{\tau^*}{\alpha^2}, \quad \alpha = 1, 2, \ldots \ll \left(\frac{1}{\chi}\right)^{1/2}. \tag{9.22}$$

In this case, the auxiliary quantities b_{ik}, d_{ik} and f_{ik} are defined, in limits of applicability of the equations ($\alpha^2 \ll 1/\chi$, $\psi \ll 1$), in terms of the anisotropy tensors β_{jl} and ϵ_{kl} as

$$b_{ik} = \beta_{ik}^{-1}, \quad d_{ik} = \beta_{ij}^{-1}\epsilon_{kj}, \quad f_{ik} = \epsilon_{ik}. \tag{9.23}$$

Relaxation Equations for Strongly Entangled Systems

This is a case, when the parameter χ has values less than a certain critical value χ^*, while additionally one requires that values of the parameter ψ are big, that is

$$\chi < \chi^* < 0.3, \quad \psi > 1.$$

The internal variables for this case are governed by relaxation equations (7.29) and (7.40) which are valid for the small mode numbers $\alpha^2 \ll \psi/\chi$. This is a case of very concentrated solutions and melts of polymers. Keeping only the zero-order terms with respect to the ratio B/E, the set of relaxation equations for the internal variables can be written in the simpler form

$$\frac{\mathrm{d}x_{ik}^\alpha}{\mathrm{d}t} - \nu_{ij}x_{jk}^\alpha - \nu_{kj}x_{ji}^\alpha$$

$$= -\frac{1}{2\tau_\alpha^\mathrm{rep}}\left(\left(x_{ij}^\alpha - \frac{1}{3}\delta_{ij}\right)b_{jk} + \left(x_{kj}^\alpha - \frac{1}{3}\delta_{kj}\right)b_{ji}\right), \tag{9.24}$$

$$\frac{\mathrm{d}u_{ik}^\alpha}{\mathrm{d}t} - \omega_{ij}u_{jk}^\alpha - \omega_{kj}u_{ji}^\alpha$$

$$= -\frac{1}{\tau}u_{ik}^\alpha - \frac{1}{\tau}\left(x_{ik}^\alpha - \frac{1}{3}\delta_{ik} - 2B\tau_\alpha^\mathrm{R}x_{il}^\alpha\gamma_{lj}f_{jk}\right) + \frac{B}{E}e_{ij}\gamma_{jl}u_{lk}^\alpha, \tag{9.25}$$

where the set of relaxation times is defined as

$$\tau, \quad \tau_\alpha^{\text{rep}} = \frac{\pi^2}{\chi} \tau_\alpha^{\text{R}}, \quad \tau_\alpha^{\text{R}} = \frac{\tau^*}{\alpha^2}, \quad \alpha = 1, 2, \ldots \ll \left(\frac{\psi}{\chi}\right)^{1/2}. \tag{9.26}$$

The auxiliary quantities b_{ik}, e_{ik} and f_{ik} are introduced in Chapter 7 to take into account the effect of the induced anisotropy of medium on the dynamics of a single macromolecule in the system. In limits of applicability of the above equations ($\alpha^2 \ll \psi/\chi$, $\psi \gg 1$), the quantities are defined in terms of the anisotropy tensors β_{jl} and ϵ_{kl} as

$$b_{ik} = \epsilon_{ik}^{-1}, \quad e_{ik} = \epsilon_{ij}^{-1}\beta_{jk}, \quad f_{ik} = \epsilon_{ij}^{-1}\beta_{jl}\epsilon_{kl}. \tag{9.27}$$

Let us remind that equation (9.24), describing the relaxation of macromolecular conformation, can be considered only as an assumed results of accurate derivation of the relaxation equation from the macromolecular dynamics.

Tensor of Anisotropy

Thus, two sets of constitutive relations are formulated. The systems of equations both (9.19)–(9.22), applicable to the weakly entangled systems, and (9.19) and (9.24)–(9.26), applicable to the strongly entangled systems, include, through equations (9.23) and (9.27), the tensors of global anisotropy

$$\beta_{ik} = (\delta_{jk} + \kappa a_{ll}\delta_{jk} + 3\beta a'_{jk})^{-1}, \quad \epsilon_{ik} = (\delta_{jk} + \nu a_{ll}\delta_{jk} + 3\epsilon a'_{jk})^{-1},$$

$$a_{ij} = \sum_\nu \left(x_{ij}^\nu - \frac{1}{3}\delta_{ij} + u_{ij}^\nu\right), \quad a'_{ij} = a_{ij} - \frac{1}{3}a_{ll}\delta_{ij}.$$

The set of equations both for weakly and strongly entangled systems contains only two relaxation branches and describe viscoelastic behaviour in the region of small frequencies (One can look at Fig. 17 to be convinced that essential contributions to the modulus are given by the first and the second branches in the region of small frequencies). These sets of equations are the basic constitutive relations which allow us to develop a reliable theory of non-linear effects in viscoelasticity of non-dilute polymer systems following the works by Pokrovskii and Pyshnograi (1990, 1991) and Pyshnograi (1994, 1996).

9.2.2 Linear Approximation

To calculate characteristics of linear viscoelasticity, one can consider linear approximation of constitutive relations derived in the previous section. The expression (9.19) for stress tensor has linear form in internal variables x_{ik}^ν and u_{ik}^ν, so that one has to separate linear terms in relaxation equations for the internal variables. This has to be considered separately for weakly and strongly entangled system.

Weakly Entangled Systems

In the cases, when

$$\chi^* < \chi < 0.3, \quad \psi < 1,$$

the relaxation equations (9.20) and (9.21) reduce to the simpler form

$$\frac{\mathrm{d}x_{ik}^\alpha}{\mathrm{d}t} = -\frac{1}{\tau_\alpha}\left(x_{ik}^\alpha - \frac{1}{3}\delta_{ik}\right) + \frac{2}{3}\gamma_{ik},$$

$$\frac{\mathrm{d}u_{ik}^\alpha}{\mathrm{d}t} = -\frac{1}{\tau_\alpha^*}u_{ik}^\alpha - \frac{1}{\tau}\psi\left(x_{ik}^\alpha - \frac{1}{3}\delta_{ik} - \frac{2}{3}B\tau_\alpha^{\mathrm{R}}\gamma_{ik}\right)$$

(9.28)

where the set of relaxation times is defined as

$$\tau, \quad \tau_\alpha^* = \frac{2\tau\tau_\alpha}{\tau + 2\tau_\alpha}, \quad \tau_\alpha = B\,\tau_\alpha^{\mathrm{R}}, \quad \tau_\alpha^{\mathrm{R}} = \frac{\tau^*}{\alpha^2}, \quad \alpha = 1, 2, \ldots \ll \left(\frac{1}{\chi}\right)^{1/2}.$$

Equations (9.28) have the following solutions for oscillatory motion

$$x_{ik}^\alpha = \frac{1}{3}\delta_{ik} + \frac{2}{3}\frac{\tau_\alpha}{1 - i\omega\tau_\alpha}\gamma_{ik},$$

$$u_{ik}^\alpha = \frac{2}{3}\frac{1}{\tau}\psi\left(B\,\tau_\alpha^{\mathrm{R}} - \frac{\tau_\alpha}{1 - i\omega\tau_\alpha}\right)\frac{\tau_\alpha^*}{1 - i\omega\tau_\alpha^*}\gamma_{ik}.$$

Then, one can make use of the expression (9.19) for the stress tensor to obtain the coefficient of dynamic modulus

$$G(\omega) = nT \sum_\alpha^{1/\sqrt{\chi}} \left[\frac{i\omega\tau_\alpha}{1 - i\omega\tau_\alpha} + \frac{1}{\tau}\psi\left(B\,\tau_\alpha^{\mathrm{R}} - \frac{i\omega\tau_\alpha}{1 - i\omega\tau_\alpha}\right)\frac{\tau_\alpha^*}{1 - i\omega\tau_\alpha^*}\right].$$

This expression can be written in standard form

$$G(\omega) = nT \times \sum_\alpha^{1/\sqrt{\chi}} \left[\left(1 + \psi\frac{\tau_\alpha\tau_\alpha^*}{\tau(\tau_\alpha^* - \tau_\alpha)}\right)\frac{-i\omega\tau_\alpha}{1 - i\omega\tau_\alpha}\right.$$

$$\left. + \frac{1}{\tau}\psi\left(B\,\tau_\alpha^{\mathrm{R}} - \frac{\tau_\alpha\tau_\alpha^*}{\tau_\alpha^* - \tau_\alpha}\right)\frac{-i\omega\tau_\alpha^*}{1 - i\omega\tau_\alpha^*}\right]. \quad (9.29)$$

The terms of the first and the second orders give the coefficients of viscosity and elasticity

$$\eta = nT \sum_\alpha^{1/\sqrt{\chi}} \left(\tau_\alpha - \psi\frac{\tau_\alpha\tau_\alpha^*}{\tau}\right) \approx \frac{\pi^2}{6}nT\tau^* B, \quad (9.30)$$

$$\nu = nT \sum_\alpha^{1/\sqrt{\chi}} \left(\tau_\alpha^2 - \psi\frac{\tau_\alpha\tau_\alpha^*}{\tau}(\tau_\alpha^* + \tau_\alpha)\right) \approx \frac{\pi^4}{90}nT(\tau^* B)^2. \quad (9.31)$$

Value of the dynamic modulus on the plateau can be found as $G_e = \lim_{\omega \to \infty} G(\omega)$ which gives

$$G_e = nT \sum_{\alpha}^{1/\sqrt{\chi}} 1 \approx nT\chi^{-\frac{1}{2}}. \qquad (9.32)$$

It is natural that estimates (9.30)–(9.32) practically coincide with estimates (6.39) and (6.40), at $\chi \ll 1$, for corresponding quantities for a system of macromolecules in viscoelastic liquid.

Strongly Entangled Systems

In the cases, when

$$\chi < \chi^* \approx 0.1, \quad \psi > 1,$$

the internal variables are governed by relaxation equations (9.24) and (9.25) which are valid for the small mode numbers $\alpha^2 \ll \psi/\chi$. Keeping only the zero-order terms with respect to velocity gradient, the set of relaxation equations for the internal variables can be written in the simpler form

$$\frac{dx_{ik}^\alpha}{dt} = -\frac{1}{\tau_\alpha^{\text{rep}}}\left(x_{ik}^\alpha - \frac{1}{3}\delta_{ik}\right) + \frac{2}{3}\gamma_{ik},$$

$$\frac{du_{ik}^\alpha}{dt} = -\frac{1}{\tau}u_{ik}^\alpha - \frac{1}{\tau}\left(x_{ik}^\alpha - \frac{1}{3}\delta_{ik} - \frac{2}{3}B\tau_\alpha^{\text{R}}\gamma_{ik}\right) \qquad (9.33)$$

where the set of relaxation times is defined as

$$\tau, \quad \tau_\alpha^{\text{rep}} = \frac{\pi^2}{\chi}\tau_\alpha^{\text{R}}, \quad \tau_\alpha^{\text{R}} = \frac{\tau^*}{\alpha^2}, \quad \alpha = 1, 2, \dots \ll \left(\frac{\psi}{\chi}\right)^{1/2}.$$

Equations (9.33) have the following solutions for oscillatory motion

$$x_{ik}^\alpha = \frac{1}{3}\delta_{ik} + \frac{2}{3}\frac{\tau_\alpha^{\text{rep}}}{1 - i\omega\tau_\alpha^{\text{rep}}}\gamma_{ik},$$

$$u_{ik}^\alpha = \frac{2}{3}\left(B\tau_\alpha^{\text{R}} - \frac{\tau_\alpha^{\text{rep}}}{1 - i\omega\tau_\alpha^{\text{rep}}}\right)\frac{1}{1 - i\omega\tau}\gamma_{ik}.$$

Then, one can make use of the expression (9.19) for the stress tensor to obtain the dynamic modulus

$$G(\omega) = nT \sum_{\alpha=1}^{\pi/\chi}\left[\frac{-i\omega\tau_\alpha^{\text{rep}}}{1 - i\omega\tau_\alpha^{\text{rep}}} + \left(B\tau_\alpha^{\text{R}} - \frac{\tau_\alpha^{\text{rep}}}{1 - i\omega\tau_\alpha^{\text{rep}}}\right)\frac{-i\omega}{1 - i\omega\tau}\right]. \qquad (9.34)$$

This expression, after some transformations, can be written in the standard form

$$G(\omega) = nT \sum_{\alpha=1}^{\pi/\chi} \left[\left(1 + \frac{\tau_\alpha^{\mathrm{rep}}}{\tau - \tau_\alpha^{\mathrm{rep}}} \right) \frac{-i\omega\tau_\alpha^{\mathrm{rep}}}{1 - i\omega\tau_\alpha^{\mathrm{rep}}} + \left(\frac{B\,\tau_\alpha^{\mathrm{R}}}{\tau} - \frac{\tau_\alpha^{\mathrm{rep}}}{\tau - \tau_\alpha^{\mathrm{rep}}} \right) \frac{-i\omega\tau}{1 - i\omega\tau} \right].$$

$$(9.35)$$

This equation, also as equation (6.49) gives description of the frequency dependency of dynamic modulus at low frequencies (the terminal zone). Both in equation (6.49) and (9.35), the second terms present the contribution from the orientational relaxation branch, while the first ones present the contribution from the conformational relaxation due to the different mechanisms: diffusive and reptational.

The terms of the first and the second orders in expansion of expression (9.34) or (9.35) in powers of $-i\omega$ determine the coefficients of viscosity and elasticity

$$\eta = nT \sum_{\alpha=1}^{\pi/\chi} B\,\tau_\alpha^{\mathrm{R}} = \frac{\pi^2}{6} nT\tau^* B,$$

$$(9.36)$$

$$\nu = nT \sum_{\alpha=1}^{\pi/\chi} (B\,\tau\tau_\alpha^{\mathrm{R}} - \tau\tau_\alpha^{\mathrm{rep}}) = nT \left(\frac{\pi^2}{3}(B\tau^*)^2\chi - \frac{\pi^4}{3}B(\tau^*)^2 \right).$$

One can see that the last terms in the last relations can be omitted in comparison with the other, so that this equations reduce to equations (6.52), that is

$$\eta = \frac{\pi^2}{6} nT\tau^* B, \qquad \nu = \frac{\pi^2}{3} nT\,(B\tau^*)^2\chi.$$

$$(9.37)$$

Value of the dynamic modulus on the plateau can be found as $G_e = \lim_{\omega\to\infty} G(\omega)$ which gives

$$G_e = nT \sum_{\alpha=1}^{\pi/\chi} \left(1 + \frac{B\,\tau_\alpha^{\mathrm{R}}}{\tau} \right) \approx nT \left(\frac{\pi}{\chi} + \frac{\pi^2}{12}\frac{1}{\chi} \right).$$

$$(9.38)$$

The contribution from the first term (reptation branch) has the same order of magnitude as the contribution from the second term at very high frequencies. However, one has to take into account that, due to distribution of relaxation times, the limit value of the first term is reached at higher frequencies than the limit value of the second term. At lower frequencies the plateau value of the dynamic modulus is determined by the second term and coincides with expression (6.52).

One can see that introduction of the reptation mechanism of conformational relaxation, instead of diffusive mechanism, does not affect the values of the terminal quantities, but, one can expect, improves the situation in the region of the minima of the loss modulus G'': reptation branch fill the gap between the orientational and the second conformational branches of relaxation times. Thus, the description with help of two relaxation branches is valid in the terminal zone and for higher frequencies close to it.

9.2.3 Steady-State Simple Shear Flow

To demonstrate the consistency of constitutive relations with experimental evidence for entangled systems, some particular cases, when the velocity gradients are known and can be assumed to be independent of time, have been investigated (Pyshnograi and Pokrovskii, 1988; Pyshnograi, 1994, 1996; Altukhov and Pyshnograi, 1995, 1996). We shall consider here steady-state shear flow of both weakly entangled system and strongly entangled system. The stress tensor is given by equation (9.19), that is

$$\sigma_{ik} = -p\delta_{ik} + 3nT \sum_{\nu} \left(x_{ik}^{\nu} - \frac{1}{3}\delta_{ik} + u_{ik}^{\nu} \right).$$

For the case of small velocity gradients, the variables x_{ik}^{α} and u_{jk}^{α} can be found in the form of an expansion in powers of velocity gradients. The first terms are defined by equations (7.28) and (7.39) for the case of weakly entangled systems $(\chi > \chi^* \approx 0.1)$ and by equations (7.32) and (7.43) for the case of strongly entangled systems $(\chi < \chi^* \approx 0.1)$.

Further on, we shall consider the case of shear stress when one of the components of the velocity gradient tensor has been specified and is constant, namely $\nu_{12} \neq 0$. This situation occurs in experimental studies of polymer solutions (Ferry 1980). In order to achieve such a flow, it is necessary that the stresses applied to the system should be not only the shear stress σ_{12}, as in the case of a linear viscous liquid, but also normal stresses, so that the stress tensor is

$$\left\| \begin{array}{ccc} \sigma_{11} & \sigma_{12} & 0 \\ \sigma_{21} & \sigma_{22} & 0 \\ 0 & 0 & \sigma_{33} \end{array} \right\|.$$

The shear stress σ_{12} and the differences between the normal stresses $\sigma_{11} - \sigma_{22}$ and $\sigma_{22} - \sigma_{33}$ are usually measured in the experiment (Meissner et al. 1989). The results of calculation of the stresses up to the third-order terms with respect to the velocity gradient will be demonstrated further on. For simplicity, we shall neglect the effect of anisotropy of the environment when the case of strongly entangled systems will be considered.

Shear Viscosity

In steady-state shear, when the only component of the velocity gradient tensor differs from zero is ν_{12}, equation (9.19) is followed by

$$\sigma_{12}^0 = \eta_0 \nu_{12}, \qquad \eta_0 = \frac{\pi^2}{6} nTB\tau^*, \qquad \chi < 0.5. \tag{9.39}$$

The third-order terms in shear stress give us the expression for the effective shear viscosity

$$\eta = \eta_0 \times \begin{cases} 1 - \left(\frac{2\pi^4}{315}\psi + \frac{2\pi^2}{15}\chi + \frac{52\pi^4}{4725}\beta + \frac{4\pi^4}{945}\kappa\right)(B\tau^*\nu_{12})^2, & \chi > \chi^*, \\ 1 - \left(4\chi^2 - \frac{\pi^2}{15}\frac{\chi}{B} - \frac{2\pi^6}{315}\frac{1}{(\chi B)^2}\right)(B\tau^*\nu_{12})^2, & \chi < \chi^*. \end{cases} \quad (9.40)$$

One can see that two factors lead to non-linear effects in shear, namely, the relaxation response of the surrounding (χ and ψ) and the effects associated with the change in dimensions and the shape of the macromolecular coils (β and κ). Though comparative influence of these effects does not investigated enough, one can suggest that the influence of χ and ψ is small as compared with the influence of the other parameters at $\chi > \chi^*$, while at $\chi < \chi^*$ the change of the environment (β and κ) can be neglected. The first term in the parentheses in (9.40) at $\chi < \chi^*$ dominates for the long macromolecules. The above relation, in agreement with the experimental evidence (Schreiber et al. 1963; Ito and Shishido 1972), shows that the deviation of the behaviour of the system from Newtonian starts at the shear stress which is the lesser, the larger the length of the macromolecule is. For very long macromolecules the deviation does not depend on the length of the macromolecules.

Normal Stresses

Calculation of terms of the second order reveals that normal stresses are

$$\sigma_{11} + p = \begin{cases} nT\left(\frac{\pi^4}{45} + \frac{\pi^2}{6}\chi - \frac{\pi^4}{90}\beta\right)(B\tau^*\nu_{12})^2, & \chi > \chi^*, \\ nT\left(\frac{\pi^6}{90}\frac{1}{B\chi} + \frac{\pi^2}{6}\chi\frac{B}{E} + \frac{\pi^2}{3}\chi\right)(B\tau^*\nu_{12})^2, & \chi < \chi^*, \end{cases}$$

$$\sigma_{22} + p = \begin{cases} -nT\left(\frac{\pi^2}{6}\chi + \frac{\pi^4}{90}\beta\right)(B\tau^*\nu_{12})^2, & \chi > \chi^*, \quad (9.41) \\ nT\left(\frac{\pi^6}{90}\frac{1}{B\chi} + \frac{\pi^2}{6}\chi\frac{B}{E} - \frac{\pi^2}{3}\chi\right)(B\tau^*\nu_{12})^2, & \chi < \chi^*, \end{cases}$$

$$\sigma_{33} + p = 0.$$

Specific characteristics of viscoelastic medium are differences of the normal stresses

$$\sigma_{11} - \sigma_{22} = \begin{cases} nT\left(\frac{\pi^4}{45} + \frac{\pi^2}{3}\chi\right)(B\tau^*\nu_{12})^2, & \chi > \chi^*, \\ nT\frac{2\pi^2}{3}\chi(B\tau^*\nu_{12})^2, & \chi < \chi^*, \end{cases}$$

$$\sigma_{22} - \sigma_{33} = \begin{cases} -nT\left(\frac{\pi^2}{6}\chi + \frac{\pi^4}{90}\beta\right)(B\tau^*\nu_{12})^2, & \chi > \chi^*, \\ nT\left(\frac{\pi^6}{90}\frac{1}{B\chi} + \frac{\pi^2}{6}\chi\frac{B}{E} - \frac{\pi^2}{3}\chi\right)(B\tau^*\nu_{12})^2, & \chi < \chi^*. \end{cases} \quad (9.42)$$

The ratio of the first difference of the normal stresses $\sigma_{11} - \sigma_{22}$ to the square of the shear stress is an important characteristic quantity. The expressions for the steady-state modulus in the region of low velocity gradients are defined as

$$\frac{2\sigma_{12}^2}{\sigma_{11} - \sigma_{22}} = \begin{cases} \frac{5}{2}nT\left(1 - \frac{15}{\pi^2}\chi\right), & \chi > \chi^*, \\ \frac{\pi^2}{12}nT\chi^{-1}, & \chi < \chi^*. \end{cases} \tag{9.43}$$

For the weakly entangled system, the steady-state modulus depends on the molecular weight of polymer as M^{-1}, while for strongly entangled system, the steady-state modulus does not depend on the molecular weight of polymer, which is consistent with typical experimental data for concentrated polymer systems (Graessley 1974). The expression for the modulus is exactly the same as for the plateau value of the dynamic modulus (equations (6.52) and (6.58))

Expressions (9.42) lead to the following relation for the ratio of the normal stresses differences

$$\frac{\sigma_{22} - \sigma_{33}}{\sigma_{11} - \sigma_{22}} = \begin{cases} -\frac{15}{2\pi^2}\chi - \frac{1}{2}\beta, & \chi > \chi^*, \\ \frac{\pi^4}{60}\frac{1}{\chi^2 B} + \frac{B}{4E} - \frac{1}{2}, & \chi < \chi^*. \end{cases} \tag{9.44}$$

This ratio depends on the molecular weight of polymer M and predicted to be negative for typical values of parameters. For the strongly entangled systems, according to equations (3.30), $B \sim M^{\delta}$, $\chi \sim M^{-1}$, so that the ratio (9.44) approaches $-1/2$ for very long macromolecules. According to experimental evidence, the second difference of the normal stresses $\sigma_{22} - \sigma_{33}$ is negative and less than the first. The ratio has generally been reported to be in the range of -0.15 to -0.3 (Brown et al. 1995), though Faitelson (1995) found the values of the quantity for polybutadiene with narrow molecular-weight distribution to lie in the range from -0.3 to -0.45. It would be desirable to measure the ratio of differences of the normal stresses for well-characterised systems to test the validity of relation (9.44).

9.3 Single-Mode Description of Entangled System

Notwithstanding the simplifying assumptions in the dynamics of macromolecules, the sets of constitutive relations derived in Section 9.2.1 for polymer systems, are rather cumbersome. Now, it is expedient to employ additional assumptions to obtain reasonable approximations to many-mode constitutive relations. It can be seen that the constitutive equations are valid for the small mode numbers α, in fact, the first few modes determines main contribution to viscoelasticity. The very form of dependence of the dynamical modulus in Fig. 17 in Chapter 6 suggests to try to use the first modes to describe low-frequency viscoelastic behaviour. So, one can reduce the number of modes to minimum, while two cases have to be considered separately.

It is clear that at transition from many modes to a single mode, weight coefficients for mode contributions into the stress tensor have to be introduced. One has to require correspondence of some specified quantities to the same ones calculated within the many-mode theory. The procedure eliminates the arbitrariness to the choice of the weights. One can see that the following form of the stress tensor

$$\sigma_{ik} = -p\delta_{ik} + \frac{\pi^2}{2}\, nT \left(x_{ik}^1 - \frac{1}{3}\delta_{ik} + u_{ik}^1 \right) \qquad (9.45)$$

provides the correct form for initial coefficient of viscosity both for the weakly and strongly entangled systems.

9.3.1 Weakly Entangled Systems

Constitutive Relations

First, we refer to constitutive relations (9.19)–(9.22) which describe the behaviour of the system with moderate concentration of polymer and/or the systems with shorter macromolecules, when the characteristic parameters of the system are satisfied to conditions

$$\chi^* < \chi < 0.5, \quad \psi < 1.$$

Every mode contains two relaxation processes, described by the relaxation equations (9.20) and (9.21). One can retain one relaxation equation from each relaxation branch only, so that the two relaxation equations have to be considered

$$\frac{\mathrm{d}x_{ik}^1}{\mathrm{d}t} - \nu_{ij}x_{jk}^1 - \nu_{kj}x_{ji}^1$$

$$= -\frac{1}{2\tau_1}\left[\left(x_{ij}^1 - \frac{1}{3}\delta_{ij} \right) b_{jk} + \left(x_{kj}^1 - \frac{1}{3}\delta_{kj} \right) b_{ji} \right], \qquad (9.46)$$

$$\frac{\mathrm{d}u_{ik}^1}{\mathrm{d}t} - \omega_{ij}u_{jk}^1 - \omega_{kj}u_{ji}^1$$

$$= -\left(\frac{1}{\tau}\delta_{ij} + \frac{1}{2\tau_1}b_{ij} \right) u_{jk}^1 - \frac{1}{\tau}\psi \left[\left(x_{ij}^1 - \frac{1}{3}\delta_{ij} \right) d_{jk} - 2B\tau_1^{\mathrm{R}}x_{il}^1\gamma_{lj}f_{jk} \right]$$

$$+ \gamma_{il}u_{lk}^1, \qquad (9.47)$$

where the relaxation time $\tau_1 = \tau^* B$ and the auxiliary quantities are

$$b_{ik} = \beta_{ik}^{-1}, \qquad d_{ik} = \beta_{ij}^{-1}\epsilon_{kj}, \qquad f_{ik} = \epsilon_{ik},$$

$$\beta_{ik} = [(1 - \kappa + (\kappa - \beta)a_{ll})\delta_{ik} + 3\beta a_{ik}]^{-1},$$

$$\epsilon_{ik} = [(1 - \nu + (\nu - \epsilon)a_{ll})\delta_{ik} + 3\epsilon a_{ik}]^{-1}, \qquad a_{ik} = x_{ik}^1 - \frac{1}{3}\delta_{ij} + u_{ij}^1.$$

The relaxation equations (9.46) and (9.47) describe the joint non-linear relaxation of the two variables which appear to be weakly connected with each other through the term with the small quantity ψ in equation (9.47).

It is convenient to introduce new variables, so that expression (9.45) for the stress tensor can be written in the form

$$\sigma_{ik} = -p\delta_{ik} + 3\frac{\eta_0}{\tau_0}\left(\xi_{ik} - \frac{1}{3}\delta_{ik}\right), \quad \xi_{ik} = x_{ik}^1 + u_{ij}^1, \tag{9.48}$$

where one retains the previous definitions of the shear viscosity η_0 and define the relaxation time τ_0 as

$$\eta_0 = \frac{\pi^2}{6}nT\tau^*B, \quad \tau_0 = \tau^*B.$$

The set of relaxation equations in the single-mode approach (9.46) and (9.47) can be written in different approximations. One can see, that in zero approximation ($\psi = 0$), the relaxation equations (9.46) and (9.47) appear to be independent. The expansion of the quantity u_{ik}^1 in powers of velocity gradient begins with terms of the second order (see equation (7.39)), so that, according to equation (9.47), the variable u_{ik}^1 is not perturbed in the first and second approximations at all and, consequently, can be omitted at $\psi = 0$. In virtue of $\psi \ll 1$, the second variable has to considered to be small in any case and can be neglected with comparison to the first variable, so that the system of equations can be written in a simpler way. In the simplest case, relaxation equation (9.46) in terms of the new variables ξ_{jk} can be rewritten as

$$\frac{d\xi_{ik}}{dt} - \nu_{ij}\xi_{jk} - \nu_{kj}\xi_{ji} = -\frac{1}{\tau_0}\left[1 + (\kappa - \beta)(\xi_{ss} - 1)\right]\left(\xi_{ik} - \frac{1}{3}\delta_{ik}\right)$$

$$- \frac{1}{\tau_0}3\beta\left(\xi_{ij} - \frac{1}{3}\delta_{ij}\right)\left(\xi_{jk} - \frac{1}{3}\delta_{jk}\right). \tag{9.49}$$

One may note that the system of constitutive relations (9.48)–(9.49), which were derived and investigated by Pyshnograi et al. (1994), Pyshnograi (1996), Altukhov and Pyshnograi (1996), is a particular case of a set of the phenomenological constitutive equations (8.30)–(8.31).

Steady-State Simple Shear Flow

The expressions for stresses in simple shear flow are followed constitutive relations (9.48)–(9.49). With accuracy up to the third-order terms with respect to velocity gradient ν_{12} one has

$$\sigma_{12} = \eta\nu_{12}, \quad \frac{\eta}{\eta_0} = 1 - \frac{1}{3}(2\kappa + 7\beta)(\tau_0\nu_{12})^2,$$

$$\sigma_{11} - \sigma_{22} = 2\eta_0\tau_0\nu_{12}^2 = 2\frac{\tau_0}{\eta_0}\sigma_{12}^2, \tag{9.50}$$

$$\sigma_{22} - \sigma_{33} = -\beta\eta_0\tau_0\nu_{12}^2 = -\beta\frac{\tau_0}{\eta_0}\sigma_{12}^2.$$

In the region of the higher velocity gradient, the viscosity coefficient η and the coefficients of normal stresses as functions of velocity gradients were calculated (Golovicheva et al. 2000) for different values of the parameters β and κ. The relations for simple shear flows are typical for polymer solutions of moderate concentration.

Constitutive relations (9.48)–(9.49) determine certain amendments to known expressions for flow of viscous liquid through the long channels. The results are available (Erenburg and Pokrovskii 1981; Altukhov and Pyshnograi 1996) for flow between parallel planes with the gap h and for flow through a round tube with the radius R, correspondingly

$$Q = \frac{A}{12\eta_0} h^3 \left(1 + \frac{1}{20} \left(\frac{Ah\tau_0}{\eta_0} \right)^2 (2\kappa + 7\beta) \right),$$

$$Q = \frac{\pi A}{8\eta_0} R^4 \left(1 + \frac{1}{6} \left(\frac{AR\tau_0}{\eta_0} \right)^2 (2\kappa + \beta) \right).$$

In these expressions, Q is the volume rate and $A = p/L$ is gradient of pressure along the channels.

One can concluded that the constitutive relations (9.48)–(9.49) do indeed approximate the behaviour of systems containing long macromolecules.

9.3.2 Strongly Entangled Systems

Constitutive Relations

In this case, when

$$\chi < \chi^* < 1, \quad \psi > 1,$$

the expression (9.45) for the stress tensor can be used in the previous form

$$\sigma_{ik} = -p\delta_{ik} + \frac{\pi^2}{2} nT \left(x_{ik}^1 - \frac{1}{3}\delta_{ik} + u_{ik}^1 \right). \tag{9.51}$$

Now, one can refer to constitutive relations (9.19) and (9.24)–(9.26) and, as in previous case, one can keep one relaxation equation from each relaxation branch only, so that one has two relaxation equations in the following form to be considered

$$\frac{\mathrm{d}x_{ik}^1}{\mathrm{d}t} - \nu_{ij}x_{jk}^1 - \nu_{kj}x_{ji}^1$$

$$= -\frac{1}{2\tau_1^{\mathrm{rep}}} \left[\left(x_{ij}^1 - \frac{1}{3}\delta_{ij} \right) b_{jk} + \left(x_{kj}^1 - \frac{1}{3}\delta_{kj} \right) b_{ji} \right], \tag{9.52}$$

$$\frac{\mathrm{d}u_{ik}^1}{\mathrm{d}t} - \omega_{ij}u_{jk}^1 - \omega_{kj}u_{ji}^1$$

$$= -\frac{1}{\tau}u_{ik}^1 - \frac{1}{\tau} \left(x_{ik}^1 - \frac{1}{3}\delta_{ik} - 2B\tau^* x_{il}^1 \gamma_{lj} f_{jk} \right) + \frac{B}{E} e_{ij}\gamma_{jl}u_{lk}^1 \tag{9.53}$$

where the auxiliary quantities are

$$b_{ik} = \epsilon_{ik}^{-1}, \qquad e_{ik} = \epsilon_{ij}^{-1}\beta_{jk}, \qquad f_{ik} = \epsilon_{ij}^{-1}\beta_{jl}\epsilon_{lk},$$

$$\beta_{ik} = [(1 - \kappa + (\kappa - \beta)a_{ll})\delta_{ik} + 3\beta a_{ik}]^{-1},$$

$$\epsilon_{ik} = [(1 - \nu + (\nu - \epsilon)a_{ll})\delta_{ik} + 3\epsilon a_{ik}]^{-1}, \qquad a_{ik} = x_{ik}^1 - \frac{1}{3}\delta_{ij} + u_{ij}^1.$$

In linear case, the dependence of the tensors β_{ik} and ϵ_{ik} on the anisotropy tensor can be neglected, and all the above quantities become the unit matrixes.

Equations (9.52) and (9.53) describe the non-linear relaxation processes, which are featured, in particular, by the anisotropy of relaxation which means that in a deformed system, different components of the tensors x_{ik}^1 and u_{ik}^1 relax at different rates. The change of the second variables depends on the first one, so that the two variables of each mode are closely connected with each other.

One considers the anisotropy of environment to give a small contribution to the terms of the second order and higher with respect to velocity gradient, so that it can be neglected for the beginning, and relaxation equations (9.52) and (9.53) take simpler forms

$$\frac{dx_{ik}^1}{dt} - \nu_{ij}x_{jk}^1 - \nu_{kj}x_{ji}^1 = -\frac{1}{\tau_1^{\text{rep}}}\left(x_{ij}^1 - \frac{1}{3}\delta_{ij}\right), \tag{9.54}$$

$$\frac{du_{ik}^1}{dt} - \omega_{ij}u_{jk}^1 - \omega_{kj}u_{ji}^1$$

$$= -\frac{1}{\tau}u_{ik}^1 - \frac{1}{\tau}\left(x_{ik}^1 - \frac{1}{3}\delta_{ik} - 2B\tau^* x_{il}^1\gamma_{lk}\right) + \frac{B}{E}\gamma_{il}u_{lk}^1. \tag{9.55}$$

The set of equations (9.51)–(9.53) or (9.51) and (9.54)–(9.55) makes up the set of constitutive equations of strongly entangled system in the single-mode approximation.

Linear Viscoelasticity

In linear case, one can rewrite the relaxation equations (9.54)–(9.55) in the form

$$\frac{dx_{ik}^1}{dt} = -\frac{1}{\tau_1^{\text{rep}}}\left(x_{ik}^1 - \frac{1}{3}\delta_{ik}\right) + \frac{2}{3}\gamma_{ik},$$

$$\frac{du_{ik}^1}{dt} = -\frac{1}{\tau}u_{ik}^1 - \frac{1}{\tau}\left(x_{ik}^1 - \frac{1}{3}\delta_{ik} - \frac{2}{3}\tau^* B\gamma_{ik}\right). \tag{9.56}$$

These equations have the following solutions for oscillatory motion

$$x_{ik}^1 = \frac{1}{3}\delta_{ik} + \frac{2}{3}\frac{\tau_1^{\text{rep}}}{1 - i\omega\tau_1^{\text{rep}}}\gamma_{ik},$$

$$u_{ik}^1 = \frac{2}{3}\left(\tau^* B - \frac{\tau_1^{\text{rep}}}{1 - i\omega\tau_1^{\text{rep}}}\right)\frac{1}{1 - i\omega\tau}\gamma_{ik}.$$

Then, one can make use of the expression for the stress tensor from (9.51), to obtain the coefficient of dynamic viscosity

$$\eta(\omega) = \frac{\pi^2}{6}nT\left[\frac{\tau_1^{\text{rep}}}{1 - i\omega\tau_1^{\text{rep}}} + \left(\tau^* B - \frac{\tau_1^{\text{rep}}}{1 - i\omega\tau_1^{\text{rep}}}\right)\frac{1}{1 - i\omega\tau}\right]. \qquad (9.57)$$

At $\omega = 0$, this expression reduces to the steady-state viscosity coefficient

$$\eta = \frac{\pi^2}{6}nT\tau^* B.$$

Expression (9.57) leads to an expression for the dynamic modulus $G(\omega) = -i\omega\eta(\omega)$, from which the value on the plateau can be found

$$G_e = \lim_{\omega \to \infty} G(\omega) = \frac{\pi^2}{6}nT\left(\frac{\tau^* B}{\tau} + 1\right) \approx \frac{\pi^2}{12}nT\chi^{-1}.$$

Thus, one can see that the single-mode approximation allows us to describe linear viscoelastic behaviour, while the characteristic quantities are the same quantities that were derived in Chapter 6. To consider non-linear effects, one must refer to equations (9.52) and (9.53) and retain the dependence of the relaxation equations on the anisotropy tensor.

9.3.3 Vinogradov Constitutive Relation

It is important to have a simple but reliable constitutive relations to investigate flows of polymer liquids in different appliances of complex geometrical forms. Now we can take one more step to simplify the set of constitutive equations (9.48)–(9.49), which approximate the behaviour of polymer liquid in the region of the applicability of the relation: $\chi^* < \chi < 0.5$, $\psi < 1$. Let us note that these conditions define the systems, which can easily flow in the devices.

One can assume that the anisotropy of the relaxation process can be neglected. This means that, in relaxation equation (9.49), we equate to zero the parameter β, but retain the parameter κ, so that the set of constitutive equations can be rewritten as follows

$$\sigma_{ij} = -p\delta_{ij} + 3\frac{\eta}{\tau}\left(\xi_{ij} - \frac{1}{3}\delta_{ij}\right),$$

$$\frac{d\xi_{ij}}{dt} - \nu_{il}\xi_{lj} - \nu_{jl}\xi_{li} = -\frac{1}{\tau}\left(\xi_{ij} - \frac{1}{3}\delta_{ij}\right), \qquad \tau = \frac{\tau_0}{1 + \kappa(\xi_{ss} - 1)}. \qquad (9.58)$$

The relaxation time τ can be considered to be a function of the first invariant of the tensor of additional stresses

$$D = 3(\xi_{ss} - 1) = \frac{\tau_0}{\eta_0}(\sigma_{ss} + 3p). \qquad (9.59)$$

The quantity η in set (9.58) represents the shear viscosity coefficient and depends on the invariant of the anisotropy tensor in the same way as the relaxation time

$$\frac{\eta}{\eta_0} = \frac{\tau}{\tau_0} = \left(1 + \frac{1}{3}\kappa D\right)^{-1} = \phi(D). \qquad (9.60)$$

The suffix zero signifies the initial values of the relevant quantities (at $D \to 0$).

One can see that the set of constitutive equations (9.58)–(9.60) contains two rheological parameters: the initial shear viscosity η_0 and the initial relaxation time τ_0, as well as a single non-dimensional parameter κ.

In this and in the next sections, we shall demonstrate the consequences of the simplified description for shear and extension motions in order to understand the applicability of the approach. We shall deal with uniform steady-state motions for which we have, from (9.58), the expression for the stress tensor

$$\sigma_{ij} + p\delta_{ij} = 3\eta(\nu_{il}\xi_{lj} + \nu_{jl}\xi_{li}). \qquad (9.61)$$

For the simple shear flow, the only one component of the velocity gradient tensor differs from zero, namely, $\nu_{12} \neq 0$. The shear stress and the differences of the normal stresses are defined by equation (9.61) as

$$\sigma_{12} = \eta\nu_{12},$$

$$\sigma_{11} - \sigma_{22} = 2\eta\tau\nu_{12}^2, \qquad (9.62)$$

$$\sigma_{22} - \sigma_{33} = 0.$$

The first equation of the set (9.62) confirms that η is the coefficient of shear viscosity, which can be estimated according to the rule

$$\eta = \frac{\sigma_{12}}{\nu_{12}}. \qquad (9.63)$$

We may note that the function $\phi(D)$, which is introduced by relation (9.60), can be excluded from expressions (9.62) which leads to a relation between normal and shear stresses

$$\sigma_{11} - \sigma_{22} = 2\frac{\tau_0}{\eta_0}\sigma_{12}^2. \qquad (9.64)$$

This relation can be used to estimate the value of the shear modulus η_0/τ_0. Measurement of the first difference of normal stresses allows us to evaluate the relaxation time

$$\tau = \frac{\sigma_{11} - \sigma_{22}}{2\sigma_{12}\nu_{12}}. \qquad (9.65)$$

TABLE 3. Characteristics of the sample systems

System	T °C	c %	$\frac{\eta_0}{\tau_0} \cdot 10^{-4}$ dyn cm^{-2}	η_0 P	τ_0 s	No. on Fig. 19
Polyisobutylene	22	100	4.8	$1.7 \cdot 10^7$	354	1
$M = 7 \cdot 10^4$ *	40	100	4.6	$3.46 \cdot 10^6$	77.8	2
Blend of 62% low density polyethylene and 38% high density polyethylene**	170	100	3.1	$1.41 \cdot 10^6$	45	3
Low density polyethylene**	170	100	5.6	$4.2 \cdot 10^5$	7.5	4
Solution of poly-acrylamide in the mixture glycerine–water (1:1)***	25	1.5	0.038	$1.8 \cdot 10^4$	47	5
Solution of poly (ethylene oxide) in the mixture glycerine–water (1:2) + 11% isopropanol***	25	3.0	0.02	$1 \cdot 10^4$	50	6

* Fikhman et al. (1970); ** Weinberger and Goddard (1974)
*** Pokrovskii et al. (1973)

Equations (9.63)–(9.65) were used, in fact, to evaluate the shear viscosity coefficient and the relaxation times which reveal the nature of dependence on the velocity gradient ν_{12} or shear stress σ_{12} (Isayev 1973). It is convenient to consider the shear viscosity coefficient and the relaxation time as a generalised function of the first invariant of the tensor of additional stresses

$$D = \frac{\tau_0}{\eta_0}(\sigma_{ss} + 3p) = 2\left(\frac{\tau_0}{\eta_0}\sigma_{12}\right)^2 = 2\Gamma^2. \tag{9.66}$$

It is remarkable that the dependencies of the non-dimensional quantities η/η_0 and τ/τ_0 on the non-dimensional argument $\tau_0\nu_{12}$ or $(\tau_0/\eta_0)\sigma_{12}$ are universal. The dependencies are not essentially effected by the temperature, the molecu lar weight, and the concentration and chemical nature of the polymers (Isayev 1973).

Despite the apparent deficiency of description (9.58) when applied to a real system, we may note that the set of constitutive equations (9.58)–(9.60) represent qualitatively the behaviour of concentrated polymer solutions and melts under shear. The set of equations include two material constants which are the individual characteristics of the system, namely, the initial shear viscosity and the initial relaxation time, which depend on the temperature, the molecular weight of polymer, and its concentration in the system. As an illustration, the estimation of the material constants for some systems are shown in Table 3. The returning to the fuller approximation (equations (9.48)–(9.49)) improves the description, as has been shown by Pyshnograi et al. (1994).

The constitutive equations (9.58)–(9.60) were derived as a consequent simplification of general equations, discussed in Section 9.2.1, so that one can conclude which assumptions have to be introduced to obtain the equations. We may note that, before this consequent derivation, the considered constitutive equations (9.58)–(9.60) were formulated (Vinogradov et al. 1972b; Phan-Thien and Tanner 1977) and used for the investigation of simple (Pokrovskii and Kruchinin 1980) and complex (Altukhov et al. 1986; Erenburg and Pokrovskii 1981) flows of polymeric liquids. The constitution equations named in honour of one of the pioneer investigator of polymer rheology G.V. Vinogradov.

It is important to note that the constitutive equation (9.58)–(9.60) belong to the class of the rare equations which are Hadamard and dissipative stable (Kwon and Leonov 1995).

9.3.4 Relation between Shear and Elongational Viscosities

Consider the case of applying the system (9.58)–(9.60) to description of uniaxial deformation with the constant elongational velocity gradient ν_{11}. The elongational viscosity coefficient λ is determined as the ratio of the extensional stress σ to the elongational velocity gradient. We shall calculate, according to Pokrovskii and Kruchinin (1980), the ratio between the coefficients of elongational and shear viscosity, namely, the quantity λ/η for a polymer liquid.[1]

For uniform uniaxial elongational deformation along axis 1, the tensor of the velocity gradients, taking into account the condition of incompressibility, can be written in the form

$$\nu_{ik} = \left\| \begin{matrix} \nu_{11} & 0 & 0 \\ 0 & -\frac{1}{2}\nu_{11} & 0 \\ 0 & 0 & -\frac{1}{2}\nu_{11} \end{matrix} \right\|.$$

If we exclude the pressure from the relation for the stresses (9.58) under the considered uniaxial deformation, we can obtain an expression for the extensional stress

$$\sigma = 3\frac{\eta}{\tau}(\xi_{11} - \xi_{22}) = 3\eta(2\xi_{11} - \xi_{22})\nu_{11} \tag{9.67}$$

where, in a steady-state case,

---

[1] The earlier history of the investigation of elongational flow can be found in the monograph by Petrie (1979).

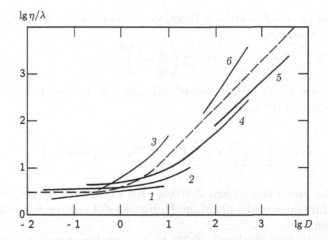

Figure 19. The ratio of elongational to shear viscosities
The theoretical dependence of the ratio of elongational to shear viscosity coefficients on the invariant of the additional stress tensor is calculated according to equation (9.71) and depicted by the dashed curve. The solid curves represent experimental data for systems listed in Table 3. Adapted from the paper of Pokrovskii and Kruchinin (1980).

$$\xi_{ii} = \frac{1}{3}\left(1 + \frac{2\tau\nu_{ii}}{1 - 2\tau\nu_{11}}\right), \quad i = 1, 2, 3. \tag{9.68}$$

The above formulae determine the elongational viscosity coefficient

$$\lambda = \frac{3\eta}{1 - \tau\nu_{11} - 2(\tau\nu_{11})^2}. \tag{9.69}$$

The viscosity coefficients η and λ are functions of the first invariant of the tensor of additional stresses

$$D = \frac{\tau_0}{\eta_0}(\sigma_{ii} + 3p). $$

When relations (9.67) and (9.68) are used, the invariant can be expressed in terms of the elongational velocity gradient or extensional stress

$$D = 2\frac{\lambda}{\eta}(\tau\nu_{11})^2 = 2\frac{\eta}{\lambda}\left(\frac{\tau_0}{\eta_0}\sigma\right)^2. \tag{9.70}$$

By eliminating the velocity gradient from relations (9.69) and (9.70) we can obtain an expression for the ratio of the coefficients of elongational and shear viscosity

$$\frac{\lambda}{\eta} = 3 + \frac{5}{4}D + \left(\frac{3}{2}D + \frac{9}{16}D^2\right)^{1/2}. \tag{9.71}$$

The ratio of the coefficients is a function of the invariant D which, for shear motion, has the form

$$D = 2 \left(\frac{\tau_0}{\eta_0} \sigma_{12} \right)^2 . \tag{9.72}$$

The expression for D in the case of uniaxial deformation is easily obtained from equations (9.70) and (9.71)

$$D = -\frac{3}{2} - \frac{1}{2} \frac{\tau_0}{\eta_0} \sigma + \frac{1}{2} \left(9 + 6 \frac{\tau_0}{\eta_0} \sigma + 9 \left(\frac{\tau_0}{\eta_0} \sigma \right)^2 \right)^{1/2} . \tag{9.73}$$

It should be noted that, when deriving this expression, we assumed that η and λ are functions of the invariant D, but we did not use a specific form of this function.

The applicability of relation (9.71) to a real polymer system was discussed in works by Pokrovskii et al. (1973); Pokrovskii and Kruchinin (1980); Pyshnograi et al. (1994). Figure 19 represents the experimental values of the ratio λ/η depending on the invariant D for the polymer systems, listed in Table 3, in comparison with the universal theoretical curve calculated according to equation (9.71). The experimental results can be seen to have a definite scatter relative to the theoretical curve; this can be ascribed to both natural experimental errors and the necessity of improving the theoretical calculation by appealing to the fuller set of constitutive relations (9.48)–(9.49). In the former case a variation of β in (9.49) leads to a set of λ/η vs D curves (Pyshnograi et al. 1994).

However, the observed consistency of the experimental and theoretical appraisals can be considered as surprisingly satisfactory. Both these results and the results of the previous section point to the possibility of employing the Vinogradov constitutive equations (9.58)–(9.60) for qualitative investigations of non-uniform flows of polymer liquids.

9.3.5 Recoverable Strain

One of the prominent features of polymeric liquids is the property to recover partially the pre-deformation state. Such behaviour is analogous to a rubber band snapping back when released after stretching. This is a consequence of the relaxation of macromolecular coils in the system: every deformed macromolecular coil tends to recover its pre-deformed equilibrium form. In the considered theory, the form and dimensions of the deformed macromolecular coil are connected with the internal variables x_{ij}^{α} which have to be considered when the tensor of recoverable strain is to be calculated. Further on, we shall consider the simplest case, when the form and dimensions of macromolecular coils are determined by the only internal tensor ξ_{ij}. In this case, the behaviour of the polymer liquid is considered to describe by one of the constitutive equations (9.48)–(9.49) or (9.58).

To determine the tensor of recoverable strains, we have to equate the stress tensor for a deformed polymer network (given in the simplest case by equation (1.43)) with the elastic part of the stress tensor for a polymer liquid, given in the general case by equation (9.19) or, in the simplified case by equation (9.48). The latter case leads to the relation

$$G\lambda_{ij}\lambda_{kj} = 3\frac{\eta_0}{\tau_0}\xi_{ik} - \left(\frac{\eta_0}{\tau_0} - G\right)\delta_{ik} \qquad (9.74)$$

where G is the shear modulus and λ_{ij} is the tensor of recoverable displacement, such that $\Lambda_{ik} = \lambda_{ij}\lambda_{kj}$ is the tensor of recoverable strains. The latter quantities are discussed in Appendix B.

To determine the shear modulus and the tensor of recoverable strains, we calculate the determinants of the left-hand and right-hand sides of equation (9.74). Taking into account the incompressibility of the polymer liquid, i.e. relation $|\lambda_{ij}\lambda_{kj}| = 1$, we obtain

$$G^3 = 27\left(\frac{\eta_0}{\tau_0}\right)^3 \left[\Xi_3 - \frac{1}{3}\left(1 - \frac{G\tau_0}{\eta_0}\right)\Xi_2 + \frac{1}{9}\left(1 - \frac{G\tau_0}{\eta_0}\right)^2\Xi_1\right]$$

where the invariants of the tensor ξ_{ij} are introduced as follows

$$\Xi_1 = \sum_{i=1}^{3}\xi_{ii}, \qquad \Xi_2 = \frac{1}{2}\sum_{i,j}(\xi_{ij}\xi_{ji} - \xi_{ii}\xi_{jj}), \qquad \Xi_3 = |\xi_{ij}|.$$

We consider these invariants independent of each other, so that we can determine the shear modulus and the tensor of recoverable strains

$$G = 3|\xi_{ls}|^{1/3}\frac{\eta_0}{\tau_0}, \qquad \lambda_{ij}\lambda_{kj} = \delta_{ik} + |\xi_{ls}|^{-1/3}\left(\xi_{ik} - \frac{1}{3}\delta_{ik}\right). \qquad (9.75)$$

As the expansion of the invariants into series with respect to the velocity gradients do not contain terms of even order, one can directly see the correctness of the above expressions with accuracy at least up to third-order terms with respect to the velocity gradient.

As an example, we shall consider simple shear when $\nu_{12} \neq 0$, and find components of the tensor of the recoverable displacement gradients $\lambda_{12}, \lambda_{11}, \lambda_{22}, \lambda_{33}$; the components of the tensor ξ_{il} are calculated from the relaxation equations (9.49) or (9.58). In this case the matrix of the deformation tensor is determined as follows

$$\Lambda = \left\| \begin{matrix} \lambda_{11}^2 + \lambda_{12}^2 & \lambda_{22}\lambda_{12} & 0 \\ \lambda_{22}\lambda_{12} & \lambda_{22}^2 & 0 \\ 0 & 0 & \lambda_{33}^2 \end{matrix} \right\|. \qquad (9.76)$$

Further on we shall consider the simple case when the relaxation equation is given by equation (9.58) and we shall assume the shear motion to be a steady-state one. So, we have the expressions

$$\xi_{11} = \frac{1}{3}(1 + 2\Gamma^2), \qquad \xi_{12} = \frac{1}{3}\Gamma, \qquad \xi_{22} = \xi_{33} = \frac{1}{3},$$

$$\xi_{13} = \xi_{23} = 0$$
(9.77)

where $\Gamma = \tau\nu_{12} = \frac{\tau_0}{\eta_0}\sigma_{12}$.

Now, equations (9.75) allow us to calculate the shear modulus and the deformation tensor. With approximation up to the third-order terms with respect to the velocity gradient, we obtain

$$G = \frac{\eta_0}{\tau_0}\left(1 + \frac{1}{3}\Gamma^2\right),$$
(9.78)

$$\Lambda = \begin{Vmatrix} 1 + \frac{5}{3}\Gamma^2 & \Gamma\left(1 - \frac{1}{3}\Gamma^2\right) & 0 \\ \Gamma\left(1 - \frac{1}{3}\Gamma^2\right) & 1 - \frac{1}{3}\Gamma^2 & 0 \\ 0 & 0 & 1 - \frac{1}{3}\Gamma^2 \end{Vmatrix}.$$
(9.79)

After comparing expressions (9.76) and (9.79), we obtain the components of the recoverable displacement tensor

$$\lambda_{11} = 1 + \frac{1}{3}\Gamma^2, \qquad \lambda_{22} = \lambda_{33} = 1 - \frac{1}{6}\Gamma^2, \qquad \lambda_{12} = \Gamma\left(1 - \frac{1}{6}\Gamma^2\right).$$
(9.80)

In accordance with the experimental data (Ferry 1980), the shear modulus increases as the velocity gradient increases and the recoverable shear deformation λ_{12} deviates from proportionality to the velocity gradient.

We may note here that the sets of constitutive equations (9.48)–(9.49) or (9.58) can be reformulated, taking into account the established connection between the internal variable ξ_{ij} and the recoverable deformation tensor (equation (9.75)), so that the constitutive equations would include the tensor of recoverable deformation as an internal variable. In fact, such constitutive equations were obtained independently (Godunov and Romenskii 1972; Leonov 1976; Prokunin 1989). Therefore, two interpretations of the internal variables and two formalisms are equivalent, but, nevertheless, one of them appears to be simpler.

Chapter 10
Optical Anisotropy

Abstract Macromolecular coils are deformed in flow, while optically anisotropic parts (and segments) of the macromolecules are oriented by flow, so
that polymers and their solutions become optically anisotropic. This is true
for a macromolecule whether it is in a viscous liquid or is surrounded by other
chains. The optical anisotropy of a system appears to be directly connected
with the mean orientation of segments and, thus, it provides the most direct
observation of the relaxation of the segments, both in dilute and in concentrated solutions of polymers. The results of the theory for dilute solutions
provide an instrument for the investigation of the structure and properties of
a macromolecule. In application to very concentrated solutions, the optical
anisotropy provides the important means for the investigation of slow relaxation processes. The evidence can be decisive for understanding the mechanism of the relaxation.

10.1 The Relative Permittivity Tensor

In order to examine the optical anisotropy, we begin with the relative permittivity tensor for the system ε_{ik}, which is defined (see, for example, Born and
Wolf 1970; Landau et al. 1987) by the relation

$$\varepsilon_{ik} E_k = E_i + 4\pi P_i \qquad (10.1)$$

where E_k is the average electric field strength acting in the medium and P_i
is the polarisation per unit volume of the system expressed in terms of the
polarisabilities of the constituent elements of the system.

One can make use of the heuristic model mentioned previously, in Section 1.1: each macromolecule consists of z segments and is surrounded by
solvent molecules. It is not essential now to know whether the segments in the
chain are connected or independent; the results of this section are applicable
in both cases.

When considering the system consisting of solvent molecules and segments,
the simple old-fashion (Vleck 1932; Fröhlich 1958) speculations allow us to

V.N. Pokrovskii, *The Mesoscopic Theory of Polymer Dynamics*,
Springer Series in Chemical Physics 95,
DOI 10.1007/978-90-481-2231-8_10, © Springer Science+Business Media B.V. 2010

determine the relative permittivity tensor of polymeric system in terms of the mean orientation of anisotropic segments of the macromolecules. The solvent molecules have an isotropic polarisability α, while the segment has an anisotropic polarisability α_{ik}. In the co-ordinate system connected with the segment, the anisotropy tensor is assumed to be diagonal. In any other co-ordinate system, the polarisability tensor of the segment has the form

$$\alpha_{ik} = c_{is}c_{ks}\alpha_{ss}$$

where c_{is} is the cosine of the angle between the ith axis of the laboratory system and the sth axis of the molecule. One can assume that the segment has axial symmetry, so that $\alpha_{22} = \alpha_{33}$, and introduce the unit vector e in direction of the axis. It allows us to rewrite the expression for the polarisability tensor of the segment in the form

$$\alpha_{ik} = \bar{\alpha}\delta_{ik} + (\alpha_{11} - \alpha_{22})\left(e_i e_k - \frac{1}{3}\delta_{ik} \right) \qquad (10.2)$$

where $\bar{\alpha} = (\alpha_{11} + \alpha_{22} + \alpha_{33})/3$. In case that is more general, we have to introduce two unit vectors $e^{\parallel}$ and $e^{\perp}$ – along the direction of the axis of the segment and in perpendicular direction, respectively. In this case

$$\alpha_{ik} = \bar{\alpha}\delta_{ik} + (\alpha_{11} - \alpha_{33})\left(e_i^{\parallel} e_k^{\parallel} - \frac{1}{3}\delta_{ik} \right) + (\alpha_{22} - \alpha_{33})\left(e_i^{\perp} e_k^{\perp} - \frac{1}{3}\delta_{ik} \right). \qquad (10.3)$$

The time of relaxation of the mean orientation of the lateral vector $e^{\perp}$ is considered to be much less than the time of relaxation of the mean orientation of the axial vector $e^{\parallel}$, so that the last term in (10.3) can be neglected for rather low frequencies and one can continue with the simpler case (10.2).

The true molecular field F acting both on the segment and on molecules of solvent differs from the average field E because the scale of the dimensions of the segments is molecular. Each solvent molecule makes an isotropic contribution to the polarisability vector; the contribution of each segment of the macromolecule is anisotropic and is expressed by the formula

$$\beta_s = c_{si}c_{ki}\alpha_{ii}F_k = \left[\bar{\alpha}\delta_{sk} + \Delta\alpha\left(e_s e_k - \frac{1}{3}\delta_{sk} \right) \right]F_k, \quad \Delta\alpha = \alpha_{11} - \alpha_{22}.$$

By taking into account all the molecules and segments and by designating with nz and n_s the densities of the number of segments and of the number of solvent molecules (n being the density of the number of macromolecules), we obtain, after averaging with respect to the orientations of the segments,

$$P_j = (nz\bar{\alpha}\delta_{jk} + nz\Delta\alpha s_{jk} + n_s\alpha\delta_{jk})F_k \qquad (10.4)$$

where a symbol has been introduced for the mean values of the directing cosines of the segment relative to the laboratory co-ordinate system – the orientation tensor

$$s_{jk} = \langle e_j e_k \rangle - \frac{1}{3}\delta_{jk}.$$

The internal field F_k is assumed to be the same for the segments and the solvent molecules.

Next, use is made of the simple hypothesis that all the positions of the molecules and segments are equally probable, and, following tradition, we shall formulate an expression for the internal field as a field within a spherical cavity (Vleck 1932; Fröhlich 1958)

$$F_i = E_i + \frac{4\pi}{3}P_i. \tag{10.5}$$

The internal field can be eliminated from relations (10.4) and (10.5), so that we have a set of equations for the components of the vector of polarisation

$$(A\delta_{sj} + as_{sj})P_j = -(B\delta_{sj} + bs_{sj})E_j$$

where the following notations are introduced

$$A = \frac{4\pi}{3}(nz\bar{\alpha} + n_s\alpha) - 1, \quad a = \frac{4\pi}{3}nz\Delta\alpha,$$
$$B = nz\bar{\alpha} + n_s\alpha, \quad b = nz\Delta\alpha.$$

The written set of equations has a simple solution for the components of the polarisation vector. We use them to write, in accordance to equation (10.1), the relative permittivity tensor

$$\varepsilon_{ik} = \delta_{ik} + \frac{4\pi}{D}\left[-A^2B\delta_{ik} + (AaB - A^2b)s_{ik} + \frac{1}{2}a^2Bs_{jl}s_{jl}\delta_{ik}\right.$$
$$\left. + (Aab - a^2B)s_{il}s_{lk} + \frac{1}{2}a^2bs_{jl}s_{jl}s_{lk} - a^2bs_{ij}s_{jl}s_{lk}\right],$$
$$D = A^3 - \frac{1}{2}Aa^2s_{ik}s_{ik} + a^3|s_{ik}|.$$

In the case when there is no preferred orientation, that is $s_{ik} = 0$, the considered system is isotropic and is characterised by the relative permittivity constant

$$\varepsilon_0 = 1 + 4\pi\frac{nz\bar{\alpha} + n_s\alpha}{1 - \frac{4\pi}{3}(nz\bar{\alpha} + n_s\alpha)}$$

from which one can find the relations

$$\frac{4\pi}{3}(nz\bar{\alpha} + n_s\alpha) = \frac{\varepsilon_0 - 1}{\varepsilon_0 + 2}, \quad A = -\frac{3}{\varepsilon_0 + 2}.$$

The written relations define the relative permittivity tensor for the system, which is formulated below to within second-order terms in the orientation tensor

$$\varepsilon_{ik} = \varepsilon_0 \delta_{ik} + 4\pi nz\Delta\alpha \left(\frac{\varepsilon_0 + 2}{3}\right)^2 s_{ik}$$

$$+ \frac{1}{2}\left[1 - \varepsilon_0 + 4\pi\left(\frac{\varepsilon_0 + 2}{3}\right)^3\right](4\pi nz\Delta\alpha)^2 \left(\frac{\varepsilon_0 + 2}{3}\right)^2 s_{jl}s_{lj}\delta_{ik}$$

$$+ \frac{1}{3}(4\pi nz\Delta\alpha)^2 \left(\frac{\varepsilon_0 + 2}{3}\right)^3 s_{il}s_{lk}. \tag{10.6}$$

One can see that, to a first approximation, as it is well known (Vleck 1932; Fröhlich 1958), allowance for the internal field by the Lorentz procedure is equivalent to multiplication by the factor

$$\left(\frac{\varepsilon_0 + 2}{3}\right)^2.$$

In conformity with the significance of the terms employed by investigators of anisotropy (Tsvetkov et al. 1964), the effects associated with the first-order terms in equation (10.6) may be called the effects of intrinsic anisotropy, while the second-order effects may be referred to as the effects of mutual interaction. In the second approximation, the principal axes of the relative permittivity tensor do not coincide, generally speaking, with the principal axes of the orientation tensor. It is readily seen that interesting situations may arise when $\Delta\alpha < 0$; in this case, the coefficients of the first- and second-order terms have different signs.

Let us note that the contribution from anisotropy due to the difference in the isotropic part of the polarisability between segments and solvents molecules, $\bar{\alpha} - \alpha_0$, ought to be added to expression (10.6). This is a first-order term in the orientational tensor (Tsvetkov et al. 1964). We shall not consider this contribution to the anisotropy, as it is not so important for the very concentrated solutions under consideration.

10.2 The Permittivity Tensor for Polymer Systems

Now, we have to return to the subchain model of macromolecule, which was used to calculate the stresses in the polymeric system, and express the tensor of the mean orientation of the segments of the macromolecule in terms of the subchain model.

Equation (10.6), formulated in the previous section, defines the relative permittivity tensor in terms of the mean orientation of certain uniformly distributed anisotropic elements, which we shall interpret here as the Kuhn segments of the model of the macromolecule described in Section 1.1. We shall now discuss the characteristic features of a polymer systems, in which the segments of the macromolecule are not independently distributed but are concentrated in macromolecular coils.

10.2.1 Dilute Solutions

In the equilibrium situation, at a given end-to-end distance R of a macro-molecule, the tensor of mean orientation of segments of a chain is determined (Flory 1969) as

$$\langle e_i e_k \rangle - \frac{1}{3}\delta_{ik} = \frac{3}{5(zl)^2}\left(R_i R_k - \frac{1}{3}R^2\delta_{ik}\right). \qquad (10.7)$$

As before, we shall consider each macromolecule to be divided into N sub-chains and assume that every subchain of the macromolecule is in the equilibrium. So, using the above formula relating the tensor of the mean orientation of the segments of the macromolecules $\langle e_j e_k \rangle$ to the distance between the ends of the subchains, we arrive from relation (10.6), taken in the first approximation, at Zimm's (1956) expression for the relative permittivity tensor

$$\varepsilon_{ik} = \varepsilon_0\delta_{ik} + n\Gamma\left(\langle r_i^\alpha A_{\alpha\gamma} r_k^\gamma\rangle - \frac{1}{3}\langle r_j^\alpha A_{\alpha\gamma} r_j^\gamma\rangle\delta_{ik}\right) \qquad (10.8)$$

where n is the density of the number of macromolecules in the solution, and the matrix A has the form specified by formula (1.8), while the coefficient of the anisotropy of the macromolecular coil Γ, for the macromolecule as a freely-jointed chain of Kuhn segments, is given by the following expression

$$\Gamma = 4\pi\Delta\alpha\left(\frac{\varepsilon_0 + 2}{3}\right)^2\frac{3N}{5zl^2}$$

where z is the number of Kuhn segments in the macromolecule, and $\Delta\alpha$ is the anisotropy of the polarisability of a Kuhn segment.

The anisotropy of the coil has been calculated for other models of the macromolecule. Expressions for the anisotropy coefficient are known in the case where the macromolecule has been represented schematically by a con-tinuous thread (the persistence length model) (Gotlib 1964; Zgaevskii and Pokrovskii 1970) and also in the case where the microstructure of the macro-molecules has been specified. In the latter case, the anisotropy coefficient of the macromolecule is expressed in terms of the bond polarisabilities and other microcharacteristics of the macromolecule (Flory 1969).

When account is taken of the excluded volume effects, one has also to take into account the possible effect of the shielding of the inner segments of the macromolecular coil, the latter effect being the greater the longer the macromolecule, so that the expression for the anisotropy coefficient, which has to be covariant in relation to subdivisions into subchain, assumes the form

$$\Gamma = 4\pi\Delta\alpha\left(\frac{\varepsilon_0 + 2}{3}\right)^2\frac{3N^{2\nu}}{5\langle R^2\rangle}. \qquad (10.9)$$

The dependence of the polarisability coefficient on the length of the macro-molecule follows from equation (10.9) as

$$\Gamma \sim M^{-2\nu}.$$

Expression (10.8) for the relative permittivity tensor in terms of the normal co-ordinates introduced by means of equations (1.13), assumes the form

$$\varepsilon_{ik} = \varepsilon_0 \delta_{ik} + n\Gamma \sum_{\alpha=1}^{N} \lambda_\alpha \left(\langle \rho_i^\alpha \rho_k^\alpha \rangle - \frac{1}{3} \langle \rho_j^\alpha \rho_j^\alpha \rangle \delta_{ik} \right)$$

or in terms of the ratios $x_{ik}^\nu = \langle \rho_i^\nu \rho_k^\nu \rangle / \langle \rho^\nu \rho^\nu \rangle_0$

$$\varepsilon_{ik} = \varepsilon_0 \delta_{ik} + \frac{3n\Gamma}{2\mu} \sum_{\alpha=1}^{N} \left(x_{ik}^\alpha - \frac{1}{3} x_{jj}^\alpha \delta_{ik} \right). \tag{10.10}$$

The last equation can be compared with equations (9.1) and (9.3) for the stresses in dilute solutions. On can see that, when internal viscosity is neglected ($\varphi_\nu = 0$), there is a relation between the permittivity tensor and stress tensor in the form

$$\varepsilon_{ij} - \varepsilon_0 \delta_{ij} = 2\bar{n}C \left(\sigma_{ij} + p\delta_{ij} - 2\eta_s \gamma_{ij} \right),$$

$$C = \frac{\Gamma}{4\bar{n}\mu T} = \frac{2\pi}{45\bar{n}T}(\varepsilon_0 + 2)^2 \Delta\alpha \tag{10.11}$$

where $\bar{n}$ is an isotropic value of the refractive index ($\bar{n}^2 = \varepsilon_0$) and C is the stress-optical coefficient, which is universally expressed through the segment anisotropy $\Delta\alpha$. The stress-optical law (10.11) reflects the fact that both the stresses and the optical anisotropy of a polymer solution under motion are determined by the mean orientation of segments of the chains.

Expression (10.10) for the relative permittivity tensor is valid only to a first approximation as regards the orientation of the segments and describes the anisotropy of the system associated with the intrinsic anisotropy of the segments. Apart from it, it was assumed that distribution of orientation of the segments inside every subchain are considered to be in equilibrium though under deformation. However, this expression has appeared to be very well applicable to dilute polymer solutions at low frequencies and small velocity gradient (Tsvetkov et al. 1964; Janeschitz-Kriegl 1983). In more general situations, one has to take into account that the mean orientation of segments under deformation of the macromolecular coil deviates from equilibrium value (10.7). One can believe that the stress-optical law (10.11) is valid in this case, so that an expression for the permittivity tensor can be found as combination of equations (9.1) and (10.11), whereby the internal viscosity is taking into account. However, an independent calculation of the tensor of orientation and the permittivity tensor in non-equilibrium situations is much desirable.

10.2.2 Entangled Systems

The situation is different for very concentrated polymer solutions. Though equation (10.6) is applicable for this case, formula (10.7) is not valid neither for the entire macromolecule nor for a separate subchain. The subchain of a macromolecule in the deformed entangled system is not in equilibrium even in the first approximation, and the problem about distribution of orientation of the interacting, connected in chains, segments apparently is not solved yet.

In this situation, which is also discussed in Section 7.5, we refer to experimental evidence according to which components of the relative permittivity tensor are strongly related to components of the stress tensor. It is usually stated (Doi and Edwards 1986) that the stress-optical law, that is proportionality between the tensor of relative permittivity and the stress tensor, is valid for an entangled polymer system, though one can see (for example, in some plots of the paper by Kannon and Kornfield (1994)) deviations from the stress-optical law in the region of very low frequencies for some samples. In linear approximation for the region of low frequencies, one can write the following relation

$$\varepsilon_{ij} - \varepsilon_0 \delta_{ij} = 2\bar{n} C \left(\sigma_{ij} + p\delta_{ij} \right) \tag{10.12}$$

where $\bar{n}$ is a value of the refractive index ($\bar{n}^2 = \varepsilon_0$) and C is the stress-optical coefficient, which is assumed to be universally expressed through the segment anisotropy $\Delta\alpha$ by formula (10.11). Relation (10.12) reflects the fact that both the stresses and the optical anisotropy of a polymeric liquid under motion are determined by the mean orientation of the interacting segments. One can use expression (9.19) for the stress tensor to write

$$\varepsilon_{ij} = \varepsilon_0 \delta_{ij} + 6nT\,\bar{n}C \sum_{\alpha=1}^{N} \left\{ x_{ij}^{\alpha} - \frac{1}{3}\delta_{ij} + u_{ij}^{\alpha} \right\}. \tag{10.13}$$

One admits that the relative permittivity tensor of the system is determined by the mean orientation of the segments, so that we consider expression (10.13) to be equivalent to the first-order terms of relation (10.6) and, at comparison, obtain the expression for the mean orientation of segments of macromolecules in an entangled system

$$\langle e_i e_k \rangle - \frac{1}{3}\delta_{ik} = \frac{3}{5z} \sum_{\alpha=1}^{N} \left\{ x_{ij}^{\alpha} - \frac{1}{3}\delta_{ij} + u_{ij}^{\alpha} \right\} \tag{10.14}$$

where z is number of segments in a macromolecules. The set of the variables $x_{ij}^{\alpha} = \frac{\langle \rho_i^{\alpha} \rho_k^{\alpha} \rangle}{\langle \rho_i^{\alpha} \rho_k^{\alpha} \rangle_0}$ represents the conformation of the macromolecular coil, while the variables u_{ij}^{α} are mainly connected with the mean orientation of the segments s_{ij}. The variables x_{ij}^{α} and u_{ij}^{α} appear to be independent from each other and can be found as a solutions of the relaxation equations (7.25) and (7.38) for weakly entangled systems and equations (7.29) and (7.40) for strongly entangled systems.

Relaxation equations for the mean orientation can be restored (see also Section 7.5). In the case of strongly entangled system, in linear approximation, assuming that $E/B \ll B$, we have

$$\frac{\mathrm{d}\langle e_i e_k \rangle}{\mathrm{d}t} = -\frac{1}{\tau}\left(\langle e_i e_k \rangle - \frac{1}{3}\delta_{ik}\right) + \frac{\pi^2}{15}\frac{1}{z}B\frac{\tau^*}{\tau}\gamma_{ik}. \tag{10.15}$$

One can see that, in this approximation, disturbed conformation of macromolecules does not affect the mean orientation of segments in the steady state, that can be found from equation (10.15) as

$$\langle e_i e_k \rangle = \frac{1}{3}\delta_{ik} + \frac{\pi^2}{15}\frac{1}{z}\tau^* B\gamma_{ik}. \tag{10.16}$$

In contrast to the case of dilute polymer solutions (relation (10.7)), mean orientation of segments does not depend (to the first approximation) on the large-scale conformation of the macromolecule. However, an independent calculation of the tensor of orientation in non-equilibrium situations is much desirable.

10.3 Optical Birefringence

The value of the refractive index n of light in the anisotropic medium depends on the direction of propagation s and on the direction of the polarisation of the light. For the given relative permittivity tensor ε_{jl}, the refractive index can be determined from the relation (Born and Wolf 1970; Landau et al. 1987)

$$\varepsilon_{jl}E_l = n^2[E_j - s_j(s \cdot E)]. \tag{10.17}$$

It follows from (10.17) that the refractive index for an isotropic medium is determined by the permittivity constant only

$$n^2 = \varepsilon_0.$$

In the case of an anisotropic system, it is convenient to consider particular cases. Further on, expressions for characteristics of optical birefringence in two typical cases will be shown.

Methods for the experimental estimation of birefringence can be found in the monograph by Tsvetkov et al. (1964), Janeschitz-Kriegl (1983) and in papers by Lodge and Schrag (1984), Inoue et al. (1991), and Kannon and Kornfield (1994).

10.3.1 Simple Elongation

In the simplest cases, the optical anisotropy of polymer systems is studied under the conditions of simple elongation, when the elongation velocity gradient ν_{11} is given. The system investigated then becomes, generally speaking, a "triaxial dielectric crystal" with components of the relative permittivity tensor

$$\left\| \begin{matrix} \varepsilon_{11} & 0 & 0 \\ 0 & \varepsilon_{22} & 0 \\ 0 & 0 & \varepsilon_{33} \end{matrix} \right\| .$$

For a system under elongational deformation along direction 1, for a beam of light propagating in direction 3, according to (10.17) one obtains different refractive indices for different polarisation of the beam, so that, for polarisation in directions 1 and 2, one has a difference of refractive indices

$$\Delta n = \frac{1}{2\bar{n}} (\varepsilon_{11} - \varepsilon_{22}) \qquad (10.18)$$

where $\bar{n}$ is the average refractive index. This relation is written on the assumption that the difference between refractive indices is small, so that non-linear terms are omitted.

10.3.2 Simple Shear

For a system undergoing simple shear, when the velocity gradient $\nu_{12} \neq 0$, the relative permittivity tensor is non-diagonal

$$\left\| \begin{matrix} \varepsilon_{11} & \varepsilon_{12} & 0 \\ \varepsilon_{12} & \varepsilon_{22} & 0 \\ 0 & 0 & \varepsilon_{33} \end{matrix} \right\| .$$

However, the tensor can be turned to diagonal form by rotating the co-ordinate frame round axis 3 by an angle χ (the extinction angle), defined by the formula

$$\tan 2\chi = \frac{2\varepsilon_{12}}{\varepsilon_{11} - \varepsilon_{22}}. \qquad (10.19)$$

The differences between the refractive indices (the extent of double refraction) in the different principal directions can be determined from equation (10.17). For a beam propagated in direction 3, we find that

$$\Delta n = \frac{1}{2\bar{n}} \sqrt{(\varepsilon_{11} - \varepsilon_{22})^2 + 4\varepsilon_{12}^2}. \qquad (10.20)$$

This relation as well as relation (10.18) is valid in linear approximation and can be therefore rewritten as

$$\Delta n = \frac{1}{\bar{n}} \varepsilon_{12}, \qquad (10.21)$$

while the extinction angle $\chi = \pi/4$.

A little bit more complicated situation appears, if one considers a beam propagating across the flow in direction characterised by the unit vector

$$s_1 = \sin \theta, \quad s_2 = \cos \theta, \quad s_3 = 0.$$

This experimental situation is described, for example, in the work of Brown et al. (1995). It is convenient to choose the electric vector of the beam in plane (1–2) or in direction 3, whereas the differences between the refractive indices can be easily found from equation (10.17)

$$\Delta n = \frac{1}{2\bar{n}} \left(\frac{\varepsilon_{11}\varepsilon_{22} - \varepsilon_{12}^2 - \varepsilon_{22}\varepsilon_{33}(1 - \sin^2\theta) - \varepsilon_{11}\varepsilon_{33}(1 - \cos^2\theta)}{\varepsilon_{22}(1 - \sin^2\theta) - \varepsilon_{11}(1 - \cos^2\theta) - 2\varepsilon_{12}\sin\theta\cos\theta} \right.$$
$$\left. - \frac{2\varepsilon_{12}\varepsilon_{33}\sin\theta\cos\theta}{\varepsilon_{22}(1 - \sin^2\theta) - \varepsilon_{11}(1 - \cos^2\theta) - 2\varepsilon_{12}\sin\theta\cos\theta} \right).$$

For $\theta = 0$, this formula reduces to

$$\Delta n = \frac{1}{2\bar{n}} \left(\varepsilon_{11} - \varepsilon_{33} - \frac{\varepsilon_{12}^2}{\varepsilon_{22}} \right).$$

10.3.3 Oscillatory Deformation

One frequently deals with the linear effects of anisotropy which are induced by oscillatory velocity gradients or by oscillatory strains

$$u_{ik}(t) = -i\omega\gamma_{ik}(t) \sim e^{-i\omega t}.$$

In this case, it is convenient to characterise the behaviour of the system by the dynamo-optical coefficient

$$S(\omega) = S'(\omega) + iS''(\omega)$$

due to Lodge and Schrag (1984), or by the strain-optical coefficient

$$O(\omega) = O'(\omega) - iO''(\omega)$$

due to Inoue et al. (1991). These quantities are introduced by relations

$$\varepsilon_{ik} = \varepsilon_0\delta_{ik} + 4\bar{n}S(\omega)\gamma_{ik},$$
$$\varepsilon_{ik} = \varepsilon_0\delta_{ik} + 4\bar{n}O(\omega)u_{ik}. \tag{10.22}$$

It is easy to find, from the above-written formulae, that the components of dynamic characteristics are connected by relations

$$O'(\omega) = \omega S''(\omega), \qquad O''(\omega) = \omega S'(\omega). \tag{10.23}$$

Relations (10.22) are quite similar to the definitions of dynamic viscosity $\eta(\omega)$ and dynamic modulus $G(\omega)$, so that relations (10.23) are similar to the relations between the components of dynamic modulus and dynamic viscosity (equations (6.10)).

Dynamo-optical and strain-optical coefficients can be estimated from measurements of birefringence Δn under elongational flow or shear flow, correspondingly

$$\Delta n = 3S(\omega)\nu_{11} = 3O(\omega)u_{11}, \tag{10.24}$$
$$\Delta n = 2S(\omega)\nu_{12} = 2O(\omega)u_{12}. \tag{10.25}$$

Note that a frequency-dependent stress-optical coefficient $C(\omega)$ can be introduced by comparing the stress tensor and the relative permittivity tensor

$$\varepsilon_{ij} - \varepsilon_0\delta_{ij} = 2\bar{n}C(\omega)(\sigma_{ij} + p\delta_{ij})$$

where

$$C(\omega) = \frac{S(\omega)}{\eta(\omega)} = \frac{O(\omega)}{G(\omega)}.$$

10.4 Anisotropy in a Simple Steady-State Shear Flow

Let us consider the anisotropy of polymer system undergoing simple steady-state shear. This situation can be realised experimentally in a simple way (Tsvetkov et al. 1964). The quantity measured in experiment are the birefringence Δn and the extinction angle χ which are defined by formulae (10.19) and (10.20), correspondingly, through components of the relative permittivity tensor.

10.4.1 Dilute Solutions

One can turn to equation (10.10) to find the components of the relative permittivity tensor. Using expressions for the moments (2.42), one determines the gradient dependence of the quantities for dilute polymer solutions to within second-order terms

$$\Delta n = 2CnT \sum_{\nu=1}^{N} \tau_\nu^\perp \nu_{12},$$
$$\tan 2\chi = \frac{1}{2A\nu_{12}}, \qquad \chi = \frac{\pi}{4} - A\nu_{12} \tag{10.26}$$

where two non-dimensional quantities, the stress-optical coefficient C and the characteristic angle A, have been introduced as

$$C = \frac{\Gamma}{4\bar{n}\mu T} = \frac{2\pi}{45\bar{n}T}(\varepsilon_0 + 2)^2\Delta\alpha, \tag{10.27}$$

$$A = \frac{1}{2}\sum_{\nu=1}^{N}(1 + \varphi_\nu)(\tau_\nu^\perp)^2 \cdot \left(\sum_{\nu=1}^{N}\tau_\nu^\perp\right)^{-1}. \tag{10.28}$$

We can see from equation (10.27) that the stress-optical coefficient depends neither on the molecular weight of the polymer nor on the number of subchains and is proportional to the polarisability of the Kuhn segment $\Delta\alpha$. The stress-optical coefficient can be estimated by investigation of the shear motion of a very dilute polymer solution, as the ratio of the characteristic extent of double refraction $[\Delta n]$ to the initial characteristic (intrinsic) viscosity defined by equation (6.23)

$$2C = \frac{[\Delta n]}{[\eta]_0}, \quad [\Delta n] = \lim_{\substack{c \to 0 \\ \nu_{12} \to 0}} \frac{n_1 - n_2}{c \eta_s \nu_{12}}. \tag{10.29}$$

As far as the characteristic angle (10.28) is concerned, taking into account the dependence of the relaxation time and of the internal viscosity on the number of the mode (formulae (2.27) and (2.31)), one can write, with the aid of the zeta-function $\zeta(x)$,

$$A = \frac{1}{2} \frac{\tau_1}{\zeta(z\nu)} \left[\zeta(2z\nu) + \varphi_1 \zeta(2z\nu - \theta) \right].$$

The first term of the expression is proportional to the solvent viscosity η_s and the second to the internal viscosity (kinetic rigidity) of the macromolecule, so that measurement of the anisotropy of solutions in different solvents makes it possible to estimate the quantity

$$\tau_1 \varphi_1 \sim M^{z\nu - \theta}.$$

The experimental results (Tsvetkov et al. 1964) for macromolecules of different lengths shows that

$$\tau_1 \varphi_1 \sim M^{1 \to 1.2}$$

and one can write an approximate empirical relation

$$\theta - z\nu + 1 = 0. \tag{10.30}$$

An independent empirical confirmation of this relation was discussed in Section 6.2.4. The relation was mention in Section 2.5 and used at the choice of specific values of the parameters for calculation of dynamic properties of dilute solutions in Section 6.2.2.

Of course, all the derived relations are valid for velocity gradients which are not too large. Otherwise, second-order terms of equation (10.6) should be taken into account, when equation (10.10) is being written, which complicates the situation. We may note that very interesting phenomena may occur, for example, at high velocity gradients. If $\Gamma < 0$, the so-called anomalous dependencies (discovered in Tsvetkov's laboratory and discussed, in particular, by Gotlib and Svetlov 1964a, 1964b) of the extent of double refraction and of the extinction angle on the velocity gradient are observed in experiments, indicating that the principal axes of the tensor of the average orientation of optical anisotropy do not coincide. In order to interpret these phenomena, one has to turn firstly to equations of type (10.6) for the relative permittivity tensor that are non-linear as regards orientation.

10.4.2 Entangled Systems

Now we refer to formula (10.13) for the relative permittivity tensor to determine the characteristic quantities in this case of strongly entangled linear polymers. We use expansions (7.32) and (7.43) for the internal variables to obtain the expression for the components of the tensor through velocity gradients

$$\varepsilon_{12} = \frac{\pi^2}{3} nT \, \bar{n} C \, \tau^* B \, \nu_{12},$$

$$\varepsilon_{22} - \varepsilon_{11} = \frac{2\pi^2}{3} nT \, \bar{n} C \, \tau\tau^* B \, \nu_{12}^2.$$

Then we can write the characteristic quantities

$$\Delta n = \frac{\pi^2}{3} nT \, C \, \tau^* B \, \nu_{12}, \qquad \tan\chi = \frac{1}{\tau\nu_{12}}. \tag{10.31}$$

Of course, these relations are trivial consequences of the stress-optical law (equation (10.12)). However, it is important that these relations would be tested to confirm whether or not there is any deviations in the low-frequency region for a polymer system with different lengths of macromolecules and to estimate the dependence of the largest relaxation time on the length of the macromolecule. In fact, this is the most important thing to understand the details of the slow relaxation behaviour of macromolecules in concentrated solutions and melts.

10.5 Oscillatory Birefringence

10.5.1 Dilute Solutions

One can turn to discussion of the dynamo-optical coefficient, defined by equation (10.22). The expression for the relative permittivity tensor (10.10) and equation (2.41) for the moments allow one to write

$$S(\omega) = nTC \sum_{\alpha=1}^{N} \frac{\tau_\alpha^\perp}{1 - i\omega\tau_\alpha^\parallel}.$$

The stress-optical coefficient C is defined by equation (10.27) and the relaxation times $\tau_\alpha^\perp$ and $\tau_\alpha^\parallel$ are defined by relations (2.30). One can see that the dynamo-optical coefficient of dilute polymer solutions depends on the non-dimensional frequency $\tau_1\omega$, the measure of internal viscosity φ_1 and indices $z\nu$ and θ

$$S(\omega) = nTC\tau_1 f(\tau_1\omega, \varphi_1, z\nu, \theta).$$

For the components of dynamo-optical coefficient, one can find the equations, established by Thurston and Peterlin (1967),

$$S'(\omega) = nTC \sum_{\alpha=1}^{N} \frac{\tau_\alpha^\perp}{1 + (\tau_\alpha^\parallel \omega)^2},$$

$$S''(\omega) = nTC \sum_{\alpha=1}^{N} (1 + \varphi_\alpha) \frac{(\tau_\alpha^\perp)^2 \omega}{1 + (\tau_\alpha^\parallel \omega)^2}. \qquad (10.32)$$

One can see that, in the case when the intramolecular viscosity is neglected ($\varphi_1 = 0$), the frequency dependence of the components of the dynamo-optical coefficient (10.32) agrees with the analogous dependence of the shear viscosity (see equation (6.20) and Fig. 14). The stress-optical law can be written in the form

$$\frac{S'(\omega)}{\eta'(\omega) - \eta_s} = C, \qquad \frac{S''(\omega)}{\eta''(\omega)} = C. \qquad (10.33)$$

Pokrovskii and Kokorin (1987) extended the results to the more general case where the internal viscosity parameter assume arbitrary values and the excluded-volume effects are taken into account.

Of course, equations (10.32) and (10.33) are valid in linear approximation for velocity gradients which are not too large and for low frequencies. Is the stress-optical law valid, at the higher frequencies, when the intramolecular relaxation processes have to be taken into account? Deviations from the stress-optical law can emerge, if one assumes the equilibrium distribution of segment orientation, when the expression for the relative permittivity tensor was written, whereas the internal viscosity in dynamic viscosity is included in proper way. At correct consideration, the deviations from the stress-optical law do not appear in the theory. At very high frequencies, the real part of the dynamo-optical coefficient is zero, while the real part of dynamic viscosity remains finite. By investigating optical anisotropy and stresses at high frequencies, one can estimate from the experimental data the importance of intramolecular relaxation processes in the dynamics of the macromolecule.

The work by Lodge et al. (1982) contains the experimental data on the frequency dependencies of the dynamo-optical coefficient for infinitely dilute solutions of polymer, which are represented as frequency dependence of the magnitude and the phase angle, respectively

$$S_m = \left((S')^2 + (S'')^2 \right)^{1/2}, \qquad \tan\theta_s = \frac{S''}{S'}.$$

10.5.2 Entangled Systems

The strain-optical coefficient $O(\omega)$, defined by equation (10.22), can be corresponded to dynamic modulus calculated in Section 6.4.2. Taking all the previous speculations into account, an expression for the strain-optical coefficient can be written in general way as

$$O(\omega) = nT \sum_{a=1}^{6} \sum_{\alpha=1}^{N} (-i\omega) \frac{C_a p_\alpha^{(a)} \tau_\alpha^{(a)}}{1 - i\omega \tau_\alpha^{(a)}} \qquad (10.34)$$

where the times of relaxation $\tau_\alpha^{(1)}$, $\tau_\alpha^{(2)}$, $\tau_\alpha^{(3)}$, $\tau_\alpha^{(4)}$, $\tau_\alpha^{(5)}$, $\tau_\alpha^{(6)}$ and the corresponding weights $p_\alpha^{(1)}$, $p_\alpha^{(2)}$, $p_\alpha^{(3)}$, $p_\alpha^{(4)}$, $p_\alpha^{(5)}$, $p_\alpha^{(6)}$ are the same as calculated for dynamic modulus in Sections 6.4.2, while the stress-optical coefficients C_1, C_2, C_3, C_4, C_5, C_6 are assume can be different for different relaxation branches. It is possible that the different relaxation branches are connected with different types of motion and are characterised with different values of the stress-optical coefficient. The stress-optical coefficients are proportional to the polarisability of the structural units of the macromolecule, which can be different for different types of motion of the chain (Gao and Weiner 1994). The strain-optical coefficient of entangled system depends on the non-dimensional frequency $\tau^*\omega$ and on the non-dimensional parameters

$$O(\omega) = nT\, f(C_1, C_2, C_3, C_4, C_5, B, \chi, \tau^*\omega).$$

The contributions into dynamic modulus and, consequently, into strain-optical coefficient from the high-frequency branches are discussed in Section 6.4.2. In the case, when the all stress-optical coefficients are equal, a graphs for the components the strain-optical coefficient have the same form as the graph for the components of the dynamic modulus, which, for example, are shown in Fig. 17. In general case, expression (10.34) allows us to describe different types of the frequency dependence of the strain-optical coefficient and this can gives an explanation to the "curious behaviour" of the strain-optical coefficient of polymer solutions and melts. In fact, considering strain-optical coefficient in a great range of frequencies, scholars have to admit that the stress-optical coefficient C depends either on frequency or is different for different relaxation branches, to explain experimental data (Inoue et al. 1991; Okamoto et al. 1995). We cannot discuss comparison between the experimental and theoretical curves any more, because it is an illustration of a phenomenon which ought to be investigated carefully.

In application to very concentrated solutions, the optical anisotropy provides the important means for the investigation of slow relaxation processes. It is important to confirm whether or not there is any deviations from the stress-optical law in the low-frequency region for a polymer melt with different lengths of macromolecules. In fact, this is the most important thing to understand the details of the slow relaxation behaviour of macromolecules in concentrated solutions and melts. The evidence can be decisive for understanding the mechanism of the relaxation.

Conclusion

The mesoscopic approach gives an amazingly consistent picture of the different relaxation phenomena in very concentrated solutions and melts of linear polymers. It is not surprising: the developed theory is a sort of phenomenological (mesoscopic) description, which allows one to get a consistent interpretation of experimental data connected with dynamic behaviour of linear macromolecules in both weakly and strongly entangled polymer systems in terms of a few phenomenological (or better, mesoscopic) parameters: it does not require any specific hypotheses.

The approach is based fundamentally on a basic picture of thermal motion of macromolecules in entangled polymer systems. The simple representation of the macromolecules, as chains of Brownian particles, allows us to explain the peculiarities of diffusion and relaxation of macromolecules both in weakly and strongly entangled systems – the peculiarities, which puzzled investigators for long times. The consequent analysis in the frame of the formal mesoscopic theory justifies the suggestion about localisation of a macromolecule in the tube, assumed by Edwards (1967a), and about reptation motion of macromolecules, guessed by de Gennes (1971), while restricting the regions of emerging these effects to the region of the systems of long macromolecules – strongly entangled systems. The formal mesoscopic theory justifies intuitive introduction of an internal intermediate length – a tube – into the consideration, though defines the radius of the tube more precisely in terms of the fundamental parameters. The motion of any Brownian particle of the chain in the system of strongly entangled macromolecules is confined. A very long macromolecule appears, in fact, to behave exactly as if confined in a tube, though no other restrictions than mesoscopic dynamic equations exist (Chapter 5). The analysis justifies the very existence of reptation mobility in the strongly entangled systems, but discovers that the role of the reptation mechanism in interpretation of viscoelasticity was exaggerated. Conformational relaxation in the strongly entangled systems is realised through the reptation mechanism of motion of the macromolecule inside the tube, but it is not only relaxation process. The reptation relaxation exists but practically does not affect lin-

V.N. Pokrovskii, *The Mesoscopic Theory of Polymer Dynamics*,
Springer Series in Chemical Physics 95,
DOI 10.1007/978-90-481-2231-8, © Springer Science+Business Media B.V. 2010

ear viscoelasticity and optical birefringence at low frequencies (Chapters 6 and 10). However, the phenomena of diffusion of long macromolecules cannot be understood without taking into account the reptation mobility. At $M > 10M_e$, the reptation mechanism of displacement predominates (Chapter 5). The analysis of the non-linear effects of viscoelasticity also confirms that the reptation relaxation has to be included in consideration to explain correctly the observed dependencies of rheological characteristics on molecular weight of polymer (Chapter 9).

Apart of empirical justification, the mesoscopic effective-field approach itself is needed in proper microscopic justification. One can anticipate that the more detailed theories could be very helpful in understanding the thermal behaviour of macromolecules and explaining the introduced mesoscopic parameters. One can imagine that a theory of a deeper level based on the heuristic model of rigid, connected in chains and interacting with each other, segments, we can say microscopic theory as compared with the exploited mesoscopic approach, could help elucidate the meaning of mesoscopic parameters and bring answers to some questions, in particular, to give us a description of a few relaxation branches, including orientation and reptation branches. The peculiarities of dynamics of macromolecules can be also deduced from geometrical and topological aspects of macromolecular dynamics (Kholodenko 1996), so that the parameters of the theory eventually could be linked with details of structure of entangled systems. One can believe that the developing methods (Schweizer 1989a, 1989b; Vilgis and Genz 1994; Guenza 1999; Rostiashvili et al. 1999; Fatkullin et al. 2000) can be helpful to bring a microscopic justification of the mesoscopic approach to the entangled systems, which will help to formulate the correct answer for the problem.

Although the microscopic theory remains to be the real foundation of the theory of relaxation phenomena in polymer systems, the mesoscopic approach has and will not lose its value. It will help to understand the laws of diffusion and relaxation of polymers of various architecture. The information about the microstructure and microdynamics of the material can be incorporated in the form of constitutive relation, thus, allowing to relate different linear and non-linear effects of viscoelasticity to the composition and chemical structure of polymer liquid.

Appendices

A The Random Walk Problem

The conformation of a macromolecule consisting of N independent subchains (or segments) can be considered as the result of a random walk of a Brownian particle after N independent steps (Flory 1953).

One can assume that the displacement r of the particle is random and a probability distribution function

$$f(r), \qquad \int_{-\infty}^{\infty} f(r)dr = 1$$

exists, so that $f(r)dr$ is the probability that the particle is displaced to a distance r and is in the volume dr.

Since the space is isotropic, we consider the distribution function to be a spherically symmetrical one, so that

$$f(r)dr = f(r)r^2 \sin\theta \, dr \, d\theta \, d\phi.$$

The mean square one-step displacement is written as

$$b^2 = \int f(r)r^2 dr = 4\pi \int_0^\infty f(r)r^4 \, dr.$$

The situation can be simplified by proposing that the one-step displacement is constant and equal to b, so that the probability distribution function takes the form

$$f(r) = \frac{1}{2\pi b}\delta(r^2 - b^2), \qquad 4\pi \int_0^\infty f(r)r^2 \, dr = 1.$$

It is easy to see that the function is normalised to unity and the mean square displacement is equal to b^2. Indeed

V.N. Pokrovskii, *The Mesoscopic Theory of Polymer Dynamics*,
Springer Series in Chemical Physics 95,
DOI 10.1007/978-90-481-2231-8, © Springer Science+Business Media B.V. 2010

$$\int_{\infty}^{-\infty} \delta(r^2 - b^2)\, \mathbf{dr} = 4\pi \int_0^\infty \delta(r^2 - b^2) r^2 \, dr$$

$$= 2\pi \int_0^\infty \delta(r^2 - b^2) r \, dr^2 = 2\pi \int_0^\infty \delta(x - b^2)\sqrt{x}\, dx = 2\pi b,$$

$$\frac{1}{2\pi b} \int_{-\infty}^\infty \delta(r^2 - b^2) r^2 \, \mathbf{dr} = \frac{2}{b} \int_0^\infty \delta(r^2 - b^2) r^4 \, dr$$

$$= \frac{1}{b} \int_0^\infty \delta(r^2 - b^2) r^3 \, dr^2 = \frac{1}{b} \int_0^\infty \delta(x - b^2) x^{3/2} \, dx = b^2.$$

Now, one can calculate a mean displacement R of the particle after N steps

$$\mathbf{R} = \sum_{\alpha=1}^{N} \mathbf{r}_\alpha.$$

First of all, we can write down the probability distribution function for R

$$W_N(\mathbf{R}) = \int_{-\infty}^\infty \delta\left(\sum_{\alpha=1}^N \mathbf{r}_i^\alpha - \mathbf{R}\right) f(\mathbf{r}^1) f(\mathbf{r}^2) \cdots f(\mathbf{r}^N)\, \mathbf{dr}^1 \mathbf{dr}^2 \cdots \mathbf{dr}^N.$$

Remembering the representation of the δ-function

$$\delta\left(\sum_{\alpha=1}^N \mathbf{r}_i^\alpha - \mathbf{R}\right) = \frac{1}{(2\pi)^3} \int_{-\infty}^\infty \exp\left[i\mathbf{q}\left(\mathbf{R} - \sum_{\alpha=1}^N \mathbf{r}^\alpha\right)\right] \mathbf{dq},$$

we rewrite distribution function in the form

$$W_N(\mathbf{R}) = \frac{1}{(2\pi)^3} \int_{-\infty}^\infty \left[\int_{-\infty}^\infty f(\mathbf{r}) \exp(-i\mathbf{qr}) \mathbf{dr}\right]^N \exp(i\mathbf{qR})\, \mathbf{dq}.$$

It is convenient to calculate separately the integral inside the square brackets. Subsequent transformations determine the expression

$$\int_{-\infty}^\infty f(\mathbf{r}) \exp(-i\mathbf{qr})\, \mathbf{dr} = \frac{1}{2\pi b} \int_{-\infty}^\infty \delta(r^2 - b^2) \exp(-i\mathbf{qr})\, \mathbf{dr}$$

$$= \frac{1}{2\pi b} \int_0^\infty \int_0^{2\pi} \int_0^\pi \delta(r^2 - b^2) e^{-iqr\cos\theta} r^2 \sin\theta \, d\theta \, d\phi \, dr$$

$$= \frac{1}{b} \int_0^\infty \int_1^{-1} \delta(r^2 - b^2) e^{-iqr\cos\theta} r^2 d\cos\theta \, dr$$

$$= \frac{1}{b} \int_0^\infty \delta(r^2 - b^2) r^2 \frac{e^{iqr} - e^{-iqr}}{iqr} \, dr$$

$$= \frac{2}{b} \int_0^\infty \delta(r^2 - b^2) r^2 \frac{\sin qr}{qr} dr = \frac{1}{bq} \sin bq.$$

One can be interested in the situation when the one-step distance is small but the number of steps is large, so that we can calculate distribution function under the conditions

$$bq \ll 1, \quad N \to \infty.$$

In this case

$$\left(\frac{1}{bq}\sin bq\right)^N \approx \left(1 - \frac{1}{6}b^2q^2\right)^N \approx \exp\left(-\frac{1}{6}b^2q^2N\right).$$

Therefore, the distribution function takes the form

$$W_N(R) = \frac{1}{(2\pi)^3}\int_{-\infty}^{\infty}\exp\left(iqR - \frac{Nq^2b^2}{6}\right)dq. \tag{A.1}$$

To calculate the integral, we divide the expression (A.1) into real and imaginary parts, taking into account that

$$e^{iqR} = \cos qR + i\sin qR.$$

The imaginary part of (A.1) is equal to zero identically. The calculation of the real part determines the distribution function

$$W_N(R) = \left(\frac{3}{2\pi Nb^2}\right)^{3/2}\exp\left(-\frac{3R^2}{2Nb^2}\right). \tag{A.2}$$

The mean square displacement for N steps can easily be calculated with the help of (A.2)

$$\langle R^2\rangle = \int W_N(R)R^2 d\mathbf{R} = Nb^2.$$

Now the distribution function (A.2) can be represented in the form 1.5.

As has been noted already, the results are valid under the assumption of the independence of the separate steps and at large N.

B Equilibrium Deformation of a Non-Linear Elastic Body

In an equilibrium state, the stress tensor is determined by the form and the volume of a deformed body. To determine the stress tensor of the deformed body at arbitrary (not small) deformation, we follow the method demonstrated by Landau and Lifshitz (1987b) for the calculation of the stress tensor for small deformation.

Thermodynamic Relations for a Deformed Body

Let us denote the Cartesian co-ordinates of an arbitrary point of the body before deformation as x_i^0. The co-ordinates of the same point after deformation x_k are functions of the original co-ordinates

$$x_k = x_k(\boldsymbol{x}^0).$$

In the case of arbitrary deformation, one can determine the relations between some small quantities in deformed and non-deformed states, namely, the relations for the co-ordinate, length and volume, respectively

$$dx_i = \frac{\partial x_i}{\partial x_k^0}\, dx_k^0 = \lambda_{ik}\, dx_k^0,$$

$$(dl)^2 = \lambda_{ik}\lambda_{il}\, dx_k^0\, dx_l^0, \qquad\qquad (B.1)$$

$$dV = |\lambda| dV_0.$$

The last relation can easily be checked if the original volume dV_0 is taken as the volume of a parallelepiped whose sides are situated along the co-ordinate axes

$$dV_0 = dx_1^0 dx_2^0 dx_3^0.$$

The tensor of the displacement gradients

$$\lambda_{ij} = \frac{\partial x_i}{\partial x_j^0}$$

appears in (B.1). One can notice that, when the body as a whole is rotated round a point at some angle, the displacement gradient tensor appears to be unequal to zero. So, this tensor cannot be a measure of the body's deformation. It is convenient, following Murnaghan (1954), to choose the symmetrical tensor $\Lambda_{kj} = \lambda_{ij}\lambda_{ik}$ as a measure of arbitrary deformation.

Let us introduce the stress tensor σ_{ik} referred to the deformed body, so that the force per unit of the deformed volume can be written as follows

$$F_i = \frac{\partial \sigma_{ik}}{\partial x_k}. \qquad\qquad (B.2)$$

Now, one can define the work done at the virtual transfer between two deformed states of the body as

$$\delta R dV = \frac{\partial \sigma_{ik}}{\partial x_k}\delta x_i\, dV = \frac{\partial \sigma_{ik}}{\partial x_j^0}\lambda_{jk}^{-1}\delta x_i\, dV.$$

The last relation can be integrated over a volume. We shall assume that the stresses disappear on the borders of the integrating volume and, after some simple transformation, obtain

$$\int_V \delta R\, dV = -\int_V \sigma_{ik}\lambda_{jk}^{-1}\delta\lambda_{ij}\, dV.$$

This relation determines the work per unit of deformed body

$$dR = -\frac{1}{2}\lambda_{ji}^{-1}\sigma_{ik}\lambda_{sk}^{-1}\, d\Lambda_{js}. \qquad\qquad (B.3)$$

The strain tensor Λ_{js} describes both a change of form and a change of volume of a body. It is convenient to separate these parts by introducing a strain tensor λ'_{ij} such that the determinator of matrix $|\lambda'_{ij}| = 1$. Then we can write

$$\lambda_{ij} = \left(\frac{V}{V_0}\right)^{1/3} \lambda'_{ij}, \qquad \Lambda_{ik} = \left(\frac{V}{V_0}\right)^{2/3} \lambda'_{li}\lambda'_{lk},$$
$$\mathrm{d}\Lambda_{ik} = \frac{2}{3}\frac{1}{V_0}\left(\frac{V}{V_0}\right)^{-1/3} \Lambda'_{ik}\,\mathrm{d}V + \left(\frac{V}{V_0}\right)^{2/3} \mathrm{d}\Lambda'_{ik}. \tag{B.4}$$

We can note that the change of volume during the deformation of a body is usually small, so we can assume further that the original strain tensor can be used instead of the newly introduced one.

Then, the relation (B.3) can be rewritten in the form

$$\mathrm{d}R = \frac{1}{V}p\,\mathrm{d}V - \frac{1}{2}\lambda_{ji}^{-1}\sigma_{ik}\lambda_{sk}^{-1}\,\mathrm{d}\Lambda_{js} \tag{B.5}$$

where the notation for isotropic pressure is introduced.

$$p = -\frac{1}{3}\sigma_{jj}.$$

The free energy F of the body depends on the deformation and is determined (Landau and Lifshitz 1969) by the general expression

$$\mathrm{d}F = -S\,\mathrm{d}T - V\,\mathrm{d}R. \tag{B.6}$$

This relation together with expressions (B.3) and (B.5) determines the stress tensor and pressure in the deformed body

$$\sigma_{ik} = \frac{2}{V}\lambda_{kj}\lambda_{is}\left(\frac{\partial F}{\partial \Lambda_{sj}}\right)_T, \qquad p = -\left(\frac{\partial F}{\partial V}\right)_T. \tag{B.7}$$

Free Energy and Stress Tensor of Deformed Body

The free energy of the deformed isotropic body depends on the strain tensor, it is a function of three invariants of the strain tensor. The volume of the deformed body

$$V = V_0|\Lambda_{ik}|^{3/2}$$

can be taken as one of the invariants. The others can be defined as

$$I_1 = \Lambda_{ii} - 3; \qquad I_2 = \Lambda_{ii}^2 - \Lambda_{ik}\Lambda_{ik}.$$

The first of the expressions in (B.7) can now be rewritten as follows

$$\sigma_{ik} = \frac{2}{V}\lambda_{kj}\lambda_{is}\left(\frac{\partial F}{\partial I_1}\frac{\partial I_1}{\partial \Lambda_{sj}} + \frac{\partial F}{\partial I_2}\frac{\partial I_2}{\partial \Lambda_{sj}} + \frac{\partial F}{\partial V}\frac{\partial V}{\partial \Lambda_{sj}}\right). \tag{B.8}$$

The derivatives of the invariants with respect to the components of the strain tensor can easily be calculated

$$\frac{\partial I_1}{\partial \Lambda_{sj}} = \delta_{sj}, \qquad \frac{\partial I_2}{\partial \Lambda_{sj}} = 2(\Lambda_{qq}\delta_{sl}\delta_{jl} - \Lambda_{sj}), \qquad \frac{\partial V}{\partial \Lambda_{sj}} = \frac{3}{2}(V_0 V)^{1/2} A_{sj}$$

where A_{sj} is the algebraic additive to components s, j of the matrix of the strain tensor.

So, one needs to know the three undefined functions of the three invariants of the strain tensor

$$C_1 = \frac{1}{V}\frac{\partial F}{\partial I_1}, \qquad C_2 = \frac{2}{V}\frac{\partial F}{\partial I_2}, \qquad p = -\frac{\partial F}{\partial V}$$

to determine, according to expression (B.8), the stress tensor for an elastic body

$$\sigma_{ik} = -p\delta_{ik} + 2C_1\lambda_{ij}\lambda_{kj} + 2C_2(\lambda_{is}\lambda_{ks}\lambda_{jl}\lambda_{jl} - \lambda_{is}\lambda_{kj}\lambda_{is}\lambda_{lj}). \tag{B.9}$$

Quantities C_1 and C_2 are functions of the two invariants of the stress tensor I_1 and I_2 for incompressible material.

C The Tensor of Hydrodynamic Interaction

To determine the perturbation of fluid velocity under the influence of volume forces $\boldsymbol{\sigma}$, we shall begin with the equations of motion of a viscous liquid at low Reynolds numbers (Landau and Lifshitz 1987a)

$$\eta_s \nabla^2 \boldsymbol{v} - \nabla p + \boldsymbol{\sigma} = 0, \qquad \operatorname{div} \boldsymbol{v} = 0, \tag{C.1}$$

where η_s is coefficient of viscosity, p is pressure, $\boldsymbol{v} = \boldsymbol{v}(\boldsymbol{x}, t)$ is the velocity and σ_i is the density of the outer forces.

Henceforth, it is convenient to use the Fourier transforms of the velocity, pressure and volume force, correspondingly

$$\boldsymbol{v}(\boldsymbol{k}) = \int \boldsymbol{v}(\boldsymbol{x}) \exp i\boldsymbol{k}\boldsymbol{x} \, d\boldsymbol{x},$$

$$p(\boldsymbol{k}) = \int p(\boldsymbol{x}) \exp i\boldsymbol{k}\boldsymbol{x} \, d\boldsymbol{x}, \tag{C.2}$$

$$\boldsymbol{\sigma}(\boldsymbol{k}) = \int \boldsymbol{\sigma}(\boldsymbol{x}) \exp i\boldsymbol{k}\boldsymbol{x} \, d\boldsymbol{x}$$

which obey a set of equations

$$-\eta_s k^2 \boldsymbol{v}(\boldsymbol{k}) - ik p(\boldsymbol{k}) + \boldsymbol{\sigma}(\boldsymbol{k}) = 0, \qquad \boldsymbol{k}\boldsymbol{v}(\boldsymbol{k}) = 0. \tag{C.3}$$

It is easy to see that the solution of the last set of equations takes the form

$$p(\mathbf{k}) = -i\frac{\mathbf{k}\boldsymbol{\sigma}(\mathbf{k})}{k^2},$$

$$v_i(\mathbf{k}) = \frac{1}{\eta_s k^2}\left(\delta_{ij} - \frac{k_i k_j}{k^2}\right)\sigma_j(\mathbf{k}).$$

The last relation determines the velocity as a function of co-ordinate

$$\mathbf{v}(\mathbf{x}) = \int d\mathbf{x}' \mathbf{H}(\mathbf{x} - \mathbf{x}')\boldsymbol{\sigma}(\mathbf{x}') \qquad (C.4)$$

where $\mathbf{H}$ is a tensor of the second rank with components

$$H_{ij}(\mathbf{r}) = \frac{1}{(2\pi)^3 \eta_s}\int \frac{d\mathbf{k}}{k^2}\left(\delta_{ij} - \frac{k_i k_j}{k^2}\right)\exp(-i\mathbf{k}\mathbf{r}) \qquad (C.5)$$

which, after the calculation of the integral, can be rewritten as

$$H_{ij}(\mathbf{r}) = \frac{1}{8\pi\eta_s r}\left(\delta_{ij} + \frac{r_i r_j}{r^2}\right). \qquad (C.6)$$

Let us note that the only assumption used here is that of slow motion: it was assumed that the Reynolds number is small.

For a suspension of Brownian particles which are far from each other, the volume force can be represented as a sum of point influences

$$\boldsymbol{\sigma}(\mathbf{x}) = -\sum_{\alpha} \mathbf{F}^{\alpha}(\mathbf{x}^{\alpha})\delta(\mathbf{x} - \mathbf{x}^{\alpha}) \qquad (C.7)$$

where $\mathbf{F}^{\alpha}(\mathbf{x}^{\alpha})$ is the resistance force acting on the particle. It is easy to see that, in this case, expression (C.4) is followed by formula (2.5) where the tensor of hydrodynamic interaction is defined by (2.6).

D Resistance Force of a Particle in a Viscoelastic Fluid

Let us find the resistance force acting on a spherical particle of radius a which moves slowly with velocity u in an incompressible viscoelastic fluid. It means that the Reynolds number of the problem is small, the convective terms are negligibly small, and the equations of fluid motion are

$$\rho\frac{\partial v_i}{\partial t} = \frac{\partial \sigma_{ij}}{\partial x_j}, \qquad \frac{\partial v_i}{\partial x_i} = 0,$$

$$\sigma_{ij} = -p\delta_{ij} + 2\int_0^{\infty}\eta(s)\gamma_{ij}(t - s)ds, \qquad \gamma_{ij} = \frac{1}{2}\left(\frac{\partial v_i}{\partial x_j} + \frac{\partial v_j}{\partial x_i}\right) \qquad (D.1)$$

where ρ is a constant density, $\mathbf{v} = \mathbf{v}(\mathbf{x}, t)$ is the velocity of the liquid and σ_i is the density of the outer forces.

A fading memory function $\eta(s)$ can be represented as a sum of exponential functions

$$\eta(s) = \sum_\alpha \frac{\eta_\alpha}{\tau_\alpha} \exp\left(-\frac{t}{\tau_\alpha}\right).$$

The coefficients of partial viscosity η_α and relaxation times τ_α are the characteristics of the liquid.

It is convenient to apply the Fourier transforms of the variables

$$v_i(\omega) = \int_{-\infty}^{\infty} v_i(t)e^{i\omega t}\,dt,$$

$$p(\omega) = \int_{-\infty}^{\infty} p(t)e^{i\omega t}\,dt$$

to transform the equations of motion (D.1) to the form

$$i\omega p v_i(\omega) + \frac{\partial p(\omega)}{\partial x_j} = \eta[\omega]\frac{\partial^2 v_i}{\partial x_i \partial x_j}, \qquad \frac{\partial v_i}{\partial x_i} = 0 \qquad \text{(D.2)}$$

where

$$\eta[\omega] = \int_0^{\infty} \eta(t)e^{i\omega t}\,dt.$$

It can be easily seen that, if we consider $\eta[\omega]$ as a constant viscosity coefficient, the written equations are identical to the equations of motion of a viscous liquid. The solution of the problem of motion of a sphere in a viscous liquid is well known, so that the resistance force in our problem can be written as follows

$$\boldsymbol{F}(\omega) = -6\pi a \eta[\omega]\boldsymbol{u}(\omega)$$

or

$$\boldsymbol{F}(t) = -6\pi a \int_0^{\infty} \eta(s)\boldsymbol{u}(t-s)\,ds. \qquad \text{(D.3)}$$

These results are valid for an arbitrary memory function $\eta(s)$, which can be represented as the sum of exponential functions. In the simplest case

$$\eta(s) = \frac{\eta}{\tau}\exp\left(-\frac{t}{\tau}\right), \qquad \eta[\omega] = \frac{\eta}{1 - i\omega\tau}, \qquad \text{(D.4)}$$

$$\boldsymbol{F}(t) = -\frac{\zeta}{\tau}\int_0^{\infty}\exp\left(-\frac{s}{\tau}\right)\boldsymbol{u}(t-s)\,ds \qquad \text{(D.5)}$$

where $\zeta = 6\pi a\eta$ is the friction coefficient of a sphere in a viscous liquid.

One can note that the resistance force (D.5) is a solution of the linear equation

$$\tau\frac{dF_i}{dt} = -F_i - \zeta u_i. \qquad \text{(D.6)}$$

To extend result (D.5) to non-linear cases, it is convenient to begin with equation (D.6). The derivative with respect to time ought to be replaced by the Yaumann derivative (see, Section 8.4), so that, in the simplest case, the covariant equation for the resistance force has the form

$$\frac{dF_i}{dt} - \omega_{il}F_l + \frac{1}{\tau}F_i = -\frac{\zeta}{\tau}u_j. \tag{D.7}$$

A matrix satisfying the following equations

$$\frac{dC_{ij}}{dt} = -C_{ik}\omega_{kj}, \qquad \frac{dC_{ij}^{-1}}{dt} = -C_{lj}^{-1}\omega_{il} \tag{D.8}$$

has to be introduced, in order to write the solution of equation (D.7) as

$$F_i = -\frac{\zeta}{\tau}\int_0^\infty C_{ik}^{-1}(t)C_{kj}(t-s)\exp\left(-\frac{s}{\tau}\right)u_j(t-s)ds. \tag{D.9}$$

Indeed, the last equation can be differentiated with respect to time to obtain equation (D.7) consequently

$$\frac{dF_i}{dt} = -\frac{\zeta}{\tau}\int_0^\infty \exp\left(-\frac{s}{\tau}\right)\frac{d}{dt}\left[C_{ik}^{-1}(t)C_{kj}(t-s)u_j(t-s)\right]ds$$

$$= -\frac{\zeta}{\tau}\int_0^\infty \exp\left(-\frac{s}{\tau}\right)\frac{dC_{ik}^{-1}(t)}{dt}C_{kj}(t-s)u_j(t-s)ds$$

$$- \frac{\zeta}{\tau}\int_0^\infty \exp\left(-\frac{s}{\tau}\right)C_{ik}^{-1}(t)\frac{d}{ds}\left[C_{kl}(t-s)u_l(t-s)\right]ds$$

$$= \omega_{iq}\frac{\zeta}{\tau}\int_0^\infty \exp\left(-\frac{s}{\tau}\right)C_{qk}^{-1}(t)C_{kj}(t-s)u_j(t-s)ds$$

$$- \frac{\zeta}{\tau}u_i(t) - \frac{\zeta}{\tau^2}\int_0^\infty \exp\left(-\frac{s}{\tau}\right)C_{ik}^{-1}(t)C_{kl}(t-s)u_l(t-s)ds$$

$$= \omega_{iq}F_q - \frac{\zeta}{\tau}u_i - \frac{1}{\tau}F_i.$$

At low velocity gradients, expression (D.9) can be expanded in a series in powers of the antisymmetrical gradient ω_{ij}. The first term of the series has the form of (D.5).

E Resistance Coefficient of a Particle in Non-Local Fluid

The motion of a spherical particle in a non-local fluid was considered by Pokrovskii and Pyshnograi (1988). We reproduce the calculation of the resistant coefficient here.

We consider the viscous liquid to be incompressible and the motion of the particle to be slow. It means that the Reynolds number of the problem is small, the convective terms are negligibly small, and the equations of motion of the fluid can be written as follows

$$\rho \frac{\partial v_i}{\partial t} = \frac{\partial \sigma_{ij}}{\partial x_j} + \sigma_i, \qquad \frac{\partial v_i}{\partial x_i} = 0,$$

$$\sigma_{ij}(\boldsymbol{r}) = -p\delta_{ij} + 2 \int \eta(\boldsymbol{r} - \boldsymbol{r}')\gamma_{ij}(\boldsymbol{r}')\,\mathrm{d}\boldsymbol{r}', \quad \gamma_{ij} = \frac{1}{2}\left(\frac{\partial v_i}{\partial x_j} + \frac{\partial v_j}{\partial x_i}\right) \tag{E.1}$$

where ρ is a constant density, $v = v(x, t)$ is the velocity of the liquid and σ_i is the density of the outer forces. The stress tensor σ_{ij} defines non-local incompressible viscous fluid and contains a decreasing influence function $\eta(\boldsymbol{r})$, which can be represented as a sum of exponential functions. Integrating over the whole volume is assumed in (E.1).

As well as for the local viscous fluid (Landau and Lifshitz 1987a), we consider the spherical particle to be immovable and to be situated at the beginning of the co-ordinate frame, so that the flux of the fluid moves around the particle with constant velocity u at infinity.

The equation of motion of the fluid takes the form

$$\frac{\partial}{\partial r_j}\sigma_{ij} = 0 \quad \text{at } |r| > a. \tag{E.2}$$

It is convenient to rewrite the equation of motion as follows

$$\frac{\partial}{\partial r_j}\sigma_{ij} = -f_i(\boldsymbol{r}). \tag{E.3}$$

Here an induced force $f_i(\boldsymbol{r})$ is introduced such that equation (E.2) can be determined for all values of the variable r. We shall assume that $f_i(\boldsymbol{r}) = 0$ for $|r| > a$.

The force acting on the particle can be calculated by integrating over the surface of the sphere or over the volume of the sphere

$$F_i = -\int_S \sigma_{ij}(\boldsymbol{r}) n_j \,\mathrm{d}S = -\int_V \frac{\partial}{\partial r_j}\sigma_{ij}\,\mathrm{d}\boldsymbol{r}.$$

Taking equation (E.3) into account, the expression for the force can be rewritten as

$$F_i = \int_V f_i(\boldsymbol{r})\,\mathrm{d}\boldsymbol{r}. \tag{E.4}$$

Then, we turn to the Fourier transforms of the quantities, which can be defined, for example, as

$$f_i(\boldsymbol{k}) = \int \exp(-i\boldsymbol{k}\boldsymbol{r}) f_i(\boldsymbol{r})\,\mathrm{d}\boldsymbol{r}.$$

So the equations of motion (E.1) and (E.3) for a non-local fluid take the form

$$ip(\mathbf{k})k_i + k_j\eta(\mathbf{k})\Big(v_i(\mathbf{k})k_j + v_j(\mathbf{k})k_i\Big) = f_i(\mathbf{k}), \qquad k_i v_i(\mathbf{k}) = 0$$

or

$$k^2\eta(\mathbf{k})v_i(\mathbf{k}) = -ip(\mathbf{k})k_i + f_i(\mathbf{k}), \qquad k_i v_i(\mathbf{k}) = 0.$$

The pressure in the last relations can be excluded so that we have an equation for the Fourier transform of velocity

$$k^2\eta(\mathbf{k})v_i(\mathbf{k}) = \left(\delta_{ij} - \frac{k_i k_j}{k^2}\right) f_j(\mathbf{k})$$

which has a solution

$$v_i(\mathbf{k}) = \frac{1}{k^2\eta(\mathbf{k})}\left(\delta_{ij} - \frac{k_i k_j}{k^2}\right) f_j(\mathbf{k}). \tag{E.5}$$

The mean velocity of the fluid taken over the surface of the sphere is equal to the velocity of the sphere. In the system of co-ordinates, where the liquid is immovable at infinity we have

$$\frac{1}{4\pi a^2}\int v_i(\mathbf{r})\delta(r-a)\,\mathbf{dr} = -u_i. \tag{E.6}$$

Relation (E.6) is followed by the relation for the Fourier transform of velocity

$$\frac{1}{(2\pi)^3}\int \frac{\sin ka}{ak}v_i(\mathbf{k})\,\mathbf{dk} = -u_i. \tag{E.7}$$

Then, we can return to expression (E.5) for the velocity transform and can rewrite relation (E.7) as

$$-u_i = \frac{1}{(2\pi)^3}\int \left(\delta_{ij} - \frac{k_i k_j}{k^2}\right)\frac{\sin ka}{\eta(\mathbf{k})ak^3}f_i(\mathbf{k})\mathbf{dk}.$$

To calculate the integral, it is convenient to refer to polar co-ordinates and write

$$-u_i = \frac{1}{3\pi a}\frac{3}{8\pi^2}\int (\delta_{ij} - \Omega_i\Omega_j)\mathrm{d}\Omega \int_0^{\infty} \frac{\sin ka}{k}\frac{f_i(k\Omega)}{\eta(k\Omega)}\mathrm{d}k$$

where $\Omega_i = k_i/k$ is the direct cosine of vector $\mathbf{k}$, and $\mathrm{d}\Omega$ is the differential of the surface of the sphere of unit radius.

Since the value of the integral does not change when we replace k by $-k$, we can also write

$$-u_i = \frac{1}{6\pi a}\frac{3}{8\pi}\int (\delta_{ij} - \Omega_i\Omega_j)\mathrm{d}\Omega\frac{1}{\pi}\int_{-\infty}^{+\infty} \frac{\sin ka}{k}\frac{f_i(k\Omega)}{\eta(k\Omega)}\mathrm{d}k.$$

The last integral can be calculated with the use of the Cauchy theorem about integral values. It results in

$$-u_i = \frac{1}{6\pi a} \frac{1}{8\pi} \int (\delta_{ij} - \Omega_i \Omega_j) \frac{f_i(\boldsymbol{k}=0)}{\eta(\boldsymbol{k}=0)} d\Omega$$

or, eventually,

$$-u_i = \frac{1}{6\pi\eta(\boldsymbol{k}=0)a} f_i(\boldsymbol{k}=0).$$

So, since $f_i(\boldsymbol{k}=0) = F_i$, we obtain the following formula for the force acting on a spherical particle in a non-local viscous fluid

$$F_i = -6\pi\eta(\boldsymbol{k}=0)au_i = -6\pi a \int \eta(\boldsymbol{r}) \, d\boldsymbol{r} \, u_i. \tag{E.8}$$

F Dynamics of Suspension of Dumbbells

In the simplest case, at $N = 1$, the considered subchain model of a macro-molecule reduces to the dumbbell model consisting of two Brownian particles connected with an elastic force. It can be called relaxator as well. The relaxator is the simplest model of a macromolecule. Moreover, the dynamics of a macromolecule in normal co-ordinates is equivalent to the dynamics of a set of independent relaxators with various coefficients of elasticity and internal viscosity. In this way, one can consider a dilute solution of polymer as a suspension of independent relaxators which can be considered here to be identical for simplicity. The latter model is especially convenient for the qualitative analysis of the effects in polymer solutions under motion.

Beginning with pioneering works by Kuhn and Kuhn (1945), the relaxator attracted the attention of researchers (Bird et al. 1987b). Further, on, we shall consider the results concerning the dynamics of the dilute suspension of the dumbbell while the hydrodynamic interaction between particles inside each dumbbell is taken into account in correct form.

The Dynamics of a Dumbbell in a Flow

Equation (1.10) is followed by the expression for elastic forces acting on the zeroth and first particles of the dumbbell

$$\left\| \begin{matrix} \boldsymbol{K}^0 \\ \boldsymbol{K}^1 \end{matrix} \right\| = -2\mu T \left\| \begin{matrix} \boldsymbol{r}^0 - \boldsymbol{r}^1 \\ \boldsymbol{r}^1 - \boldsymbol{r}^0 \end{matrix} \right\|. \tag{F.1}$$

We shall assume that the dumbbell is situated in the stream of viscous fluid characterised by the mean velocity gradient tensor ν_{ij}. According to (2.8), the resistance force for every particle of the dumbbell can be written as

$$F_i^\alpha = -\zeta B_{il}^{\alpha\gamma}(u_l^\gamma - \nu_{lj} r_j^\gamma). \tag{F.2}$$

Here and henceforth in this appendix, Greek labels take the values 0 and 1.

The matrix of the hydrodynamic resistance for two particles can exactly be determined based on results of Section 2.2. The components of the matrix are as follows

$$B_{ij}^{00} = B_{ij}^{11} = \frac{1}{1 - l^2}\delta_{ij} + \frac{3l^2}{(1 - l^2)(1 - 4l^2)}e_i e_j,$$

$$B_{ij}^{01} = B_{ij}^{10} = -\frac{l}{1 - l^2}\delta_{ij} - \frac{l(1 + 2l^2)}{(1 - l^2)(1 - 4l^2)}e_i e_j$$

(F.3)

where e is the unit vector in the direction from the first particle to the zeroth one

$$e = \frac{r^0 - r^1}{|r^0 - r^1|}.$$

The parameter of the hydrodynamic interaction is determined by the radius of the particle and by the distance between the particles

$$l = \frac{\zeta}{8\pi\eta^0|r^0 - r^1|} = \frac{3a}{4|r^0 - r^1|}.$$

We shall also consider intramolecular viscosity. According to equation (2.20), we can write the resistance force in the form

$$\left\| \begin{matrix} G_i^0 \\ G_i^1 \end{matrix} \right\| = -\frac{\lambda}{2} \left\| \begin{matrix} (u_j^0 - u_j^1)e_j e_i \\ (u_j^1 - u_j^0)e_j e_i \end{matrix} \right\|.$$

(F.4)

Further on, we can ignore the inertia forces and introduce instead of the stochastic thermal forces, the mean diffusion forces for each of the particles of the dumbbell

$$-T \left\| \begin{matrix} \frac{\partial \ln W}{\partial r^0} \\ \frac{\partial \ln W}{\partial r^1} \end{matrix} \right\|.$$

(F.5)

In this situation every particle is characterised by the mean diffusion velocity in the co-ordinate space

$$w^\alpha = \langle u^\alpha \rangle.$$

Now we can write down the balance of all the forces acting on each particle of the dumbbell

$$-\zeta B_{ji}^{00}(w_i^0 - \nu_{il}r_l^0) - \zeta B_{ji}^{01}(w_i^1 - \nu_{il}r_l^1)$$

$$-\frac{\lambda}{2}e_j e_i(w_i^0 - w_i^1) - 2\mu T(r_j^0 - r_j^1) - T\frac{\partial \ln W}{\partial r_j^0} = 0,$$

(F.6)

$$-\zeta B_{ji}^{10}(w_i^0 - \nu_{il}r_l^0) - \zeta B_{ji}^{11}(w_i^1 - \nu_{il}r_l^1)$$

$$-\frac{\lambda}{2}e_j e_i(w_i^1 - w_i^0) - 2\mu T(r_j^1 - r_j^0) - T\frac{\partial \ln W}{\partial r_j^1} = 0.$$

The Modes of Motion of the Dumbbell

Transformation (1.14) and (1.16) determine the normal co-ordinates of the dumbbell

$$\rho^0 = \frac{1}{\sqrt{2}}(r^0 + r^1), \qquad \rho = \frac{1}{\sqrt{2}}(r^0 - r^1),$$

$$\psi^0 = \frac{1}{\sqrt{2}}(w^0 + w^1), \qquad \psi = \frac{1}{\sqrt{2}}(w^0 - w^1). \tag{F.7}$$

To transform the set of equations (F.6), it is also necessary to take into account that

$$\frac{\partial}{\partial r^0} = \frac{1}{\sqrt{2}}\left(\frac{\partial}{\partial \rho^0} + \frac{\partial}{\partial \rho}\right), \qquad \frac{\partial}{\partial r^1} = \frac{1}{\sqrt{2}}\left(\frac{\partial}{\partial \rho^0} - \frac{\partial}{\partial \rho}\right).$$

Then the set of equations (F.6) can be rewritten in normal co-ordinates

$$-\zeta\left(\frac{1}{1+l}\delta_{ij} - \frac{l}{(1+l)(1+2l)}e_je_i\right)(\psi_i^0 - \nu_{il}\rho_l^0) - T\frac{\partial \ln W}{\partial \rho_j^0} = 0, \quad \text{(F.8)}$$

$$-\zeta\left(\frac{1}{1-l}\delta_{ij} + \frac{l}{(1-l)(1-2l)}e_je_i\right)(\psi_i - \nu_{il}\rho_l)$$

$$-\lambda e_je_i\psi_i - 4\mu T\rho_j - T\frac{\partial \ln W}{\partial \rho_j} = 0. \tag{F.9}$$

Relation (F.8) determines the mean velocity of the centre of mass of the dumbbell

$$\psi_j^0 = \nu_{jl}\rho_l^0 - \left[(1+l)\delta_{ji} + le_je_i\right]\frac{T}{\zeta}\frac{\partial \ln W}{\partial \rho_i^0}. \tag{F.10}$$

To obtain the velocity of the relaxation mode, we shall transform vector equation (F.9). By multiplying it by the unit vector in two optional different ways (scalar and vector), we obtain the relations

$$\psi_je_j = \frac{1}{1+(1-2l)\gamma}\nu_{il}\rho_ie_l - \frac{4\mu T}{\zeta}\frac{1-2l}{1+(1-2l)\gamma}\rho$$

$$-\frac{T}{\zeta}\frac{1-2l}{1+(1-2l)\gamma}e_i\frac{\partial \ln W}{\partial \rho_i},$$

$$\psi_je_k - \psi_ke_j = \nu_{jl}\rho_le_k - \nu_{kl}\rho_le_j - \frac{T}{\zeta}(1-l)\left(e_k\frac{\partial \ln W}{\partial \rho_j} - e_j\frac{\partial \ln W}{\partial \rho_k}\right)$$

where the coefficient of relative internal viscosity $\gamma = \lambda/\zeta$ is introduced.

These relations define the velocity of the relative motion of the particles of the dumbbell

$$\psi_j = \nu_{jl}\rho_l - \frac{(1-2l)\gamma}{1+(1-2l)\gamma}\nu_{il}e_i e_l \rho_j - \frac{4\mu T}{\zeta}\frac{1-2l}{1+(1-2l)\gamma}\rho_j$$

$$- \left[(1-l)\delta_{ji} - \frac{l+(1-l)(1-2l)\gamma}{1+(1-2l)\gamma}e_j e_i\right]\frac{T}{\zeta}\frac{\partial \ln W}{\partial \rho_i}. \qquad \text{(F.11)}$$

One can see that the relaxation mode does not depend on the diffusion mode and can be considered separately, whereas the diffusion mode cannot.

Diffusion Equation

The equation for distribution function $W(t,\boldsymbol{\rho}^0,\boldsymbol{\rho})$ has the form

$$\frac{\partial W}{\partial t} + \sum_{j=1}^{3}\left(\frac{\partial(\psi_j^0 W)}{\partial \rho_j^0} + \frac{\partial(\psi_j W)}{\partial \rho_j}\right) = 0 \qquad \text{(F.12)}$$

where velocities ψ_j^0 and ψ_j are defined by relations (F.10) and (F.11).

To separate space diffusion, we represent the distribution function as

$$W(t,\boldsymbol{\rho}^0,\boldsymbol{\rho}) = n(t,\boldsymbol{q})\,W(t,\boldsymbol{\rho}), \qquad \boldsymbol{q} = \frac{1}{2}(\boldsymbol{r}^0 + \boldsymbol{r}^1) = \frac{1}{\sqrt{2}}\boldsymbol{\rho}^0,$$

where $\boldsymbol{q}$ is the centre of mass of the dumbbell particles.

The distribution function $W(t,\boldsymbol{\rho})$ is a function normalised to unity, so that the number density function $n(t,\boldsymbol{q})$ can be calculated as

$$n(t,\boldsymbol{q}) = \int W(t,\boldsymbol{\rho}^0,\boldsymbol{\rho})\mathrm{d}\boldsymbol{\rho}.$$

The mean value of a quantity $A(\boldsymbol{\rho})$ depends on co-ordinate $\boldsymbol{q}$ and has to be calculated according to the rule

$$\langle A(\boldsymbol{\rho})\rangle = \frac{1}{n}\int A(\boldsymbol{\rho})W(t,\boldsymbol{\rho}^0,\boldsymbol{\rho})\mathrm{d}\boldsymbol{\rho}.$$

After having integrated equation (F.12), one can obtain the equation of diffusion

$$\frac{\partial n}{\partial t} - \frac{T}{\zeta}\frac{\partial^2}{\partial q_i \partial q_j}[(1+\langle l\rangle)\delta_{ij} + \langle le_i e_j\rangle]n = 0. \qquad \text{(F.13)}$$

In the case of non-homogeneous flows, equation (F.13) determines the effects of orientation on the diffusion of the particles. One can notice that the equation for the diffusion of the centre of mass of the dumbbell, that is an equation for $W(t,\boldsymbol{\rho}^0)$, cannot be written down separately, without reference to the equation for relaxation mode $W(t,\boldsymbol{\rho})$.

In the case when mean values $\langle l\rangle$ and $\langle le_i e_j\rangle$ do not depend on co-ordinates, equation (F.13) is reduced to the known diffusion equation

$$\frac{\partial n}{\partial t} - D_{ij}\frac{\partial^2 n}{\partial q_i \partial q_j} = 0 \tag{F.14}$$

with the anisotropic diffusion coefficient

$$D_{ij} = \frac{T}{2\zeta}[(1 + \langle l \rangle)\delta_{ij} + \langle le_i e_j \rangle]. \tag{F.15}$$

One can use the relations (1.23) to obtain the mean diffusion coefficient of the centre of mass of the relaxators in equilibrium (Öttinger 1989a)

$$D_{ij} = \frac{T}{2\zeta}\left(1 + \sqrt{\frac{6}{\pi}\frac{a}{b}}\right)\delta_{ij}. \tag{F.16}$$

Distribution Function

Equations (F.12) and (F.13) are followed an equation for the distribution function of the distance between the centres of resistance of the relaxator

$$\frac{\partial W}{\partial t} + \sum_{j=1}^{3}\frac{\partial(\psi_j W)}{\partial \rho_j} = 0 \tag{F.17}$$

where velocity ψ_j is determined by relation (F.11).

For the cases, when hydrodynamic interaction is neglected, that is $l = 0$, the equation for distribution function was found by Pokrovskii (1978) and was confirmed later by Schieber (1992)

$$\frac{\partial W}{\partial t} - \frac{T}{\zeta}\frac{\partial^2 W}{(\partial \rho_i)^2} + \frac{T}{\zeta}\frac{\gamma}{1+\gamma}\left(\frac{2\rho_i}{\rho_j \rho_j}\frac{\partial W}{\partial \rho_i} + e_s e_j\frac{\partial^2 W}{\partial \rho_j \partial \rho_s}\right)$$

$$- \frac{4\mu T}{\zeta}\frac{1}{1+\gamma}\left(3W + \rho_j\frac{\partial W}{\partial \rho_j}\right) + \nu_{js}\rho_s\frac{\partial W}{\partial \rho_j}$$

$$- \frac{\gamma}{1+\gamma}e_s e_i \nu_{si}\left(3W + \rho_j\frac{\partial W}{\partial \rho_j}\right) = 0. \tag{F.18}$$

Exact solutions of equation (F.18) can be found in some particular cases. For example, if the tensor of velocity gradients is symmetrical, equation (F.18) has an exact simple solution

$$W = C(\gamma_{ik})\exp\left[-2\mu(\rho_j \rho_j - 2\tau\gamma_{ik}\rho_i \rho_k)\right]. \tag{F.19}$$

When there are no velocity gradients, the solution of equation (F.18) normalised with respect to unity has the form

$$W(\boldsymbol{\rho}) = \left(\frac{2\mu}{\pi}\right)^{\frac{3}{2}}\exp(-2\mu\boldsymbol{\rho}\boldsymbol{\rho}). \tag{F.20}$$

In the case of low velocity gradients, the time-independent distribution function may be found in the form of an expansion in terms of the invariant combinations of vector ρ and the symmetrical and anti-symmetrical velocity-gradient tensors γ_{ik} and ω_{ik}. In the steady-state case, one has, to within the second-order terms in the velocity gradients,

$$W = W_0 \left\{ 1 + 4\mu\tau\gamma_{ik}\rho_i\rho_k + 8(\mu\tau)^2\gamma_{ik}\gamma_{sj}\rho_i\rho_k\rho_s\rho_j \right.$$
$$\left. - \tau^2\gamma_{ik}\gamma_{ik} - 64G(\mu\tau)^2\gamma_{si}\omega_{sk}\rho_i\rho_k \right\} \tag{F.21}$$

where W_0 is defined by equation (F.20) and the following notation for the relaxation time has been introduced

$$\tau = \frac{\zeta}{8T\mu} \tag{F.22}$$

which is the characteristic relaxation time of the dumbbell.

In equation (F.21), the coefficient of the last term G depends on the scalar $\rho_j\rho_j$ and the internal viscosity. It is not difficult to find the relations for two asymptotic cases

$$G = \begin{cases} \frac{1}{8\mu}, & \gamma = 0, \\ \frac{\rho^2}{6} + \frac{\rho^2}{9\gamma}(5 - 4\mu\rho^2) + \frac{\rho^2}{27\gamma^2}[35 - 112\mu\rho^2 + 48(\mu\rho^2)^2], & \gamma \gg 1. \end{cases} \tag{F.23}$$

Let us note that in the particular case when the macromolecule has no internal viscosity, equation (F.18) is followed by the equation

$$\frac{\partial W}{\partial t} - \frac{T}{\zeta}\frac{\partial^2 W}{(\partial\rho_i)^2} + \nu_{ik}\rho_k\frac{\partial W}{\partial\rho_i} - \frac{2\mu\lambda_\alpha T}{\zeta}\left(3W + \rho_i\frac{\partial W}{\partial\rho_i}\right) = 0. \tag{F.24}$$

This particular form of diffusion equation was used in earlier works (Cerf 1958; Peterlin 1967; Zimm 1956).

In the steady-state case, the solution of equation (F.24) for simple shear ($\nu_{12} \neq 0$) was deduced by Peterlin (see Zimm 1956).

$$W = C\exp\left\{ -\frac{2\mu}{1 + (\tau\nu_{12})^2}\left[\rho_j\rho_j - 2\tau\nu_{12}\rho_1\rho_2 + 2(\tau\nu_{12}\rho_2)^2 + (\tau\nu_{12}\rho_3)^2\right] \right\}.$$

The first terms of the expansion of the written functions are identical to expression (F.21) taken for the appropriate cases.

Equation (F.18) and equations that are more general which can be obtained in the case when $l \neq 0$, can be used to calculate mean quantities $\langle l \rangle$, $\langle le_ie_j \rangle$ and others in non-equilibrium situations, which is needed to consider macroscopic phenomena.

Equations of Relaxation

The distribution function (F.21) considered in the previous section makes it possible to calculate the moments in the stationary case, for example the second-order moments

$$\langle e_i e_k \rangle = \int W e_i e_k \{ \mathbf{d}\boldsymbol{\rho} \}, \qquad \langle \rho_i \rho_k \rangle = \int W \rho_i \rho_k \{ \mathbf{d}\boldsymbol{\rho} \}$$

as expansions in powers of velocity gradient.

In general, it is more convenient to determine the moments from equations which can be derived directly from the diffusion equation (F.18). For example, on multiplying equation (F.18) by $\rho_i \rho_k$ and integrating with respect to all the variables, we find the relaxation equation

$$\frac{\mathrm{d}\langle \rho_i \rho_k \rangle}{\mathrm{d}t} = -\frac{1}{\tau} \frac{3}{2\mu\lambda} \left(\langle e_i e_k \rangle - \frac{1}{3}\delta_{ik} \right) - \frac{1}{\tau'} \left(\langle \rho_i \rho_k \rangle - \frac{3}{2\mu\lambda} \langle e_i e_k \rangle \right)$$

$$+ \nu_{ij}\langle \rho_j \rho_k \rangle + \nu_{kj}\langle \rho_j \rho_i \rangle - \frac{2\gamma}{1+\gamma} \langle \rho_i \rho_k e_j e_s \rangle \nu_{js}. \qquad (F.25)$$

Two relaxation times appear here: the first time τ, defined by equation (F.22), refers to the orientation processes; the second time

$$\tau' = (1 + \gamma) \cdot \tau \qquad (F.26)$$

refers to the deformation processes.

Indeed, by multiplying equation (F.18) by ρ^2 and integrating with respect to all the variables, or, by carrying out a direct summation of equation (F.25) with identical indices, we find

$$\frac{\mathrm{d}\langle \rho^2 \rangle}{\mathrm{d}t} = -\frac{1}{\tau'} \left(\langle \rho^2 \rangle - \frac{3}{2\mu\lambda} \right) + \frac{2}{1+\gamma} \langle \rho_s \rho_j \rangle \gamma_{sj}. \qquad (F.27)$$

This equation describes only the deformation of the macromolecular coil and therefore τ' is the relaxation time of the deformation process. In order to isolate the orientation process, we now formulate the moments in the form

$$\langle \rho_i \rho_k \rangle = \langle \rho^2 \rangle \langle e_i e_k \rangle, \qquad \langle \rho_i \rho_k e_j e_s \rangle = \langle \rho^2 \rangle \langle e_i e_k e_j e_s \rangle.$$

Then, equation (F.25) gives rise to the relaxation equation for the orientation process

$$\frac{\mathrm{d}\langle e_i e_k \rangle}{\mathrm{d}t} = -\frac{1}{\tau} \left(\langle e_i e_k \rangle - \frac{1}{3}\delta_{ik} \right) + \nu_{ij}\langle e_j e_k \rangle$$

$$+ \nu_{kj}\langle e_j e_i \rangle - \frac{2\gamma}{1+\gamma} \langle e_i e_k e_j e_s \rangle \gamma_{js}. \qquad (F.28)$$

Thus τ is the relaxation time for the orientation process and τ' is the relaxation time for the deformation process.

We may note that, for a nonzero internal viscosity, the system of equations for the moments is found to be open: the equations for the second-order moments contain the fourth-order moments, etc. This situation is encountered in the theory of the relaxation of the suspension of rigid particles (Pokrovskii 1978). Incidentally, for $\gamma \to \infty$, equation (F.28) becomes identical to the relaxation equation for the orientation of infinitely extended ellipsoids of rotation (Pokrovskii 1978, p. 58).

In contrast to the situation described above, the system (F.25) for the moments is closed for the case, when the internal viscosity may be neglected. This factor makes it possible to find the moments in the form of a series expansion for small values of the internal viscosity and the velocity gradients.

Second-Order Moments of Co-Ordinates

We use the expansion of the distribution function (F.21) and the relaxation equations (F.25) and (F.28) to calculate the second-order moments of coordinates in steady-state and non-steady-state cases in the form of a series expansion for low values of the velocity gradients. Calculations are simple but tedious. As a first step of calculations, we demonstrate the mean values of the products of different variables or moment of equilibrium distribution functions. They are defined, for example, as

$$\langle \rho_i \rho_k \rangle_0 = \int W_0 \rho_i \rho_k \{d\rho\}, \quad \langle e_i e_k \rangle_0 = \int W_0 \frac{\rho_i \rho_k}{\rho\rho} \{d\rho\}$$

and are calculated with the help of function (F.20)

$$\langle \rho_i \rho_k \rangle_0 = \frac{1}{2\mu\lambda}\delta_{ik},$$

$$\langle e_i e_k \rangle_0 = \frac{1}{3}\delta_{ik},$$

$$\langle \rho_i \rho_k \rho_s \rho_j \rangle_0 = \frac{1}{4(\mu\lambda)^2}(\delta_{\alpha\beta}\delta_{\gamma\epsilon})_{iksj},$$

$$\langle e_i e_k e_s e_j \rangle_0 = \frac{1}{15}(\delta_{\alpha\beta}\delta_{\gamma\epsilon})_{iksj},$$

$$\langle e_i e_k \rho_s \rho_j \rangle_0 = \frac{1}{10\mu\lambda}(\delta_{\alpha\beta}\delta_{\gamma\epsilon})_{iksj},$$

$$\langle \rho_i \rho_k \rho_s \rho_j \rho_l \rho_m \rangle_0 = \frac{1}{8(\mu\lambda)^3}(\delta_{\alpha\beta}\delta_{\gamma\epsilon}\delta_{\mu\nu})_{iksjlm},$$

$$\langle e_i e_k \rho_s \rho_j \rho_l \rho_m \rangle_0 = \frac{1}{28(\mu\lambda)^2}(\delta_{\alpha\beta}\delta_{\gamma\epsilon}\delta_{\mu\nu})_{iksjlm},$$

$$\langle e_i e_k e_s e_j \rho_l \rho_m \rangle_0 = \frac{1}{70\mu\lambda}(\delta_{\alpha\beta}\delta_{\gamma\epsilon}\delta_{\mu\nu})_{iksjlm},$$

$$\langle e_i e_k e_s e_j e_l e_m \rangle_0 = \frac{1}{105} (\delta_{\alpha\beta}\delta_{\gamma\epsilon}\delta_{\mu\nu})_{iksjlm},$$

$$\langle \rho_i \rho_k \rho_s \rho_j \rho_l \rho_m \rho_p \rho_q \rangle_0 = \frac{1}{16(\mu\lambda)^4} (\delta_{\alpha\beta}\delta_{\gamma\epsilon}\delta_{\mu\nu}\delta_{\sigma\kappa})_{iksjlmpq}.$$

In these formulae, an expression of the form $(\delta_{\alpha\beta}\delta_{\gamma\epsilon})_{iksj}$ means the sum of the similar terms in which the Greek labels inside the brackets take subsequently all the Latin labels outside the brackets. Identical terms are taken into account only once, so, for example, the last formula contains 105 terms and the formula before the last contains 15 terms.

In the steady-state case, the expansion assumes the form

$$\langle \rho_i \rho_k \rangle = \frac{1}{4\mu} \left\{ \delta_{ik} + 2\tau\gamma_{ik} + 2\tau^2 [2\gamma_{ij}\gamma_{jk} + (1+Z)(\omega_{ij}\gamma_{jk} + \omega_{kj}\gamma_{ji})] \right\}$$

$$(F.29)$$

where

$$Z = \begin{cases} \frac{2}{5}\gamma, & \gamma \ll 1, \\ \frac{4}{27}\left(9 - \frac{42}{\gamma} + \frac{245}{\gamma^2}\right), & \gamma \gg 1. \end{cases}$$

One may assume that Z is a monotonically increasing function of γ which, if necessary, may be fitted to any kind of convenient function.

In the non-steady-state case, the second-order moments of co-ordinates are calculated as solutions of equations (F.25) and (F.28). We assume that the velocity gradient and, consequently, the moments do not depend on space co-ordinates. To find the solutions, we multiply equation (F.25) by $\exp(\frac{t}{\tau'})$, equation (F.28) by $\exp(\frac{t}{\tau})$, and integrate over time from $t \to -\infty$. After some transformation, we obtain

$$\langle \rho_i \rho_k \rangle = \frac{1}{2\mu\lambda}\delta_{ik} + \int_0^\infty \exp\left(-\frac{s}{\tau'}\right)(\nu_{ij}\langle \rho_j \rho_k \rangle + \nu_{kj}\langle \rho_j \rho_i \rangle)\,\mathrm{d}s$$

$$- \frac{1}{\tau}\frac{3}{2\mu\lambda}\frac{\gamma}{1+\gamma}\int_0^\infty \exp\left(-\frac{s}{\tau'}\right)\left(\langle e_i e_k \rangle - \frac{1}{3}\delta_{ik}\right)\mathrm{d}s$$

$$- \frac{2\gamma}{1+\gamma}\int_0^\infty \exp\left(-\frac{s}{\tau'}\right)\langle e_i e_k \rho_j \rho_s \rangle \gamma_{js}\mathrm{d}s,$$

$$\langle e_i e_k \rangle = \frac{1}{3}\delta_{ik} + \int_0^\infty \exp\left(-\frac{s}{\tau}\right)(\nu_{ij}\langle e_j e_k \rangle + \nu_{kj}\langle e_j e_i \rangle - 2\nu_{js}\langle e_i e_k e_j e_s \rangle)\mathrm{d}s.$$

The moments and velocity gradients in the integrands are taken at the point $t - s$.

Now we can use the equilibrium moments, shown in the beginning of this section, to find the first terms of the expansion of the moments as a series of repeated integrals

$$\langle e_i e_k \rangle = \frac{1}{3}\delta_{ik} + \frac{2}{5}\int_0^\infty \exp\left(-\frac{s}{\tau}\right)\gamma_{ik}(t-s)ds, \tag{F.30}$$

$$\langle \rho_i \rho_k \rangle = \frac{1}{2\mu\lambda}\delta_{ik} + \frac{5+3\gamma}{5\mu\lambda(1+\gamma)}\int_0^\infty \exp\left(-\frac{s}{\tau'}\right)\gamma_{ik}(t-s)ds$$

$$-\frac{1}{\tau}\frac{1}{2\mu\lambda}\frac{\gamma}{1+\gamma}\int_0^\infty \exp\left(-\frac{s}{\tau'}\right)\int_0^\infty \exp\left(-\frac{u}{\tau}\right)\gamma_{ik}(t-s-u)du\,ds. \tag{F.31}$$

The Stress Tensor

The results for the case, when hydrodynamic interaction is taken into account (Altukhov 1986) are rather cumbersome. So, we consider here the more simple case, when hydrodynamic interaction is neglected but internal viscosity is retained. The results were obtained by Pokrovskii and Chuprinka (1973) (see also Pokrovskii 1978).

When the elastic force and the force of internal viscosity are defined, at $N = 1$, the expression for the stress tensor directly follows relation (6.7)

$$\sigma_{ik} = -nT\delta_{ik} + 2\eta_s\gamma_{ik} + \frac{1}{2}n\zeta\left[\frac{1}{\tau'}\left(\langle\rho_i\rho_k\rangle - \frac{3}{4\mu}\langle e_k e_i\rangle\right)\right.$$

$$\left. + \frac{1}{\tau}\frac{3}{4\mu}\left(\langle e_i e_k\rangle - \frac{1}{3}\delta_{ik}\right) + \frac{2\gamma}{1+\gamma}\langle\rho_k\rho_i e_j e_s\rangle\gamma_{js}\right]. \tag{F.32}$$

This equation contains two relaxation times: orientational and deformational, correspondingly

$$\tau = \frac{\zeta}{8\mu T}, \qquad \tau' = (1+\gamma)\cdot\tau. \tag{F.33}$$

One can consider relations (F.25), (F.28), and (F.32) as a constitutive set of equations which determine non-linear stresses in a suspension of relaxators. From the macroscopic point of view moments $\langle\rho_i\rho_k\rangle$, $\langle e_k e_i\rangle$ and others in equations (F.25), (F.28), and (F.32) are thermodynamic internal variables. One can see, that at proper choice of thermodynamic forces, a set of relations (F.25), (F.28), and (F.32) in linear approximation with respect to fluxes are a particular case of relations (8.28). Though the set is not closed, solutions can be found for small velocity gradients and/or for small internal viscosity.

The thermodynamic consistency of the theory was considered by Schieber and Öttinger (1994). Following the methods of Grmela (1985) and Jongschaap (1991), they considered the distribution function $W(t,\rho)$, determined by equation (F.18) to be a set of internal variables, while co-ordinate ρ serves as a label, which takes a continuous set of values, and have demonstrated that equations (F.18) and (F.32) are thermodynamically consistent. It is an important result, because there were some disagreements in the works concerning the method of calculation of the stresses. Eventually, their result confirms the method used in Chapter 6 for the calculation of stresses.

We can use expressions (F.25) and (F.28) for moments, in order to determine the stresses with accuracy within the first-order term with respect to velocity gradients

$$\sigma_{ik} = -p\delta_{ik} + 2\eta_s\gamma_{ik}$$

$$+ \frac{2}{5}nT\frac{\tau}{1+\gamma}\left[\frac{1}{\tau'}(5+3\gamma)\int_0^\infty \exp\left(-\frac{s}{\tau'}\right)\gamma_{ik}(t-s)\mathrm{d}s\right.$$

$$- \frac{3\gamma}{\tau\tau'}\int_0^\infty \exp\left(-\frac{s}{\tau'}\right)\int_0^\infty \exp\left(-\frac{s}{\tau'}\right)\gamma_{ik}(t-s-u)\mathrm{d}u\,\mathrm{d}s$$

$$+ \left. \frac{3\gamma}{\tau}\int_0^\infty \exp\left(-\frac{s}{\tau}\right)\gamma_{ik}(t-s)\mathrm{d}s + 2\gamma\gamma_{ik}\right]. \qquad (F.34)$$

The constitutive equation (F.34) contains two relaxation times, which are defined by equations (F.33).

The study of the reaction of the system in the simple case when the velocity gradients are independent of the co-ordinates, and vary in accordance with the law

$$\gamma_{ik} \sim e^{-i\omega t}$$

for different deformation frequencies ω, yields important information about the relaxation processes in the system.

In this case, equation (F.34) defines, as was shown by Pokrovskii and Chuprinka (1973), the stresses in a dilute solution of a polymers in terms of a linear approximation

$$\sigma_{ik} = -p\delta_{ik} + 2\eta(\omega)\gamma_{ik}$$

where $\eta(\omega)$ is the complex shear viscosity with components

$$\eta'(\omega) = \eta_s + nT\frac{\tau}{1+\gamma}\frac{1}{5}\left[2\gamma + \frac{3(1+\gamma)}{1+(\tau\omega)^2} + \frac{2}{1+(\tau'\omega)^2}\right],$$

$$\eta''(\omega) = nT\tau^2\omega\frac{1}{5}\left[\frac{3}{1+(\tau\omega)^2} + \frac{2}{1+(\tau'\omega)^2}\right]. \qquad (F.35)$$

It is interesting for us to understand the effect of the non-averaged hydrodynamic interaction on the stresses in a suspension of the dumbbell under deformation. The simple, but somewhat tedious calculations (Altukhov 1986), for the considered case determine the stresses at simple shear

$$\sigma_{12} = \eta \nu_{12}, \qquad \eta = \eta_s + nT\tau \left[1 + \frac{16}{5} h^2 \left(1 + \frac{2}{3\gamma} \right) \frac{\gamma^3}{(1+\gamma)^3} \right],$$

$$\sigma_{11} - \sigma_{33} = 2nT \left[1 - \frac{2}{3\gamma} + \frac{16}{105} h^2 \left(\frac{14\gamma}{(1+\gamma)^2} + \frac{\gamma^3}{(1+\gamma)^3} \right. \right.$$
$$\left. \left. + \frac{6\gamma^2}{(1+\gamma)^3} \right) \right] (\tau\nu_{12})^2,$$

$$\sigma_{22} - \sigma_{33} = -\frac{8}{35} nTh^2 \frac{\gamma^3}{(1+\gamma)^3} \left(\frac{25}{3} - \frac{8}{\gamma} \right) (\tau\nu_{12})^2$$

(F.36)

where n is the number density of the dumbbells, $\gamma = \lambda/\zeta$ is the coefficient of the internal viscosity, and $\tau = \zeta/8T\mu$ is the relaxation time.

Results (F.36) are valid with approximation to the term of second order with respect to the parameter of hydrodynamic interaction

$$h = \left(\frac{27}{32} \frac{a^2}{\langle |r^0 - r^1|^2 \rangle_0} \right)^{1/2}.$$

The expressions (F.36), calculated at the exact hydrodynamic interaction, contain the terms, which disappear, if the hydrodynamic interaction is averaged beforehand. It can, thus, be believed that, if hydrodynamic interaction is taken accurately, extra terms will also appear for the subchain model.

G Estimation of Some Series

We shall estimate the sums of the form

$$\sum_{\alpha=1}^{N} f(\alpha, \chi).$$

So as the upper limit N in the sums, which need to be estimated, is large, one can approximate it by infinity. In simple cases one comes to the known zeta-function

$$\zeta(x) = \sum_{u=1}^{\infty} \frac{1}{\alpha^x}.$$

We use values of the zeta-function, to estimate the following sums

$$\sum_{\alpha=1}^{N} \frac{1}{\alpha^2} = \frac{\pi^2}{6}, \qquad \sum_{\alpha=1}^{N} \frac{1}{\alpha^4} = \frac{\pi^4}{90}, \qquad \sum_{\alpha=1}^{N} \frac{1}{\alpha^6} = \frac{\pi^6}{945}.$$

The more complicated sums depend on parameter χ which is considered to be small. This allows one to approximate the sums by integrals. For example,

$$\sum_{\alpha=1}^{N} \frac{1}{1+\chi\alpha^2} = \int_0^{\infty} \frac{\chi^{-\frac{1}{2}}dx}{1+x^2} = \frac{\pi}{2}\chi^{-\frac{1}{2}}.$$

Estimates of the most important sums are listed below

$$\sum_{\alpha=1}^{N} \frac{1}{1+\chi\alpha^2} = \frac{\pi}{2}\chi^{-\frac{1}{2}}, \tag{G.1}$$

$$\sum_{\alpha=1}^{N} \frac{1}{(1+\chi\alpha^2)^2} = \frac{\pi}{4}\chi^{-\frac{1}{2}}, \tag{G.2}$$

$$\sum_{\alpha=1}^{N} \frac{1}{(1+\chi\alpha^2)^3} = \frac{3\pi}{16}\chi^{-\frac{1}{2}}, \tag{G.3}$$

$$\sum_{\alpha=1}^{N} \frac{\alpha^2}{(1+\chi\alpha^2)^2} = \frac{\pi}{4}\chi^{-\frac{3}{2}}, \tag{G.4}$$

$$\sum_{\alpha=1}^{N} \frac{1}{\alpha^2(1+\chi\alpha^2)} = \frac{\pi^2}{6} - \frac{\pi}{2}\chi^{\frac{1}{2}}, \tag{G.5}$$

$$\sum_{\alpha=1}^{N} \frac{1}{\alpha^2(1+\chi\alpha^2)^2} = \frac{\pi^2}{6} - \frac{3\pi}{4}\chi^{\frac{1}{2}}, \tag{G.6}$$

$$\sum_{\alpha=1}^{N} \frac{1}{\alpha^4(1+\chi\alpha^2)} = \frac{\pi^4}{90} - \frac{\pi^2}{6}\chi + \frac{\pi}{2}\chi^{\frac{3}{2}}. \tag{G.7}$$

References

Adachi K, Kotaka T (1993) Dielectric normal mode relaxation. Prog Polym Sci 18:585–622

Adelman SA, Freed KF (1977) Microscopic theory of polymer internal viscosity: Mode coupling approximation for the Rouse model. J Chem Phys 67(4):1380–1393

Aharoni SM (1983) On entanglements of flexible and rodlike polymers. Macromolecules 16(11):1722–1728

Aharoni SM (1986) Correlation between chain parameters and the plateau modulus of polymers. Macromolecules 19(2):426–434

Ahlrichs P, Dünweg B (1999) Simulation of a single polymer chain in solution by combining lattice Boltzman and molecular dynamics. J Chem Phys 111(17):8225–8239

Akcasu Z, Gurol H (1976) Quasielastic scattering by dilute polymer solutions. J Polym Sci: Polym Phys Ed 46(1):1–10

Akers LC, Williams MC (1969) Oscillatory normal stresses in dilute polymer solutions. J Chem Phys 51(9):3834–3841

Alkhimov VI (1991) Self-avoiding walk problem. Sov Phys – Usp 34(9):804–816

Allal A, Lamaison S, Leonardi F, Marin G (2002) De la microstructure du polymère à ses propriétés rhéologiques. C R Phys 3:1451–1458

Al-Naomi GF, Martinez-Mekler GC, Wilson CA (1978) Dynamical scaling exponent z for a single polymer chain by renormalization along the chain. J Phys Lett 39(21):L373–L377

Altukhov YuA (1986) Constitutive equation for polymer solutions based on the dynamics of non-interacting relaxators. Zh Prikl Mech Tekhnich Fiz 3:101–105 (in Russian)

Altukhov YuA, Pyshnograi GV (1995) The mobility anisotropy and non-linear effects in molecular theory of viscoelasticity of linear polymers. Fluid Dyn 30(4):487–495

Altukhov YuA, Pyshnograi GV (1996) Microstructural approach to the theory of flow of linear polymers and related effects. Polym Sci A 38(7):766–774

Altukhov YuA, Kekalov AN, Pokrovskii VN, Popov VI, Khabakhpasheva EM (1986) On a description of pulsating flow of solutions of polymers. In: Struktura gidrodinamicheskikh potokov (Structures of hydrodynamical flows, in Russian). ITF SO AN SSSR, Novosibirsk, pp 5–14

Altukhov YuA, Pokrovskii VN, Pyshnograi GV (2004) On the difference between weakly and strongly entangled linear polymers. J Non-Newton Fluid Mech 121(2–3):73–86

Astarita G, Marrucci G (1974) Principles of non-Newtonian fluid mechanics. McGraw-Hill, New York

Aubert JH, Tirell H (1980) Macromolecules in non-homogeneous velocity gradient fields. J Chem Phys 72:2694–2701

Baldwin PR, Helfand E (1990) Dilute solutions in steady shear flow: Non-Newtonian stress. Phys Rev A 41:6772–6785

Baumgaertel M, Schausberger A, Winter HH (1990) The relaxation of polymers with linear flexible chains of uniform length. Rheol Acta 29(5):400–408

V.N. Pokrovskii, *The Mesoscopic Theory of Polymer Dynamics*,
Springer Series in Chemical Physics 95,
DOI 10.1007/978-90-481-2231-8, © Springer Science+Business Media B.V. 2010

Baumgaertel M, De Rosa ME, Machado J, Masse M, Winter HH (1992) The relaxation time spectrum of nearly monodisperse polybutadiene melts. Rheol Acta 31(1):75–82

Baumgärter A, Muthukumar M (1996) Polymers in disordered media. Adv Chem Phys 94:625–708. Eds I. Prigogine and S.A. Rice, Wiley, New York

Berry GC, Fox TG (1968) The viscosity of polymers and their concentrated solutions. Adv Polym Sci 5:261–357

Bird RB, Armstrong RC, Hassager O (1987a) Dynamics of polymeric liquids, vol 1: Fluid mechanics, 2nd edn. Wiley, New York

Bird RB, Curtiss CF, Armstrong RC, Hassager O (1987b) Dynamics of polymeric liquids, vol 2: Kinetic theory, 2nd edn. Wiley, New York

Birshtein TM, Ptitsyn OB (1966) Conformations of macromolecules. Interscience, New York

Bixon M, Zwanzig R (1978) Optimised Rouse-Zimm theory for stiff polymers. J Chem Phys 68(4):1896–1902

Boon JP, Yip S (1980) Molecular hydrodynamics. McGraw-Hill, New York

Born M, Wolf E (1970) Principles of optics. Electromagnetic theory of propagation, interference, and diffraction of light, 4th edn. Pergamon, Elmsford

Brown EF, Burghardt WR, Kahvand H, Venerus DC (1995) Comparison of optical and mechanical measurements of second normal stress difference relaxation following step strain. Rheol Acta 34(3):221–234

Brunn PO (1984) Polymer migration phenomena based on the general bead-spring model for flexible polymers. J Chem Phys 80:5821–5826

Bueche F (1956) Viscosity of polymers in concentrated solutions. J Chem Phys 25:599–600

Chandrasekhar S (1943) Stochastic problems in physics and astronomy. Rev Modern Phys 15(1):1–89

Chang X-Y, Freed KF (1993) Test of theory for long time dynamics of floppy molecules in solution using Brownian dynamics simulation of octane. J Chem Phys 99(10):8016–8030

Chompff AJ, Duiser JA (1966) Viscoelasticity of networks consisting of crosslinked or entangled macromolecules. I Normal modes and mechanical spectra. J Chem Phys 45(5):1505–1514

Chompff AJ, Prins W (1968) Viscoelasticity of networks consisting of crosslinked or entangled macromolecules. II Verification of the theory for entanglement networks. J Chem Phys 48(1):235–243

Cerf R (1958) Statistical mechanics of chain macromolecules in a velocity field. J Phys Radium 19(1):122–134

Coleman BD, Gurtin ME (1967) Thermodynamics with internal state variables. J Chem Phys 47:597–613

Coleman BD, Nolle W (1961) Foundations of linear viscoelasticity. Rev Modern Phys 33:239–249

Cooke BJ, Matheson AJ (1976) Dynamic viscosity of dilute polymer solutions at high frequencies of alternating shear stress. J Chem Soc: Faraday Trans II 72(3):679–685

Curtiss CF, Bird RB (1981a) A kinetic theory for polymer melts. I The equation for the single-link orientational distribution function. J Chem Phys 74:2016–2025

Curtiss CF, Bird RB (1981b) A kinetic theory for polymer melts. II The stress tensor and the rheological equation of state. J Chem Phys 74(3):2026–2033

Daoud M, de Gennes PG (1979) Some remarks on the dynamics of polymer melts. J Polym Sci: Polym Phys Ed 17:1971–1981

Daoud M, Cotton JP, Farnoux B, et al (1975) Solution of flexible polymers. Neutron experiments and interpretation. Macromolecules 8(6):804–818

Dasbach TP, Manke CW, Williams MC (1992) Complex viscosity for the rigorous formulation of the multibead internal viscosity model with hydrodynamic interaction. J Phys Chem 96(10):4118–4125 (McMahon)

Dean P (1967) Atomic vibration in solids. J Inst Math Appl 3:98–165

De Gennes PG (1967) Quasi-elastic scattering of neutrons by dilute polymer solutions: I. Free-draining limit. Physics 3(1):37–45

De Gennes PG (1971) Reptation of a polymer chain in the presence of fixed obstacles. J Chem Phys 55:572–579

De Gennes PG (1977) Origin of internal viscosity in dilute polymer solution. J Chem Phys 66(12):5825–5826

De Gennes PG (1979) Scaling concepts in polymer physics. Cornell Univ. Press, Ithaca

De Gennes PG (1981) Coherent scattering by one reptating chain. J Phys (Paris) 42:735–740

Des Cloizeaux J (1993) Dynamic form function of a long polymer constrained by entanglements in a polymer melt. J Phys I France 3:1523–1539

Des Cloizeaux J, Jannink G (1990) Polymers in solution: their modelling and structure. Oxford University Press, Oxford

Doi M, Edwards SF (1978) Dynamics of concentrated polymer systems. Part 1 Brownian motion in the equilibrium state. J Chem Soc: Faraday Trans II 74:1789–1801

Doi M, Edwards SF (1986) The theory of polymer dynamics. Oxford Univ. Press, Oxford

Dubuis-Violette E, De Gennes PG (1967) Quasi-elastic scattering by dilute, ideal, polymer solutions: II Effects of hydrodynamic interactions. Physics 3(4):181–198

Dünweg B (2003) Langevin methods. In: Dünweg B, Landau DP, Milchev A (eds) Computer simulations of surfaces and interfaces. Proceedings of the NATO Advanced Study Institute/Euroconference, Albena, Bulgaria, September 2002. Kluwer, Dordrecht, pp 77–92

Dünweg B, Reith D, Steinhauser M, Kremer K (2002) Corrections to scaling in the hydrodynamic properties of dilute polymer solutions. J Chem Phys 117(2):914–924

Dušek K, Prins W (1969) Structure and elasticity of non-crystalline polymer networks. Adv Polym Sci 6:1–102

Ebert U, Baumgärtner A, Schäfer L (1996) Universal short-time motion of a polymer in a random environment: Analytical calculations, a blob picture, and Monte Carlo results. Phys Rev E 53(1):950–965

Edwards SF (1965) The statistical mechanics of polymers with excluded volume. Proc Phys Soc 85(546):613–624

Edwards SF (1967a) Statistical mechanics with topological constraints. I. Proc Phys Soc 91(3):513–519

Edwards SF (1967b) Statistical mechanics of polymerised material. Proc Phys Soc 92(1):9–16

Edwards SF (1969) Theory of crosslinked polymerised material. Proc Phys Soc: Solid State Phys [2] 2(1):1–13

Edwards SF, Freed KF (1974) Theory of the dynamical viscosity of polymer solutions. J Chem Phys 61(3):1189–1202

Edwards SF, Grant JWV (1973) The effect of entanglements on diffusion in a polymer melt. J Phys A: Math Nucl Gen 6:1169–1185

Erenburg VB, Pokrovskii VN (1981) Non-uniform shear flow of linear polymers. Inzh Fiz Zh 41(3):449–456 (in Russian)

Erman B, Flory PJ (1978) Theory of elasticity of polymer networks. II The effect of geometric constraints on junctions. J Chem Phys 68:5363–5369

Erukhimovich IYa, Irzhak VI, Rostiashvili VG (1976) On concentration dependence of swelling coefficient of weakly non-Gaussian macromolecules. Polym Sci USSR 18:1682–1089

Everaers R, Sukumaran SK, Grest GS, Svaneborg C, Sivasubramanian A, Kremer K (2004) Rheology and microscopic topology of entangled polymeric liquids. Science 303:823–826

Ewen B, Richter D (1995) The dynamics of polymer melts as seen by neutron spin echo spectroscopy. Macromol Symp 90:131–149

Ewen B, Maschke U, Richter D, Farago B (1994) Neutron spin echo studies on the segmental diffusion behaviour in the different chain sections of high molecular weight poly(dimetylsiloxane) melt. Acta Polym 45:143–147

Faitelson LA (1995) Some aspects of polymer melts rheology. Mech Compos Mater 31(1):101–116

Fatkullin NF, Kimmich R, Kroutieva M (2000) The twice-renormalised Rouse formalism of polymer dynamics: Segment diffusion, terminal relaxation, and nuclear spin-lattice relaxation. J Exp Theor Phys 91(1):150–166

Ferry JD (1980) Viscoelastic properties of polymers, 3rd edn. Wiley, London

Ferry JD (1990) Some reflections on the early development of polymer dynamics: Viscoelasticity, dielectric dispersion, and self-diffusion. Macromolecules 24:5237–5245

Ferry JD, Landel RF, Williams ML (1955) Extensions of the Rouse theory of viscoelastic properties to undilute linear polymers. J Appl Phys 26:359–362

Fikhman VD, Radushkevich BV, Vinogradov GV (1970) Reological properties of polymers under extension at constant deformation rate and at constant extension rate. In: Vinogradov GV (ed) Uspekhi reologii polimerov (Advances in polymer rheology, in Russian). Khimija, Moscow, pp 9–23

Fleisher G, Appel M (1995) Chain length and temperature dependence of the self-diffusion of polyisoprene and polybutadiene in the melt. Macromolecules 28(21):7281–7283

Flory PJ (1953) Principles of polymer chemistry. Cornell Univ. Press, New York

Flory PJ (1969) Statistics of chain molecules. Interscience, New York

Flory PJ (1977) Theory of elasticity of polymer networks. The effect of local constraints on junctions. J Chem Phys 66(12):5720–5729

Fodor JS, Hill DA (1994) Determination of molecular weight distribution of entangled *cis*-polystyrene melts by inversion of normal-mode dielectric loss spectra. J Phys Chem 98(31):7674–7684

Fox TG, Flory PJ (1948) Viscosity-molecular weight and viscosity-temperature relationships for polystyrene and polyisobutylene. J Am Chem Soc 70(7):2384–2395

Freed KF, Edwards SF (1974) Polymer viscosity in concentrated solutions. J Chem Phys 61(9):3626–3633

Freed KF, Edwards SF (1975) Huggins coefficient for the viscosity of polymer solutions. J Chem Phys 62(10):4032–4035

Fröhlich H (1958) Theory of dielectrics, dielectric constant and dielectric loss, 2nd edn. Clarendon, Oxford

Gabay M, Garel T (1978) Renormalisation along the chemical sequence of a single polymer chain. J Phys Lett 39(9):L123–L126

Gao J, Weiner JH (1994) Monomer-level description of stress and birefringence relaxation in polymer melts. Macromolecules 27(5):1201–1209

Gardiner CW (1983) Handbook of stochastic methods for physics, chemistry, and the natural sciences. Springer, Berlin

Godunov SK, Romenskii EI (1972) Non-steady state equation of non-linear theory of elasticity in Euler co-ordinates. Zh Prikl Mech Technich Phys 6:124–144 (in Russian)

Golovicheva IE, Sinovich SA, Pyshnograi GV (2000) Effect of molecular weight on shear and elongational viscosity of linear polymers. Prikl Mech Tekhnich Fiz 41(2):154–160 (in Russian)

Gotlib YuYa (1964) Theory of optical anisotropy for short or rigid chain macromolecules in terms of persistent model. Vysokomolek Soedin 6(3):389–398 (in Russian)

Gotlib YuYa, Svetlov YuE (1964a) To the theory of anomalous angles in dynamic birefringence of polymer solutions. Vysokomolek Soedin 6(5):771–776 (in Russian)

Gotlib YuYa, Svetlov YuE (1964b) On the gradient dependence of dynamic birefringence near inverse mode. Vysokomolek Soedin 6(9):1591–1592 (in Russian)

Gotlib YuYa, Darinskii AA, Svetlov YuE (1986) Fizicheskaya kinetika makromolekul (Physical kinetics of macromolecules, in Russian). Khimiya, Leningrad

Graessley WW (1974) The entanglement concept in polymer rheology. Adv Polym Sci 16:1–179

Graessley WW (1982) Entangled linear, branched and network polymer systems—Molecular theories. Adv Polym Sci 47:67–117

Grassberger P, Hegger R (1996) 'Smart' self-avoiding trails and θ collapse of chain polymers in three dimensions. J Phys A: Math Gen 29:279–288

Gray P (1968) The kinetic theory of transport phenomena in simple liquids. In: Temperley HNV, Rowlinson JS, Rushbrook GS (eds) Physics of simple liquids. North-Holland, Amsterdam, pp 507–562

Green AE, Adkins JE (1960) Large elastic deformations and non-linear continuum mechanics. Clarendon Press, Oxford

Grmela M (1985) Stress tensor in generalised hydrodynamics. Phys Lett A 111:41–44

Grossberg AYu, Khokhlov AR (1994) Statistical physics of macromolecules. AIP, New York

Guenza M (1999) Many chain correlated dynamics in polymer fluids. J Chem Phys 110(15):7574–7588

Guenza M, Freed KF (1996) Extended rotational isomeric model for describing the long time dynamics of polymers. J Chem Phys 105(9):3823–3837

Hansen JP, McDonald JR (1986) Theory of simple liquids, 2nd edn. Academic Press, London

Havriliak S (1990) Equilibrium polarization of polar polymers in a matrix of arbitrary compliance. Macromolecules 23:2384–2388

Hess W (1986) Self-diffusion and reptation in semi-dilute polymer solutions. Macromolecules 19(5):1395–1404

Hess W (1988) Generalised Rouse theory of entangled polymeric liquids. Macromolecules 21(8):2620–2632

Higgins JS, Benoit HC (1994) Polymers and neutron scattering. Oxford University Press, New York

Higgins JS, Roots JE (1985) Effect of entanglement on the single-chain motion of polymer molecules in melt samples observed by neutron scattering. J Chem Soc: Faraday Trans II 81:757–767

Hinch EJ (1994) Uncoiling a polymer molecule in a strong extensional flow. J Non-Newton Fluid Mech 54:209–230

Imanishi Y, Adachi K, Kotaka T (1988) Further investigation of the dielectric normal mode process in undiluted cis-polyisoprene with narrow distribution of molecular weight. J Chem Phys 89(12):7685–7592

Inoue T, Okamoto H, Osaki K (1991) Birefringence of amorphous polymers. 1 Dynamic measurements on polystyrene. Macromolecules 24:5670–5675

Isayev AI (1973) Generalised characterisation of relaxation properties and high elasticity of polymer systems. J Polym Sci A-2 116:2123–2133

Ito Y, Shishido S (1972) Critical molecular weight for onset of non-Newtonian flow and upper Newtonian viscosity of polydimethylsiloxane. J Polym Sci: Polym Phys Ed 10:2239–2248

Jackson JK, Winter HH (1995) Entanglement and flow behaviour of bidisperse blends of polystyrene and polybutadiene. Macromolecules 28:3146–3155

Janeschitz-Kriegl H (1983) Polymer melt rheology and flow birefringence. Springer, Berlin

Johnson RM, Schrag JL, Ferry JD (1970) Infinite-dilution viscoelastic properties of polystyrene in θ-solvents and good solvents. Polym J (Japan) 1(6):742–749

Jongschaap RJJ (1991) Towards a unified formulation of microrheological models. Lect Notes Phys 381:215–247

Kalashnikov VN (1994) Shear-rate dependent viscosity of dilute polymer solutions. J Rheol 38(5):1385–1403

Kannon RM, Kornfield JA (1994) Stress-optical manifestation of molecular and microstructural dynamics in complex polymer melts. J Rheol 38(4):1127–1150

Kargin VA, Slonimskii GL (1948) On deformation of amorphous linear polymers in fluid state. Dokl Akad Nauk SSSR 62:239–242 (in Russian)

Kholodenko AL (1996) Reptation theory: geometrical and topological aspects. Macromol Theory Simul 5:1031–1064

Kholodenko AL, Vilgis TA (1998) Some geometrical and topological problems in polymer physics. Phys Rep 298(5–6):251–370

Kehr M, Fatkullin N, Kimmich R (2007) Molecular diffusion on a time scale between nano- and milliseconds probed by field-cycling NMR relaxometry of intermolecular dipolar interactions: Application to polymer melts. J Chem Phys 126:094903

Kirkwood JG, Riseman J (1948) The intrinsic viscosity and diffusion constant of flexible macromolecules in solution. J Chem Phys 16:565–573

Klein J (1986) Dynamics of entangled linear, branched, and cyclic polymers. Macromolecules 19(1):105–118

Kokorin YuK, Pokrovskii VN (1990) Mechanism of ultra-slow relaxation in non-dilute linear polymers. Polym Sci USSR 32:2532–2540

Kokorin YuK, Pokrovskii VN (1993) New approach to the relaxation phenomena theory of the amorphous linear entangled polymers. Int J Polym Mater 20(3–4):223–237

Kostov KS, Freed KF (1997) Mode coupling theory for calculating the memory functions of flexible chain molecules: influence on the long time dynamics of oligoglycines. J Chem Phys 106(2):771–783

Kramers HA (1946) The behaviour of macromolecules in inhomogeneous flow. J Chem Phys 14(7):415–424

Kremer K, Grest GS (1990) Dynamics of entangled linear polymer melts: A molecular-dynamics simulation. J Chem Phys 92(8):5057–5086

Kremer K, Sukumaran SK, Everaers R, Grest GS (2005) Entangled polymer systems. Comput Phys Commun 169(1–3):75–81

Kuhn W (1934) Über die Gestalt fadenförmiger Moleküle in Lösung. Kolloid Z 68:2

Kuhn W, Kuhn H (1945) Bedeutung beschränk freier Drehbarkeit für die Viskosität and Strömungsdoppelbrechung von Fadenmolekullösungen I. Helv Chim Acta 28(7):1533–1579

Kwon Y, Leonov AI (1995) Stability constraints in the formulation of viscoelastic constitutive equations. J Non-Newton Fluid Mech 58:25–46

Laius LA, Kuvshinskii EV (1963) Structure and mechanical properties of "oriented" amorphous linear polymers. Fizika Tverdogo Tela 5(11):3113–3119 (in Russian)

Landau LD, Lifshitz EM (1969) Statistical physics, 2nd edn. Oxford, Pergamon

Landau LD, Lifshitz EM (1987a) Fluid mechanics, 2nd edn. Oxford, Pergamon

Landau LD, Lifshitz EM (1987b) Theory of Elasticity, 3rd edn. Oxford, Pergamon

Landau LD, Lifshitz EM, Pitaevskii LP (1987) Electrodynamics of continuous media, 2nd edn. Oxford, Pergamon

Leonov AI (1976) Non-equilibrium thermodynamics and rheology of viscoelastic polymer medium. Reol Acta 15(2):85–98

Leonov AI (1992) Analysis of simple constitutive equations for viscoelastic liquids. J Non-Newton Fluid Mech 42(3):323–350

Leonov AI (1994) On a self-consistent molecular modelling of linear relaxation phenomena in polymer melts and concentrated solutions. J Rheol 38(1):1–11

Liu B, Dünweg B (2003) Translational diffusion of polymer chains with excluded volume and hydrodynamic interactions by Brownian dynamics simulation. J Chem Phys 118(17):8061–8072

Lodge AS (1956) A network theory of flow birefringence and stress in concentrated polymer solutions. Trans Faraday Soc 52:120–130

Lodge TP (1999) Reconciliation of the molecular weight dependence of diffusion and viscosity in entangled polymers. Phys Rev Lett 83(16):3218–3221

Lodge TP, Schrag JL (1984) Oscillatory flow birefringence properties of polymer solutions at high effective frequencies. Macromolecules 17:352–360

Lodge TP, Miller JW, Schrag JL (1982) Infinite-dilution oscillatory flow birefringence properties of polystyrene and poly (α-methylstyrene) solutions. J Polym Sci: Polym Phys Ed 20:1409–1425

Lohmander U (1964) Non-Newtonian flow of dilute macromolecular solutions studied by capillary viscosimetry. Macromol Chem 72(1):159–173

Maugin GA (1999) Thermomechanics of nonlinear irreversible processes. World Scientific, Singapore

MacInnes DA (1977a) Internal viscosity in the dynamics of polymer molecules. J Polym Sci: Polym Phys Ed 15(3):465–476

MacInnes DA (1977b) Internal viscosity in the dynamics of polymer molecules. II A new model for the Lamb – Matheson – Philippoff increment. J Polym Sci: Polym Phys Ed 15(4):657–674

McLeish TCB, Milner ST (1999) Entangled dynamics and melt flow of branched polymers. Adv Polym Sci 143:195–256

Maconachie A, Richards RW (1978) Neutron scattering and amorphous polymers. Polymer 19(7):739–762

Mandelstam LI, Leontovich MA (1937) To the theory of the second sound. Zh Exper Theor Fiziki 7:438–449 (in Russian)

Marcone B, Orlandini E, Stella AL, Zonta F (2005) What is the length of a knot in a polymer? J Phys A: Math Gen 38:L15–L21

Mazars M (1996) Statistical physics of the freely jointed chain. Phys Rev 53(6):6297–6319

Mazars M (1998) Canonical partition functions of the freely jointed chains. J Phys A: Math Gen 31:1949–1964

Mazars M (1999) Freely jointed chains in external potentials: analytical computations. J Phys A: Math Gen 32:1841–1861

Meissner J, Garbella W, Hosteller J (1989) Measuring normal stress differences in polymer melt shear flow. J Rheol 33:843–864

Migliorini G, Rostiashvili VG, Vilgis TA (2003) Polymer chain in a quenched random medium: slow dynamics and ergodicity breaking. Eur Phys J B 33:61–73

Milchev A, Rostiashvili VG, Vilgis TA (2004) Dynamics of a polymer in a quenched random medium: a Monte Carlo investigation. Europhys Lett 66:384–390

Montfort JP, Marin G, Monge Ph (1984) Effects of constraint release on the dynamics of entangled linear polymer melts. Macromolecules 17(8):1551–1560

Mori H (1965) Transport, collective motion and Brownian motion. Prog Theor Phys 33:423–450

Müller-Plathe F (2002) Coarse-graining in polymer simulation: From the atomistic to the mesoscopic scale and back. J Chem Phys Phys Chem 3:754–769

Muller R, Picot C, Zang YH, Froelich D (1990) Polymer chain conformation in the melt during steady elongational flow as measured by SANS. Temporary network model. Macromolecules 23(9):2577–2582

Murnaghan FD Finite deformation of an elastic solid. New York (1954)

Noordermeer JaWM, Ferry JD, Nemoto N (1975) Viscoelastic properties of polymer solutions in high-viscosity solvents and limiting high-frequency behaviour. III Poly (2-substituted methyl acrylates). Macromolecules 8(5):672–677

Okamoto H, Inoue T, Osaki K (1995) Viscoelasticity and birefringence of polyisobutylene. J Polym Sci B Polym Phys 33:1409–1416

Oldroyd JG (1950) On the formulation of rheological equations of state. Proc R Soc A 200(1063):523–541

Oleinik EF (2003) Plasticity of semicrystalline flexible-chain polymers at the microscopic and mesoscopic levels. Polym Sci C 45(1):17–117

Onogi S, Masuda T, Kitagawa K (1970) Rheological properties of anionic polystyrenes. I Dynamic viscoelasticity of narrow-distribution polystyrenes. Macromolecules 3(2):109–116

Oono Y, Ohta T, Freed KF (1981) Application of dimensional regularisation to single chain polymer static properties: Conformational space renormalization of polymers. J Chem Phys 74(11):6458–6466

Osaki K, Mitsuda G, Johnson R, Schrag J, Ferry JD (1972a) Infinite-dilution viscoelastic properties of linear and star-shaped polybutadiene. Macromolecules 5(1):17–19

Osaki K, Schrag J, Ferry JD (1972b) Infinite-dilution viscoelastic properties of poly (α-methylstyrene). Applications of Zimm theory with exact eigenvalues. Macromolecules 5(2):144–147

Öttinger HCh (1989a) Gaussian approximation for Hookian dumbbells with hydrodynamic interaction: Translational diffusivity. Colloid Polym Sci 267(1):1–8

Öttinger HCh (1989b) Renormalization-group calculation of excluded-volume effects on the viscometric functions for dilute polymer solutions. Phys Rev A 40(5):2664–2671

Öttinger HCh (1990) Renormalisation of various quantities for dilute polymer solutions undergoing shear flow. Phys Rev A 41(8):4413–4420

Öttinger HCh (1995) Stochastic processes in polymeric fluids. Tools and examples for developing simulation algorithms. Springer, Berlin

Öttinger HCh, Beris AN (1999) Thermodynamically consistent reptation model without independent alignment. J Chem Phys 110:6593–6596

Pakula T, Geyler S, Edling T, Boese D (1996) Relaxation and viscoelastic properties of complex polymer systems. Rheol Acta 35(6):631–644

Panyukov S, Rabin Y (1996) Statistical physics of polymer gels. Phys Rep 269(1–2):1–131

Patlazhan SA (1993) Photoelastic properties of textured heterogeneous polymer materials. J Polym Sci Part B, Polym Phys 31:17

Paul W, Smith GD (2004) Structure and dynamics of amorphous polymers: computer simulations compared to experiment and theory. Rep Prog Phys 67:1117–1185

Peterlin A (1967) Frequency dependence of intrinsic viscosity of macromolecules with finite internal viscosity. J Polym Sci: A - 2 5(1):179–193

Peterlin A (1972) Origin of internal viscosity in linear macromolecules. Polym Lett 10:101–105

Petrie CJS (1979) Elongational flows. Piman, London

Phan-Thien N, Tanner RI (1977) A new constitutive equation derived from network theory. J Non-Newton Fluid Mech 2(4):353–365

Phillies GDJ (1995) Hydrodynamic scaling of viscosity and viscoelasticity of polymer solutions, including chain architecture and solvent quality effects. Macromolecules 28(24):8198–8208

Poh BT, Ong BT (1984) Dependence of viscosity of polystyrene solutions on molecular weight and concentration. Eur Polym J 20(10):975–978

Pokrovskii VN (1970) Equations of motion of viscoelastic systems as derived from the conservation laws and the phenomenological theory of non-equilibrium processes. Polym Mech 6(5):693–702

Pokrovskii VN (1978) Statisticheskaya mekhanika razbavlennykh suspenzii (Statistical mechanics of dilute suspensions, in Russian). Nauka, Moscow. Available via http://ecodynamics.narod.ru/polymer/Suspension.pdf. Cited 1 Feb 2009

Pokrovskii VN (2005) Extended thermodynamics in a discrete-system approach. Eur J Phys 26:769–781

Pokrovskii VN (2006) A justification of the reptation-tube dynamics of a linear macromolecule in the mesoscopic approach. Physica A 366:88–106

Pokrovskii VN (2008) The reptation and diffusive modes of motion of linear macromolecules. J Exper Theor Phys 106(3):604–607

Pokrovskii VN, Chuprinka VI (1973) The effect of internal viscosity of macromolecules on the viscoelastic behaviour of polymer solutions. Fluid Dyn 8(1):13–19

Pokrovskii VN, Kokorin YuK (1984) Theory of viscoelasticity of dilute blends of linear polymers. Vysokomolek Soedin B 26:573–577 (in Russian)

Pokrovskii VN, Kokorin YuK (1985) Effect of entanglement on the macromolecule mobility. Vysokomolek Soedin B 27:794–798 (in Russian)

Pokrovskii VN, Kokorin YuK (1987) The theory of oscillating birefringence of solutions of linear polymers. Dilute and concentrated systems. Polym Sci USSR 29:2385–2393

Pokrovskii VN, Kruchinin NP (1980) On the non-linear behaviour of linear polymer flow. Vysokomolek Soedin B 22:335–337 (in Russian)

Pokrovskii VN, Pyshnograi GV (1988) Dependence of viscoelasticity of concentrated polymer solutions upon polymer concentration and macromolecule length. Vysokomolek Soedin B 30:35–39 (in Russian)

Pokrovskii VN, Pyshnograi GV (1990) Non-linear effects in the dynamics of concentrated polymer solutions and melts. Fluid Dyn 25:568–576

Pokrovskii VN, Pyshnograi GV (1991) The simple forms of constitutive equation of polymer concentrated solution and melts as consequence of molecular theory of viscoelasticity. Fluid Dyn 26:58–64

Pokrovskii VN, Tonkikh GG (1988) Dynamics of dilute polymer solutions. Fluid Dyn 23(1):115–121

Pokrovskii VN, Tskhai AA (1986) Slow motion of a particle in slightly anisotropic liquid. Prikl Matem Mechanika 50(3):512–515 (in Russian)

Pokrovskii VN, Volkov VS (1978a) The theory of slow relaxation processes in linear polymers. Polym Sci USSR 20(2):288–299

Pokrovskii VN, Volkov VS (1978b) The calculation of relaxation time and dynamical modulus of linear polymers in one-molecular approximation with self-consistency. (A new approach to the theory of viscoelasticity of linear polymers). Polym Sci USSR 20:3029–3037

Pokrovskii VN, Kruchinin NP, Danilin GA, Serkov AT (1973) A relation between coefficients of shear and extensional viscosity for concentrated polymer solutions. Mechanika Polymerov 1:124–131 (in Russian)

Polverary M, van de Ven TGM (1996) Dilute aqueous poly(ethylene oxide) solutions: Clusters and single molecules in thermodynamic equilibrium. J Phys Chem 100(32):13687–13695

Priss LS (1957) On reasons of discrepancies between kinetic theory of high-elasticity and experience. Dokl Akad Nauk SSSR 116(2):225–228 (in Russian)

Priss LS (1980) The theory of high elasticity and birefringence of rubber. Int J Polym Mater 8(1):99–116

Priss LS (1981) Molecular origin of constants in the theory of rubber-like elasticity considering network chains steric interactions. J Pure Appl Chem 53:1581–1596

Priss LS, Gamlitski YuA (1983) Mechanism of conformation transitions in polymer chains. Polym Sci USSR 25:117–123

Priss LS, Popov VF (1971) Relaxation spectrum of one-dimensional model of polymer chains. J Macromol Sci B 5(2):461–472

Prokunin AN (1989) On the description of viscoelastic flows of polymer fluids. Rheol Acta 28(1):38–47

Pyshnograi GV (1994) The effect of the anisotropy of macromolecular tangles on the non-linear properties of polymer fluids stretched along a single axis. J Appl Mech Techn Phys 35(4):623–627

Pyshnograi GV (1996) An initial approximation in the theory of viscoelasticity of linear polymers and non-linear effects. J Appl Mech Techn Phys 37(1):123–128

Pyshnograi GV (1997) The structure approach in the theory of flow of solutions and melts of linear polymers. J Appl Mech Techn Phys 38(3):122–130

Pyshnograi GV, Pokrovskii VN (1988) Stress dependence of stationary shear viscosity of linear polymers in the molecular field theory. Polym Sci USSR 30:2624–2629

Pyshnograi GV, Pokrovskii VN, Yanovsky YuG, Karnet YuN, Obraztsov IF (1994) Constitutive equation on non-linear viscoelastic (polymer) media in zeroth approximation by parameter of molecular theory and conclusions for shear and extension. Phys – Doklady 39(12):889–892

Rabin Y, Öttinger HCh (1990) Dilute polymer solutions: internal viscosity, dynamic scaling, shear thinning, and frequency-dependent viscosity. Europhys Lett 13(5):423–428

Rallison JM, Hinch EJ (1988) Do we understand the physics in the constitutive equation? J Non-Newton Fluid Mech 29(1):37–55

Ren J, Urakawa O, Adachi K, Kotaka T (2003) Dielectric and viscoelastic studies of segmental and normal mode relaxations in undiluted poly(d,l-lactic acid). Macromolecules 36:210–219

Riande E, Siaz E (1992) Dipole moments and birefringence of polymer. Prentice Hall, Englewood Cliffs, NJ

Richter D, Farago B, Fetters LJ, Huang JS, Ewen B, Lartique C (1990) Direct microscopic observation of the entanglement distance in a polymer melt. Phys Rev Lett 64:1389–1392

Richter D, Farago B, Butera R, Fetters LJ, Huang JS, Ewen B (1993) On the origin of entanglement constraints. Macromolecules 26(4):795–804

Rice SA, Gray P (1965) Statistical mechanics of simple liquids. Wiley, New York

Ronca G (1983) Frequency spectrum and dynamic correlation of concentrated polymer liquids. J Chem Phys 79(2):1031–1043

Rouse PE (1953) A theory of the linear viscoelastic properties of dilute solutions of coiling polymers. J Chem Phys 21:1272–1280

Rossers RW, Schrag JL, Ferry JD (1978) Infinite-dilution viscoelastic properties of polystyrene of very high molecular weight. Macromolecules 11(5):1060–1062

Rostiashvili VG, Rehkopf M, Vilgis TA (1999) The Hartree approximation in dynamics of polymeric manyfolds in the melt. J Chem Phys 110(1):639–651

Schausberger A, Schindlauer G, Janescitz-Kriegl H (1985) Linear elastico-viscous properties of molten polysterenes. I Presentation of complex moduli; role of short range structural parameters. Rheol Acta 24:220–227

Schieber JD (1992) Do internal viscosity models satisfy the fluctuation-dissipation theorem? J Non-Newton Fluid Mech 45:47–61

Schieber JD, Öttinger HCh (1994) On consistency criteria for stress tensors in kinetic theory models. J Rheol 38(6):1909–1924

Schieber JD, Neergaard J, Gupta S (2003) A full-chain temporary network model with sliplinks, chain-length fluctuations, chain connectivity and chain stretching. J Rheol 47(1):213–233

Schrag JL, et al (1991) Local modification of solvent dynamics by polymeric solutes. J Non-Cryst Solids 131–133(2):537–543

Schreiber HP, Bagley EB, West DC (1963) Viscosity/molecular weight relation in bulk polymers–I. Polymer 4:355–374

Schweizer KS (1989a) Microscopic theory of the dynamics of polymeric liquids: General formulation of a mode-mode coupling approach. J Chem Phys 91:5802–5821

Schweizer KS (1989b) Mode-coupling theory of the dynamics of polymer liquids: Qualitative predictions for flexible chain and ring melts. J Chem Phys 91:5822–5839

Scolnick J, Kolinski A (1990) Dynamics of dense polymer systems: Computer simulations and analytic theories. Adv Chem Phys 78:223–278

Shishkin NI, Milagin NF, Gabarajeva AD (1963) Molecular network and orientational processes in amorphous polystyrene. Fizika Tverdogo Tela 5(12):3453–3462 (in Russian)

Shliomis MI (1966) To the hydrodynamics of liquid with internal rotation. Soviet Phys – JETF 51(1):259–270

Tang WH, Chang XY, Freed KF (1995) Theory for long time polymer and protein dynamics: Basic functions and time correlation functions. J Chem Phys 103(21):9492–9501

Tao H, Lodge TP, von Meerwall ED (2000) Diffusivity and viscosity of concentrated hydrogenated polybutadiene solutions. Macromolecules 33:1747–1758

Takahashi Y, Isono Y, Noda I, Nagasawa M (1985) Zero-shear viscosity of linear polymer solutions over a wide range of concentration. Macromolecules 18(50):1002–1008

Tashiro K (1993) Molecular theory of mechanical properties of crystalline polymer. Prog Polym Sci 18:377–435

Thurston GB, Peterlin A (1967) Influence of finite number of chain segments, hydrodynamic interaction, and internal viscosity on intrinsic birefringence and viscosity of polymer solutions in an oscillating laminar flow field. J Chem Phys 46(12):4881–4884

Treloar LRG (1958) The physics of rubber elasticity. Oxford University Press, London

Tsenoglou C (2001) Non-Newtonian rheology of entangled polymer solutions and melts. Macromolecules 34:2148–2155

Tskhai AA, Pokrovskii VN (1985) Rotational mobility of non-spherical particle in anisotropic liquid. Kolloid Zh 47(1):106–111

Tsvetkov VN, Eskin VE, Frenkel SYa (1964) Struktura makromolekul v rastvorakh (Structure of macromolecules in solutions, in Russian). Nauka, Moscow

Valleau JP (1996) Distribution of end-to-end length of an excluded-volume chain. J Chem Phys 104(8):3071–3074

Van Meerwall E, Grigsby J, Tomich D, Van Antverp R (1982) Effect of chain-end free volume on the diffusion of oligomers. J Polym Sci: Polym Phys Ed 20(6):1037–1053

Van Vleck JH (1932) The Theory of electric and magnetic susceptibilities. Oxford University Press, London

van Zanten JH, Rufener KP (2000) Brownian motion in a single relaxation time Maxwell fluid. Phys Rev E 62(4):5389–5396

van Zanten JH, Amin S, Abdala AA (2004) Brownian motion of colloidal spheres in aqueous PEO solutions. Macromolecules 37:3874–3880

Vilgis TA, Genz U (1994) Theory of the dynamics of tagged chains in interacting polymer liquids: General theory. J Phys I 4(10):1411–1425

Vinogradov GV, Malkin AYa, Yanovsky YuG et al (1972a) Viscoelastic properties and flow of polybutadienes and polyisoprenes. Vysokomolek Soedin A 14(11):2425–2442 (in Russian)

Vinogradov GV, Pokrovskii VN, Yanovsky YuG (1972b) Theory of viscoelastic behaviour of concentrated polymer solutions and melts in one-molecular approximation and its experimental verification. Rheol Acta 7(2):258–274

Volkov VS (1990) Theory of Brownian motion in viscoelastic Maxwellian liquid. Sov Phys – JETP 71(1):93–97

Volkov VS, Pokrovskii VN (1978) Effect of intramolecular relaxation processes on the dynamics of dilute polymer solutions. Vysokomolek Soedin B20(11):834–837 (in Russian)

Volkov VS, Vinogradov GV (1984) Theory of dilute polymer solutions in viscoelastic fluid with a single relaxation time. J Non-Newton Fluid Mech 15:29–44

Volkov VS, Vinogradov GV (1985) Relaxational interactions and viscoelasticity of polymer melts. Part I Model development. J Non-Newton Fluid Mech 18:163–172

Wang S-Q (2003) Chain dynamics in entangled polymers: diffusion versus rheology and their comparison. J Polym Sci: Part B: Polym Phys 41:1589–1604

Watanabe H (1999) Viscoelasticity and dynamics of entangled polymers. Prog Polym Sci 24:1253–1403

Watanabe H (2001) Dialectric relaxation of type-A polymers in melts and solutions. Macromol Rapid Commun 22(3):127–175

Watanabe H, Kotaka T (1984) Viscoelastic properties and relaxation mechanism of binary blends of narrow molecular weight distribution polystyrene. Macromolecules 17(11):2316–2325

Watanabe H, Sakamoto T, Kotaka T (1985) Viscoelastic properties of binary blends of narrow molecular weight distribution polystyrene. 2. Macromolecules 18(5):1008–1015

Watanabe H, Yao M-L, Osaki K (1996) Comparison of dialectric and viscoelastic relaxation behaviour of polyisoprene solutions: Coherence in subchain motion. Macromolecules 29(1):97–103

Weinberger CB, Goddard JD (1974) Extensional flow behaviour of polymer solutions and particle suspensions in a spinning motion. Int J Multiphase Flow 1:465–486

Wishnewski A, Monkenbush M, Willner L, Richter D, Likhtman AE, McLeish TCB, Farago B (2002) Molecular observation of contour-length fluctuations limiting topological confinment in polymer melts. Phys Rev Lett 88(5):058301-1–058301-4

Wittmer JP, Beckrich P, Crevel F, Huang CC, Cavallo A, Kreer T, Meyer H (2007) Are polymer melts "ideal"? Comput Phys Commun 177(1–2):146–149

Wood LC (1975) The thermodynamics of fluid systems. Clarendon, Oxford

Yanovski YuG, Vinogradov GV, Ivanova LN (1982) Viscoelasticity of blends of narrow molecular mass distribution polymers. In: Novyje aspecty v reologii polimerov (New aspects in polymer rheology, in Russian), Part 1. INKHS AN USSR, Moscow, pp 80–87

Yong CW, Clarke JHR, Freire JJ, Bishop M (1996) The theta condition for linear polymer chains in continuous space and three dimensions. J Chem Phys 105(21):9666–9673

Zgaevskii VE (1977) Theoretical description of the elastic properties of micro-inhomo-
geneous polymers. Polym Mech 13(4):624–625

Zgaevskii VE, Pokrovskii VN (1970) Calculation of polarizability of macromolecule as elastic
thread. Zh Prikl Spektrosk 12(2):312–317 (in Russian)

Zimm BH (1956) Dynamics of polymer molecules in dilute solution: viscoelasticity, flow
birefringence and dielectric loss. J Chem Phys 24:269–278

Zwanzig R (1961) Memory effects in irreversible thermodynamics. Phys Rev 124(4):983–992

Index

A

anisotropy of mobility
  global, 138
  local, 45, 139
approximation of memory functions, 47, 54

B

bead-and-spring model of macromolecule, 3
birefringence
  dilute solution flow, 209
  entangled system flow, 211

C

conformation of macromolecule
  in equilibrium, 7
  relaxation of, 34, 143, 145
constitutive equation
  differential form, 168
  for dilute solution, 172, 173
  for entangled system
    basic, 180
    Vinogradov, 191
  integral form, 168
  single-mode form
    for strongly entangled system, 190
    for weakly entangled system, 188
crystalline polymer
  model of, 19
  modulus of elasticity, 19

D

diffusion of macromolecule
  in entangled system, 87
    dependence on length and concentration, 93
    numerical results, 94

in viscous liquid, 84
dilute blend
  characteristic viscoelasticity, 131, 132
  definition, 129
  terminal time, 133
dilute solution
  characteristic viscosity
    internal viscosity dependence, 110
    molecular-weight dependence, 108
  constitutive equation, 172
    many-mode approximation, 172
    single-mode approximation, 173
  dynamic modulus, 106, 107
  dynamic viscosity, 102, 107, 110
  flow birefringence, 209
  general features, 12
  normal stresses, 174
  oscillatory birefringence, 211
  stress tensor, 104
  viscosity, 174
dissipation-fluctuation relation, 32, 65
dynamic modulus
  chains in viscoelastic liquid, 113
  dilute solution, 106
  entangled system, 119
dynamic viscosity
  dilute solution, 102, 110
dynamics of macromolecule
  in entangled system, 39
    Cerf-Rouse mode, 64
    diffusive mode, 64
    Doi-Edwards approach, 57
    linear equation, 44
    Markovian form of equation, 54, 137
    non-linear equation, 46
  in viscous liquid, 21

V.N. Pokrovskii, *The Mesoscopic Theory of Polymer Dynamics*,
Springer Series in Chemical Physics 95,
DOI 10.1007/978-90-481-2231-8, © Springer Science+Business Media B.V. 2010

dynamo-optical coefficient
  definition, 208
  dilute solution, 211

E
elastic thread model of macromolecule, 1
elasticity
  crystalline polymer, 19
  entangled system, 197
  network polymer, 18
elongational/shear viscosities ratio
  stress dependence, 195
  theory and experiment, 196
end-to-end distance
  and orientation of segments, 150
  distribution function, 3, 219
  of chain with excluded volume, 10
  of elastic thread, 2
  of freely-jointed chain, 2
entangled system
  basic constitutive equation, 180
  diffusion of macromolecule
    dependence on length and concentra-
      tion, 93
    numerical results, 94
  diffusive mobility of macromolecule, 86
  Doi-Edwards approach, 57
  dynamic modulus, 119
  dynamics of macromolecule, 39
  dynamics of macromolecule in Markovian
    form, 54
  elongational viscosity, 194, 195
  flow birefringence, 211
  general features, 13
  intermediate dynamic length and
    elasticity, 125
  many-chain approach, 39
  modulus of elasticity, 197
  molecular weight between entanglements,
    124
  neutron scattering, 93
  normal stresses, 186, 192
  oscillatory birefringence, 213
  recoverable strains, 197
  relaxation time, 193
  relaxation times of macromolecule
    dependence on length, 78
    diffusive branch, 72, 73
    reptation branch, 74
    transition point, 78
  reptation mobility of macromolecule, 90
  reptation of macromolecule, 57
  shear viscosity, 184, 193
  stress tensor, 111, 117

  terminal relaxation time, 126
  transition empirical points, 116
  Vinogradov constitutive equation, 191
  viscosity, 126
excluded-volume effect
  general features, 8
  screening of, 14

F
freely-jointed segment model, 2
frequency-temperature reduction rule, 128

H
hydrodynamic interaction
  screening of, 44
  tensor of, 23, 222

I
intermediate dynamic length
  and localisation scale, 89
  and neutron scattering, 96
  and quasi-equilibrium elasticity, 125
  and the tube diameter, 97
  dependence on concentration and
    temperature, 97
  introduction, 87
  problem of introduction, 16
internal variables in thermodynamics, 159
intramolecular resistance for bead-and-
    spring chain
  in entangled system, 52
  in viscous liquid, 28

J
Jaumann derivative, 164

L
large deformation
  stress-strain relation, 222
  thermodynamics of, 219

M
macromolecule
  bead-and-spring or subchain model, 3
  elastic thread model, 1
  freely-jointed segment model, 2
  persistence length model, 2
  radius of gyration, 7
  random walk model, 217
memory functions, 47, 54
mobility of macromolecule
  in entangled system, 90
    diffusive mechanism, 86
    reptation mechanism, 90

in viscous liquid, 84

N

network polymer
  model of, 16
  modulus of elasticity, 18
  stress-strain relation, 18
neutron scattering measurement
  intermediate length, 93
  macromolecular dimensions, 15
  macromolecular dimensions in flow, 81
normal co-ordinates, 5
normal stresses
  for entangled system, 186, 192
  for weakly entangled system, 188

O

orientation of segments
  and end-to-end distance, 150
  relaxation, 46, 150
oscillatory birefringence
  dilute solution, 211
  entangled system, 213

P

parameters of the mesoscopic approach, 53
permittivity tensor
  dependence of orientation tensor, 201
  for dilute solution, 204
  for entangled system, 205
persistence length model of macromolecule,
    2
polarisability tensor of segment, 200
principle of material objectivity, 163

R

radius of gyration of a macromolecule, 7
recoverable strains, 197
refractive index, 206
relaxation equation
  for deformation of dumb-bell, 234
  for orientation of dumb-bell, 234
  for segment orientation, 150
relaxation modulus, 103
relaxation of conformational tensors
  due to reptation, 145
  in strongly entangled systems, 145
  in viscous liquid, 34
  in weakly entangled systems, 143
relaxation of internal variables, 162
relaxation of orientational tensors
  in strongly entangled systems, 147
  in weakly entangled systems, 148
relaxation of segment orientation, 46

relaxation time of dumb-bell
  deformational, 234
  orientational, 233, 234
relaxation times of macromolecule
  in entangled system
    dependence on length, 78
    diffusive, 72, 73, 78
    numerical results, 80
    reptational, 74
  in viscous liquid, 31
resistance-drag coefficient
  in non-local liquid, 225
  in viscoelastic liquid, 224
  molecular-weight dependence, 27
  Stokes relation, 22
resistance-drag matrix, 24

S

screening of
  excluded-volume effect, 14
  hydrodynamic interaction, 44
self-consistency condition, 122, 124
strain-optical coefficient
  definition, 208
  entangled system, 213
stress tensor
  method of calculation, 100
stress tensor in moments
  for dilute solution, 104
  for entangled system, 111, 117
stress-optical coefficient, 204, 205, 209, 211
stress-optical law, 204, 205
strongly entangled system
  dynamic viscosity, 191
  relaxation times, 78
  single-mode constitutive equation, 190
  transition point, 78
subchain model of macromolecule, 3
suspension of dumb-bells
  diffusion, 231
  normal stresses, 238
  shear viscosity, 238
  stress tensor, 237

T

temperature-frequency reduction rule, 128
terminal relaxation time, 126
theta solvent, 13
theta temperature, 11
transition temperatures, 13, 19

V

viscosity

256   Index

dependence on concentration, molecular
  weight and temperature, 126
for entangled system, 184

W

weakly entangled system
  normal stresses, 188
  relaxation times, 78
  single-mode constitutive equation, 188
  transition point, 78